From Enzyme Models to Model Enzymes

From Enzyme Models to Model Enzymes

Anthony J. Kirby
University Chemical Laboratory, University of Cambridge, Cambridge, UK

Florian Hollfelder
Department of Biochemistry, University of Cambridge, Cambridge, UK

RSCPublishing

Cover picture. Graphic representation (by Dr. Jörg Harms) of the X-ray structure of a bacterial 50S ribosome (*Cell*, 2001, **107**, 679–688). This is the larger subunit of the complex "machine" that translates the genetic code into the specific primary sequence of amino-acids defining each natural protein (Figure 1.1) – enzymes included. Significantly, the ribosome combines a few RNA chains (silver ribbons) with many protein molecules (coloured), and can be thought of as the prototype model enzyme. We thank Professor Ada Yonath for supplying the graphics file.

ISBN: 978-0-85404-175-6

A catalogue record for this book is available from the British Library

Published by The Royal Society of Chemistry,
Thomas Graham House, Science Park, Milton Road,
Cambridge CB4 0WF, UK

Registered Charity Number 207890

For further information see our web site at www.rsc.org

Preface

We have been fascinated by enzymes since we were students in the 1960s and the 1990s. In the early days the mantra – "there's nothing magic about enzymes" – simply added to the fascination. Of course "*like magic*" in these postromantic times simply means (to quote the Oxford English Dictionary) "*without any apparent explanation; with incredible rapidity*": simple maybe, but it's hard to think of a more intriguing starting point for a research career. And we still would like to understand – as well as possible – how enzymes work.

At one level we can already explain enzyme catalysis: what an enzyme does is bind, and so stabilize selectively, the transition state for its specific reaction. But our current level of understanding fails the more severe, practical test – the creation of new systems with catalytic efficiencies to rival those of natural enzymes. As researchers, mostly in chemistry but at defining points in biochemistry departments, our approach has been experimental: applying the methods of contemporary organic chemistry, enzymology and more recently molecular biology, to develop systems that model or – better – mimic enzymes: and in particular their catalytic efficiency.

We settled on the title of this book only after some discussion, and some definitions are in order. An enzyme *model* simply models an aspect, sometimes more than one aspect, of the chemistry of a particular enzyme. At the next level, we expect enzyme *mimics* to catalyze reactions of substrates free in solution, by mechanisms that are demonstrably enzyme-like, involving both binding and catalysis. However, to call the most successful systems artificial enzymes seems no longer appropriate. As protein chemistry develops in countless new directions, existing protein structures can provide appropriate starting points for new catalysts, which are not exactly artificial. We propose the term model enzyme. Everyone expects that their model aeroplane should at least fly! And

From Enzyme Models to Model Enzymes
By Anthony J. Kirby and Florian Hollfelder
© Anthony J. Kirby and Florian Hollfelder 2009
Published by the Royal Society of Chemistry, www.rsc.org

ribozymes (Chapter 5), which qualify as successful systems, are serious candidates to have been Nature's models for all enzymes.

Our first intention was to write a short student text, introducing the area of enzyme models for readers from the many disciplines that offer a basic grounding in chemistry and biochemistry. The project evolved as we wrote, reflecting the ways the subject is developing, to take advantage of the availability of new and powerful biological methodologies. The resulting text is still short, and still intended to be accessible to the same readership: but attempts in a final chapter to put recent developments in the perspective of the major body of work that has gone before: to trace the underlying logic of these developments, and to identify the lessons to be learned. So that as we understand and come to terms with the limitations of classical approaches to "design-based" model systems, we learn how to complement – and perhaps even replace – them with new and emerging methods: based on the successful development of enzymes by natural evolution, and using iterative approaches to make the best use of design and selection. It is in the nature of science that the very recent work we review may well be superseded by even more remarkable results in this fast moving field.

We have been helped, encouraged and inspired by our students, colleagues and teachers, too numerous to mention, from all over the world: but we must acknowledge our special debts of gratitude to the late Bill Jencks, who could turn physical organic chemistry into biological magic.

Tony Kirby, Florian Hollfelder
Cambridge, UK

Contents

Chapter 1   From Models Through Mimics to Artificial Enzymes   **1**

    1.1   Introduction to Enzyme Chemistry   2
        1.1.1   Why are Enzymes so Big?   3
        1.1.2   Functional Groups Available to Enzymes   4
    1.2   Principles of Catalysis by Enzymes   6
        1.2.1   Dependence on pH   7
    1.3   General Acid–Base Catalysis   8
        1.3.1   Experimental Evidence   8
        1.3.2   Mechanisms   9
        1.3.3   Kinetic Equivalence   13
    1.4   Intramolecularity   14
        1.4.1   Efficiency of Intramolecular Catalysis   17
    1.5   Energetics   18
    1.6   Binding and Recognition   20
        1.6.1   Hydrophobic Binding   22
        1.6.2   The Special Environment of the Active Site   24
    1.7   Cofactors   24
    1.8   Many Enzymes Use the Same Basic Mechanisms   25
    1.9   Enzyme Models, Mimics and Pretenders   26

**Chapter 2   Evaluation of Catalytic Efficiency in Enzymes and
Enzyme Models**   **29**

    2.1   Introduction   29
    2.2   Measurement of Uncatalyzed Rate Constants, k_{uncat} 30
    2.3   Characterizing the Catalytic Reactivity of an Enzyme
        or an Enzyme Model   32
        2.3.1   Comparing Catalyzed and Uncatalyzed Rates   34

From Enzyme Models to Model Enzymes
By Anthony J. Kirby and Florian Hollfelder
© Anthony J. Kirby and Florian Hollfelder 2009
Published by the Royal Society of Chemistry, www.rsc.org

        2.3.2   Calculating Rate Accelerations   35
    2.4 Catalytic Efficiencies of Representative Enzymes:
        The Size of the Challenge   41

Chapter 3   Constructing Enzyme Models – Building up Complexity   42

    3.1 Solvents – Catalysis Without Functional Groups?   43
    3.2 Introducing Catalytic Groups – Without
        Positioning Them   46
    3.3 Positioning of Substrate and Catalytic Groups
        by Covalent Design   47
    3.4 Binding the Ground State by Noncovalent
        Interactions   47
    3.5 Binding the TS More Strongly than the GS by
        Noncovalent Interactions   50
    3.6 Existing Enzymes as Catalytic Scaffolds
        to Accommodate New Functions   52
        3.6.1   Enzymes Modified by Addition of
                Functionality   54
                3.6.1.1   Exploiting Modular Build-Up
                          of Binding and Catalytic Features:
                          Chimeras of Binding Proteins and
                          Reactive Chemical Functionality   54
                3.6.1.2   Noncovalent Introduction of
                          Reactive Cofactors: Transition-Metal
                          Catalysts in a Protein   55
                3.6.1.3   Covalent Derivatization of
                          Active-Site Residues
                          to Introduce Reactive Cofactors   56
        3.6.2   Site-Directed Mutants of Enzymes: Minimalist
                Protein Redesign   56
        3.6.3   Exploring Enzyme Promiscuity   57

Chapter 4   Enzyme Models Classified by Reaction   61

    4.1 Acyl Transfer   61
        4.1.1   The Serine Proteases. Typical Active Sites   62
                4.1.1.1   The Active-Site Environment . . .
                          and Mechanism   64
                4.1.1.2   Intramolecular Models   66
                4.1.1.3   Supramolecular Models   69
                          4.1.1.3.1 Cyclodextrins   69
                          4.1.1.3.2 Synthetic Models   72
        4.1.2   SH Hydrolases   75
                4.1.2.1   Models   76

		4.1.2.2 Intramolecular Models	78
		4.1.2.3 Supramolecular Models	79
	4.1.3	Aspartic Proteinases	79
		4.1.3.1 Intramolecular Models	80
	4.1.4	Metallopeptidases/Amide Hydrolases	83
		4.1.4.1 Enzymes	86
		4.1.4.2 Model Systems with One Metal Centre	86
		4.1.4.2.1 Intramolecular Reactions	88
		4.1.4.2.2 Supramolecular Metalloprotease/Peptidase Models	88
		4.1.4.3 Model Systems with Two Metal Centres	90
		4.1.4.3.1 Aminopeptidase and Lactamase Models	92
4.2	Phosphoryl Transfer		95
	4.2.1	Phosphoryl Group Transfer from Monoesters	96
		4.2.1.1 Enzymes I. Phosphoryl Transfer Without Metals: PTPases	98
		4.2.1.1.1 Models	100
		4.2.1.1.2 Intramolecular Models	100
		4.2.1.1.3 Supramolecular Models	101
		4.2.1.2 Enzymes II. Metalloenzymes	102
		4.2.1.2.1 Models	102
	4.2.2	Phosphoryl Group Transfer from Phosphodiesters	105
		4.2.2.1 Intramolecular Reactions	106
		4.2.2.1.1 Intramolecular Attack by OH	108
		4.2.2.2 Enzymes I. Phosphoryl Group Transfer Without Metals	111
		4.2.2.2.1 Transfer to Neighbouring OH	113
		4.2.2.2.2 Supramolecular Models	114
		4.2.2.3 Enzymes II. Metalloenzymes	117
		4.2.2.3.1 Supramolecular Models: RNA Cleavage	118
		4.2.2.3.2 Supramolecular Models: DNA-Cleavage	122
4.3	Glycosyl Transfer		126
	4.3.1	Simple Models	129
	4.3.2	Enzymes	132
		4.3.2.1 Glycoside Hydrolases	133
		4.3.2.2 Glycoside Transferases	135
	4.3.3	Intramolecular Models	138
	4.3.4	Enzyme Mimics	142

4.4   Hydrogen Transfer   145
      Introduction   145
      4.4.1   Enolization: Proton Transfer from Carbon   146
              4.4.1.1   Simple Models   148
              4.4.1.2   Intramolecular Models   149
              4.4.1.3   Catalysis by Metal Ions   151
              4.4.1.4   Enzymes Catalyzing Enolization   152
                        4.4.1.4.1 Triose Phosphate Isomerase   152
                        4.4.1.4.2 Citrate Synthase   155
                        4.4.1.4.3 The Enolase Superfamily   155
                        4.4.1.4.4 Mandelate Racemase   157
                        4.4.1.4.5 Models   157
      4.4.2   Hydride Transfer   159
              4.4.2.1   Uridine Diphosphate-galactose-4-
                        epimerase   160
              4.4.2.2   Dehydrogenases   161
              4.4.2.3   Models   161
              4.4.2.4   Intramolecular Models   165
      4.4.3   Hydrogen-Atom Transfer   165
4.5   Radical Reactions   168
      4.5.1   Coenzyme Initiators Based on
              Adenosylcobalamin   168
      4.5.2   Radicals in Enzyme-Active Sites   171
              4.5.2.1   Pyruvate-formate Lyase   171
              4.5.2.2   Ribonucleotide Reductases   172
      4.5.3   Models   174
              4.5.3.1   Initiation Stages   176
              4.5.3.2   Hydrogen-Atom Transfers   176
4.6   Pericyclic Reactions   180
      4.6.1   Chorismate Mutase   181
              4.6.1.1   Models: Catalysis by Antibodies   182
      4.6.2   Antibodies Catalyzing the Diels–Alder
              Reaction   185
              4.6.2.1   Supramolecular Catalysis of the
                        Diels–Alder Reaction   187
      4.6.3   Catalysis by RNA   191

Chapter 5   Design *vs.* Iterative Methods – Mimicking the Way Nature
            Generates Catalysts   195

      Introduction   195
5.1   Catalytic Polymers   197
      5.1.1   Synzymes   199
      5.1.2   Dendrimers   202
              5.1.2.1   Peptide Dendrimers   204
      5.1.3   Molecular Imprinting   209

	5.2	Catalytic Antibodies	212
		5.2.1 Other Approaches	214
		5.2.2 Proton Transfer from Carbon	216
		5.2.3 Conclusions	219
	5.3	Nucleic Acids as Catalysts	220
		5.3.1 Mechanisms of Nucleic-acid Catalysis	220
		5.3.2 Selection as an Alternative to Design Strategies	225
		5.3.3 Access to New Catalysts Using SELEX	227
		5.3.4 Changing the Catalyst Backbone: DNAzymes	228
		5.3.5 Nucleic-Acid Catalysis of Other Reactions	229
	5.4	Improving Protein Enzymes	231
		5.4.1 Challenges in Exploring Protein Catalysts	231
		5.4.2 What Fraction of Diversity Space is Practically Accessible?	233
		5.4.3 Mapping Enzyme Function in the Proteome: Protein Superfamilies as a Basis for Understanding Functional Links	236
		5.4.3.1 The Enolase Superfamily	236
		5.4.3.2 The Alkaline Phosphatase Superfamily	238
		5.4.4 Challenging Chance by Design and Directed Evolution	243

| **References** | **248** |

| **Subject Index** | **266** |

CHAPTER 1

From Models Through Mimics to Artificial Enzymes

Enzymes are the all-purpose catalysts that make the Chemistry of Life run smoothly and efficiently. They do the sorts of things that chemists want to do, under the mildest, "greenest" conditions – in aqueous solution near pH 7, at atmospheric pressure and temperatures close to ambient. They are wonderfully efficient catalysts, capable of handling with ease the most unreactive compounds present in biological systems, and their reactions, where necessary, are completely chemo-, regio- and stereoselective. Small wonder that an understanding of how enzymes work has been an ambition for generations of researchers in a whole range of disciplines, from pure enzymology, through almost all of chemistry to X-ray crystallography.

An understanding of the principles of enzyme catalysis is of far more than academic interest. The industrial use of enzymes is widespread and growing.[1] The food industry has always used the enzymes present in various organisms such as yeasts, but a growing trend is to use isolated enzymes at key stages to improve quality control. Medical applications have become hugely important, both in diagnostic testing and directly in therapeutic applications. And nowhere is the need to understand the fundamental principles more important than in the development of artificial enzymes, which have far-reaching potential in the fine chemicals and pharmaceutical industries. Thus, asymmetric synthesis is a major activity and growth area for organic and pharmaceutical chemistry, and chiral catalysis the most elegant – and most efficient – way of achieving it. Enzymes are chiral catalysts *par excellence*, and natural (wild-type) or specifically modified enzymes play increasingly important roles.

As proteins, enzymes do, however, have certain practical disadvantages outside their native organisms: they are often denatured inconveniently fast, by changes in pH, heat or solvent, and by surfactants and many other chemicals.

From Enzyme Models to Model Enzymes
By Anthony J. Kirby and Florian Hollfelder
© Anthony J. Kirby and Florian Hollfelder 2009
Published by the Royal Society of Chemistry, www.rsc.org

And the typical enzyme works best on just one specific substrate, in water, and at concentrations that are inconveniently low for serious synthesis. Hence the interest in developing synthetic "artificial enzymes": which can be more robust, can work in a solvent or solvents of choice; and could in principle be designed to catalyze a particular reaction, rather than a particular reaction of a specific substrate. Last but not least, a major advantage of synthetic systems is that they can in principle be designed to catalyze any reaction of interest, including non-natural reactions, for which no natural enzymes exist. Successful design in this context will inevitably be based on developing enzyme models.

Enzymes are far more than just highly evolved catalysts for specific reactions: they may also have to recognize and respond to molecules other than their specific substrate and product, as part of the control mechanisms of the cell. The evolution of artificial enzymes is at a much more primitive stage, with efficient catalysis the primary, and often the sole, objective. Systems are known that model various other functions, including potential control mechanisms. But to be useful as an industrial catalyst an artificial enzyme has no need of sophisticated built-in feedback control mechanisms or high substrate specificity: a stable molecule that is an efficient catalyst for a key target reaction in a chemical reactor will not be required to select its substrate from many hundreds in the same solution, as enzymes routinely must in the cell. So, a rational design strategy is indeed to consider simply those features of enzymes that are essential for catalytic efficiency.

In these first two chapters we discuss enzyme mechanisms in rather general terms, to identify and define these key features. We then go on to discuss the developing range of enzyme models: by which we mean systems designed to test basic ideas on enzyme mechanism by reproducing specific, key features of enzyme reactions; and attempts to develop them into enzyme-like catalysts. We reserve the term enzyme mimics for the most highly developed enzyme models, which combine successfully more than one of these key features, and catalyze reactions by mechanisms that are demonstrably enzyme-like, involving both binding and catalysis. An enzyme mimic that can do all this, *and* achieve turnover at a reasonable rate, deserves to be called an artificial enzyme.

1.1   Introduction to Enzyme Chemistry

Enzymes are proteins. The "central dogma" of biological chemistry underlines the pivotal role of enzyme catalysis (and highlights a fascinating problem in biochemical evolution!):

$$\text{DNA} \xrightarrow{\text{enzymes}} \text{RNA} \xrightarrow{\text{enzymes}} \text{Proteins} \xrightarrow{\text{enzymes}} \text{Chemistry}$$

Enzyme proteins are made up of one (sometimes more than one) polypeptide chain (Figure 1.1), each of which is folded into a flexible, more or less unique active conformation. The preferred 3-dimensional structure is determined by a complex array of physicochemical interactions between side-chains, main-chain

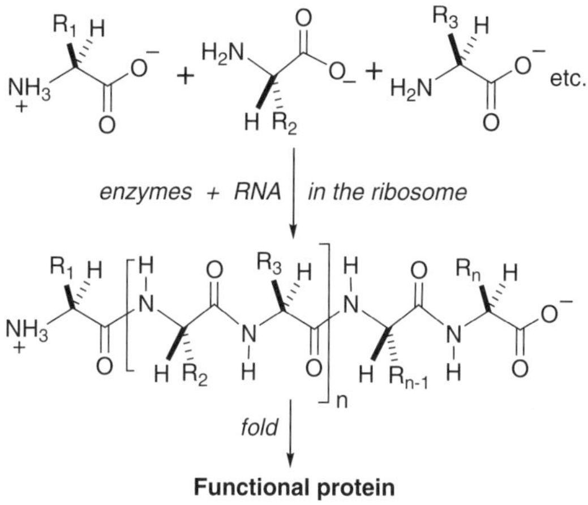

Figure 1.1 Protein structure and biosynthesis.

Proteins are composed of long chains of the 20 naturally occurring L-α-amino-acids, linked by peptide (amide) bonds. The primary sequence of the amino-acids of a natural protein is derived from the genetic code, and the tightly controlled synthesis carried out in a complex "machine" called the ribosome, one of the miracles of molecular biology (and acknowledged on the cover of this book).

amide groups and especially solvent water. Whole books and half a dozen current journals deal specifically with protein chemistry, and the basic ideas are described in many textbooks. So, only those properties of special relevance to catalysis will be introduced in this chapter, and discussed in the necessary detail later in the book. Specific suggestions for further reading are to be found at the end of each chapter.

The easiest way to a broader understanding of the 3-dimensional structures of proteins is to spend time on your computer "playing" with real structures. A good place to start is with the simple-to-use software available at http://www.umass.edu/microbio/rasmol/ or http://www.pymol.org/). While the structure of practically any enzyme of special interest is likely to be one of the many thousands accessible online from the Protein Data Base (http://www.rcsb.org/pdb/).

1.1.1 Why are Enzymes so Big?

Enzymes have evolved to operate under most of the various environments natural to living organisms. The most important of these are the cytoplasm – an aqueous solution containing hundreds of other proteins and small-molecule metabolites – and the surfaces of membranes of various sorts. So enzymes have to be "comfortable" in various operating environments, and "tunable" – to work at different, controlled levels of activity appropriate to the changing requirements of the system. They must also be capable of catalyzing specific reactions of specific substrates at rates (based on values of k_{cat} – see Section 1.2 – typically in the range 1–1000 s^{-1}) high enough to support the immediate demands of the interactive network of local control mechanisms. Substrates range in size from O_2 and CO_2 to macromolecules, and k_{cat} values between 1–1000 s^{-1} can represent accelerations of up to 10^{20} compared with rates of the corresponding uncatalyzed reactions at physiological pH in water.

So, an enzyme has to provide a highly sophisticated single-molecule "reaction vessel," to support such rapid reactions. Not surprisingly, filling and emptying this "reaction vessel" efficiently poses its own problems, because when reactions become very fast, simple diffusion processes can become rate limiting. So, important additional requirements for the active-site "reaction vessel" are rapid substrate binding and – no less important – rapid release of product. (This last is no trivial requirement, considering that the product is typically almost identical to the substrate, give or take the cleavage or formation of a single covalent bond.) Furthermore, most if not all of any water molecules present in the resting active site have to be removed for the duration of the reaction process. All this makes the simple picture of a (static) cavity complementary in structure to the substrate a highly unconvincing proposition. Fischer's imaginative lock-and-key principle[2] remains a valuable starting point, but an enzyme must provide specific binding complementarity not just to its substrate (or substrates), but to all intermediates and transition states on the reaction pathway.

Thus, the processes involved in catalysis are dynamic, and make complex demands on the enzyme protein. So, we should not be surprised that all this should require a large, sometimes a very large, molecule. An additional if less immediate factor is what might be called "protein bloat". We are all familiar with the way systems like software applications (or government legislation) that are continually being improved and extended, can grow rapidly in size and complexity over just a small number of years. Enzyme evolution is a great deal slower – but it has been going on for millions of years . . .

1.1.2 Functional Groups Available to Enzymes

What makes one protein different from another is the sequence of amino-acids, and thus the arrangement of the side-chains R_n (Figure 1.1) in the polypeptide chain. The chemical reactions involved in enzyme catalysis are implemented by the functional groups available on the amino-acid side-chains (backbone amide groups are not usually directly involved). Nine of the 20 naturally occurring amino-acids carry one of six different reactive groups (seven if phenol and alcohol OH groups count as different). These are listed in Table 1.1, which also gives the standard one- and three-letter abbreviations for the amino-acids, and approximate values for the pK_as of the side-chain groups. The pK_a value tells us how much of each ionic form of a group is present at any given pH, for the amino-acid free in solution. Though it gives only an indication of what might be the situation in the controlled environment of an active site, where in the absence of full solvation pK_as can be perturbed, sometimes by as much as 4–5 units. (Qualitatively, this perturbation generally favours neutral species, because ionized forms are stabilized more strongly by hydrogen-bonding solvation.)

Table 1.1 Amino-acids with side-chains ionizing in the pH region.

Amino acid	Side-chain functional group	Catalytic function of the group
Aspartic acid (Asp, D)	$pK_a \sim 4$	The COOH group can act as a general acid, but will be present near pH 7 only in a specially controlled environment.
Glutamic acid (Glu, E)	$pK_a \sim 5$	The carboxylate anion can act as a general base or as a nucleophile.
Histidine (His, H)	$pK_a \sim 7$	The imidazole group is about 50% protonated at pH 7, so is the most versatile side-chain group: making available general acid, general base or nucleophilic groups near pH 7.
Cysteine (Cys, C)	$pK_a \sim 9$	The thiolate group is a powerful nucleophile, and more than 1% may be present in a controlled active-site environment if its pK_a is lowered.
Lysine (Lys, K)	$pK_a \sim 10$	NH_2 is a good nucleophile, especially for C=O, and its effective pK_a can be lowered in a controlled active-site environment.
Tyrosine (Tyr, Y)	$pK_a \sim 10$	An oxyanion is a good nucleophile, especially for phosphorus.
Arginine (Arg, R)		The guanidinium group generally stays protonated. It makes strong, stabilizing double H-bonding interactions with CO_2^- and PO_2^- groups.
Serine (Ser, S) Threonine (Thr, T)	pK_a high (13-14)	The OH group is not ionized significantly in the pH region, but can act as a nucleophile as the proton is removed by a general base.

The functional groups listed in Table 1.1 are by no means an impressive selection of reagents, by the standards of the present-day organic or inorganic chemist. Nor can they be, because they must operate, and have long-term stability, in water: any strong base, acid or electrophile would immediately be neutralized by the solvent.

1.2   Principles of Catalysis by Enzymes

Most current thinking about enzyme catalysis is based on Pauling's original suggestion that enzymes work by binding and thus selectively stabilizing the transition states for their reactions. Starting from simple transition-state theory: consider the interaction between two reacting molecules **A** and **B**. The essential first step is for the two to come together. In the gas phase this involves a simple collision, but in solution molecules are separated by bulk solvent, and since each has its own solvation shell, making contact is a more complicated business. We can allow for this step by introducing into the reaction pathway the "encounter complex" $A \cdot B$: without defining it in detail. Once in contact the molecules can undergo multiple "collisions" within the encounter complex before either reacting or diffusing apart. Thus, simple geometrical requirements for reaction, *e.g.* the directionality of approach of the reacting centres on the separate molecules, are not generally critical. If the chemical reaction is faster than diffusional separation, as is typically the case for many proton-transfer reactions, the diffusion step is rate determining and the reaction is diffusion controlled. (For examples, see Section 4.1.)

$$A + B \underset{diffusion}{\overset{K_a}{\rightleftharpoons}} A \cdot B \rightleftharpoons AB^{\ddagger} \longrightarrow [products] \overset{diffusion}{\rightleftharpoons} products$$

A ball-park estimate of the equilibrium constant for the random association of small molecules in aqueous solution is $K_a \sim 0.07 \, M^{-1}$. Making the interaction between **A** and **B** "sticky" – *i.e.* by the various binding interactions of molecular recognition – will increase K_a, and thus, other things being equal, (see Figure 1.4) the overall rate of formation of products. Enzymes work by (i) binding their own particular substrate, usually very specifically from the hundreds available in solution in the cell, (ii) catalyzing a specific reaction of the bound molecule; and (iii) finally releasing the product into solution.

$$E + S \underset{(i)}{\overset{1/K_M}{\rightleftharpoons}} E \cdot S \underset{(ii)}{\overset{k_{cat}}{\rightleftharpoons}} E \cdot P \underset{(iii)}{\longrightarrow} E + products$$

This mode of catalysis – whether by an enzyme or a simpler catalyst – is characterized by "saturation" or "Michaelis–Menten" kinetics: whereby a limiting rate is reached when all catalyst molecules are "busy" – *i.e.* binding reactant, intermediates or product. The defining equation for Michaelis–Menten kinetics introduces two key parameters, k_{cat} and K_M:

$$\text{rate} = \frac{[E]_0[S]k_{cat}}{K_M + [S]}$$

In the simplest case, where the chemical step k_{cat} is clearly rate determining (*i.e.* **E.S** dissociates faster than it is converted to products), the rate becomes $k_{cat}[E]_0$ at $[S] >> K_M$, defining k_{cat} as the first-order rate constant for the conversion of **E.S** to products. And at $[S] << K_M$ (as $[S]$ becomes very small) the rate $\rightarrow k_{cat}/K_M [E]_0[S]$, defining k_{cat}/K_M as the second-order rate constant for the overall reaction. (For further detail see Section 2.3.)

We will use the parameters k_{cat} and K_M to characterize catalysis by enzyme models as well as enzymes: because they are familiar, and because they allow direct comparisons with enzyme reactions. We will also use the association constant K_a, in discussions of simple binding equilibria. (This corresponds to $1/K_M$ in the simple Michaelis–Menten mechanism.)

1.2.1   Dependence on pH

The basic mechanistic problem solved by enzyme catalysis is simply illustrated by the schematic pH–rate profiles shown in Figure 1.2. Ionic reactions between nucleophilic and electrophilic reactants, for example most hydrolysis reactions, are typically acid and/or base catalyzed; so slowest (as shown by the minimum

pH-rate profiles
Curve **I** represents the reaction *in vitro* of a typical, unreactive, compound: it shows only acid- and base-catalyzed reactions, and reaction is very slow at the minimum, near pH 7. For very reactive compounds an additional feature (curve **II**) is a pH-independent region, where the uncatalyzed reaction with water has become faster near neutrality than the acid and base-catalyzed reactions. *What happens in this region can provide clues to the mechanisms of enzyme catalysis.* Curve **III** is the pH–rate profile for a typical – much faster – enzyme-catalyzed reaction: enzymes are "designed" to operate near pH 7, and typically show pH optima in this region, with rates falling off at higher and lower pH

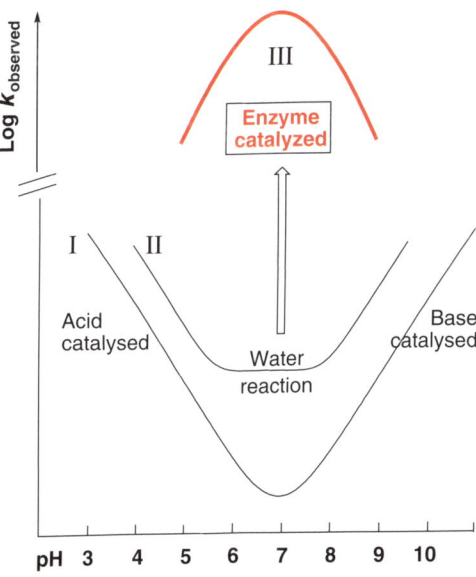

Figure 1.2   Representative pH–rate profiles for three typical reaction types in water.

in the pH–rate profile **I**) near pH 7. A sufficiently highly reactive system may react without the need for acid/base catalysis, and show also an uncatalyzed "water reaction" (curve **II**): which is pH independent over a certain range but still slowest near neutrality. Enzyme reactions, by contrast, are fastest under physiological conditions, often showing bell-shaped pH–rate profiles (curve **III**), with a rate *maximum* near pH 7.

1.3   General Acid–Base Catalysis

The reactions at high and low pH (curves **I** and **II** of Figure 1.2) are not directly relevant to catalysis by enzymes, which operate of necessity near pH 7, and use catalytic groups that are only weak acids, bases and nucleophiles. The great majority of enzyme-catalyzed reactions are ionic, involving heterolytic bond making and breaking, and thus the creation or neutralization of charge. Under conditions of constant pH this commonly requires the transfer of protons. General acid–base catalysis provides mechanisms for bringing about the necessary proton transfers without involving hydrogen or hydroxide ions, which are present in water at concentrations of the order of 10^{-7} M under physiological conditions. At pHs near neutrality relatively weak acids and bases can compete effectively with lyonium or lyate species because they can be present in much higher concentrations.

1.3.1   Experimental Evidence

Acid–base catalysis is termed *specific* if the rate of the reaction concerned depends only on the acidity (pH, *etc.*) of the medium (as in curve **I** of Figure 1.2). This is the case if the reaction involves as an intermediate a small amount of the conjugate acid or base of the reactant preformed in a rapid equilibrium process – normal behaviour if the reactant is weakly basic or acidic.

General acid–base catalysis is defined experimentally by the appearance in the rate law of acids and/or bases other than lyonium or lyate ions. Thus, the hydrolysis of enol ethers (Scheme 1.1) is general acid catalyzed: the rate depends on pH, but near neutrality depends also on the concentration of the buffer (HA + A⁻) used to maintain the pH. Measurements at different buffer ratios show that the catalytic species is the conjugate acid HA. Any "general acid" can be a catalyst: the most reactive will always be the hydronium

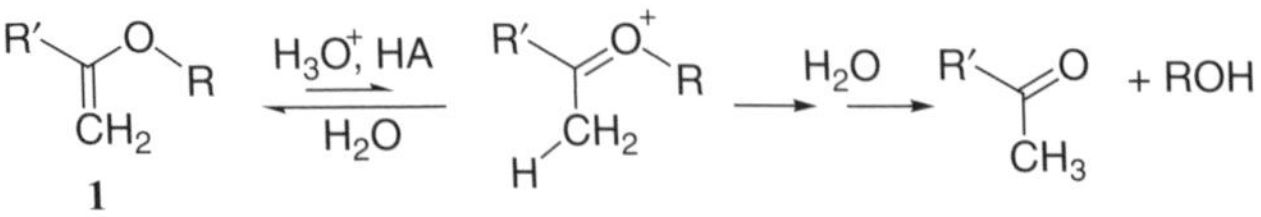

Scheme 1.1   The hydrolysis of enol ethers is a typical general acid-catalyzed reaction. The rate law is found to be $-d[\mathbf{1}]/dt = k_H[\mathbf{1}]\,[H_3O^+] + k_{HA}[\mathbf{1}]\,[HA]$.

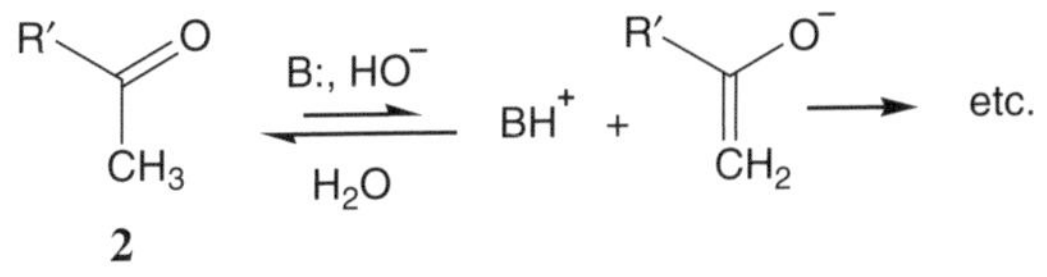

Scheme 1.2 The enolization of ketones is general base catalyzed. The rate law is found to be $-d[2]/dt = k_{OH}[2]\,[HO^-] + k_B\,[2]\,[B]$.

ion H_3O^+, which is (by definition) the strongest acid available in aqueous solution.

If the measurements at different buffer ratios show that the catalytic species is the conjugate base A^- the reaction is kinetically general base catalyzed (in which case HA and A^- will usually subsequently be referred to as BH^+ and B). Thus, the enolization of ketones is general base catalyzed (Scheme 1.2).

The rate constants k_{HA} and k_B depend on the strength of the acid or base, and for a given reaction are correlated by the Brönsted equation. This is written differently for general acid- and general base-catalyzed reactions, respectively:

$$\log k_{HA} = \alpha \log k_{HA} + \text{constant} = -\alpha\, pK_{HA} + \text{constant}$$
$$\log k_B = \beta \log k_B + \text{constant} = -\beta\, pK_B + \text{constant}$$

The pK_as used are those of the conjugate acids, HA and BH^+.

1.3.2 Mechanisms

To illustrate the fundamental mechanisms involved we use as examples the hydrolysis reactions of carboxylic ester and amide groups: though the treatment applies to any reaction involving the attack of a nucleophile bearing a potentially acidic proton on any polar single bond, or its addition to a multiple bond.

If the substrate is reactive (electrophilic) enough, activation by protonation is not necessary and the attack of water on the neutral molecule is rate determining near pH 7 (see curve **II** of Figure 1.2). Formally this generates both a positive and a negative charge, and as the reaction proceeds both will be "delocalized" into the surrounding solvent via the network of hydrogen bonds. Scheme 1.3 shows the first step in the hydrolysis of a carboxylic acid derivative, RCOX, where X^- is a potential leaving group.

The *nucleophile* in a hydrolysis reaction is of course a water molecule (labeled **nuc** in Scheme 1.3). As the new C–O bond forms, positive charge develops on the nucleophilic oxygen, and the attached OH protons become more and more acidic, until they can be transferred to solvating water, acting formally as a *general base* (**gb**). The negative charge developing on the carbonyl oxygen may similarly be transferred via a hydrogen bond to another water molecule, acting this time as a *general acid* (**ga**). This mechanism is always available, though it

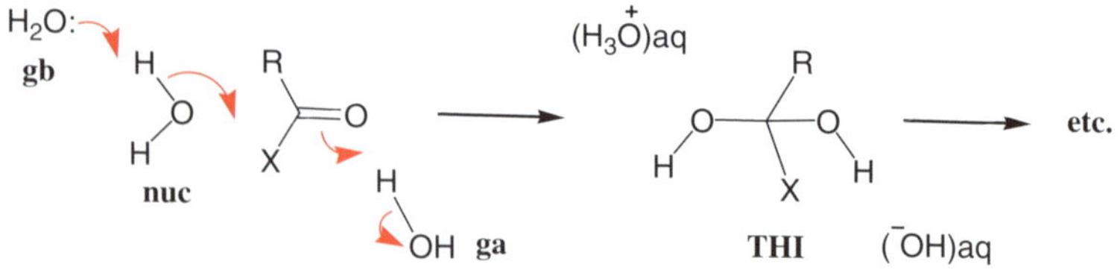

Scheme 1.3 Mechanism of spontaneous hydrolysis of a carboxylic acid derivative. The second step ("*etc.*") of the reaction is discussed in Scheme 1.6. Note that only the principal reacting solvent molecules are shown. Every lone pair on every O or X atom will also be solvated by hydrogen bonding to a water OH, and every potentially acidic proton solvated by hydrogen bonding to a water oxygen.

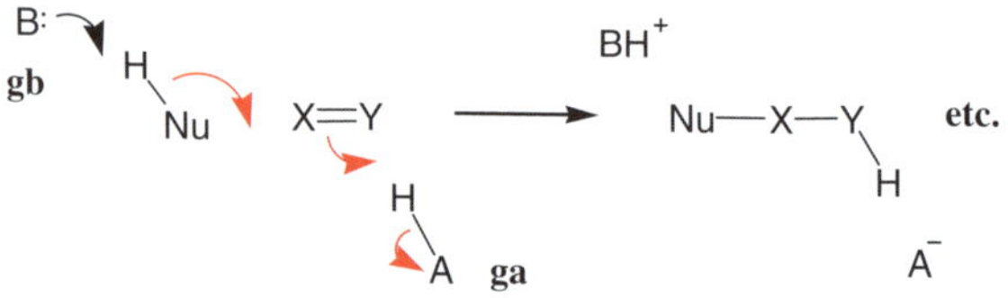

Scheme 1.4 Generalized reaction scheme for ionic bond breaking. Given the right combination of nucleophile, general base and general acid this mechanism can provide major reductions in the enthalpy of activation for the cleavage of the X–Y bond.

does not always lead to reaction at an observable rate. It accounts, for example, for the rapid hydrolysis of ethyl trifluoroacetate, CF_3COOEt, which shows a pH-independent region near neutrality (see curve **II** of Figure 1.2): but the hydrolysis of ethyl acetate, CH_3COOEt is extremely slow (half-life ~80 years at 25 °C at the pH minimum (see curve **I** of Figure 1.2), while the neutral reaction of $CH_3CONHEt$ is too slow to observe.

The mechanism outlined in Scheme 1.3 is quite general, and can be applied to the ionic cleavage of any σ- or π-bond X–Y or X=Y (Scheme 1.4).

However, H_2O is a weak nucleophile, acid and base and it is easy to find stronger examples of all three types of reagent – general base, nucleophile and general acid – among the functional groups available on amino-acid side-chains (Table 1.1). So we are not surprised to learn that the hydrolysis of CF_3COOEt, for example, is general base catalyzed by carboxylate anions (Scheme 1.5).

The direct reaction (Scheme 1.6) does have the entropic advantage of not involving a third molecule, and a carboxylate anion *can* – given the right conditions – displace better leaving groups than ethoxide (and poorer leaving groups than itself). As long as the intermediate (in this case an anhydride) is more reactive than the starting material the result is catalysis of hydrolysis.

This mechanism competes equally with general base catalysis (*i.e.* accounts for some 50% of the reaction) for esters with a leaving group of pK_a of ~7, and becomes dominant for derivatives with better leaving groups. The two

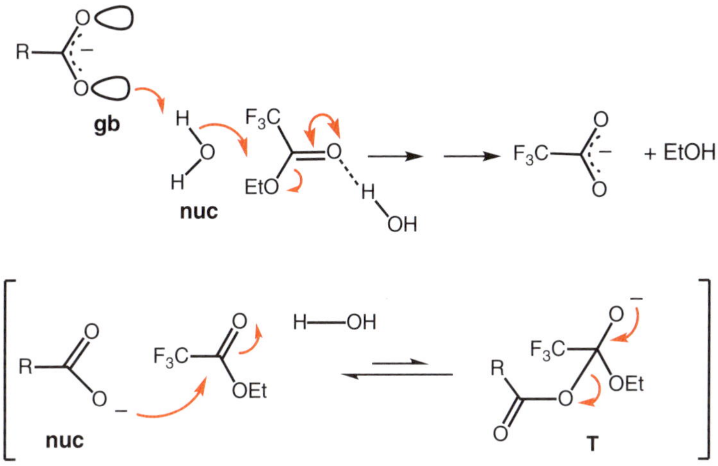

Scheme 1.5 Mechanism of spontaneous hydrolysis of ethyl trifluoroacetate. Nucleophilic catalysis [in brackets] is disfavoured because the tetrahedral addition intermediate **T** reverts too rapidly to reactants. See the text.

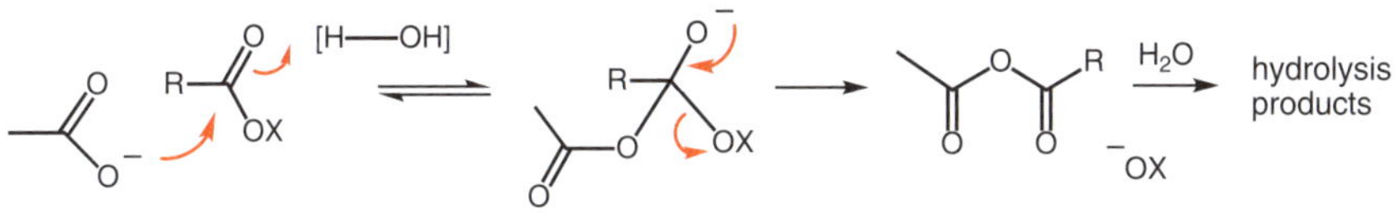

Scheme 1.6 Nucleophilic catalysis in the spontaneous hydrolysis of an acyl derivative with a good leaving group. See the text.

mechanisms are kinetically identical (the third participant in the general base-catalysis mechanism being a kinetically invisible solvent molecule). But they can be distinguished in a number of ways, most simply by identifying the intermediate, directly or by trapping, and showing that it is "kinetically competent" (reactive enough to account for the observed rate of reaction). This provides convincing positive evidence for a nucleophilic mechanism: as do irregularities in the fit of data to the Brönsted equation (Section 1.3.1) that can be ascribed to steric effects (not to be expected for general base catalysis, where the attacking nucleophile is always H_2O). The simplest positive evidence for general acid or general base catalysis is a significant solvent deuterium isotope effect, k_{H2O}/k_{D2O}, of ≥ 2 consistent with a proton transfer in the rate-determining step. Though, as always, an accumulation of several independent pieces of evidence consistent with the suggested mechanism is preferred to support a convincing conclusion.[3]

If the leaving group is poor it can be made viable by protonation: complete protonation to form the conjugate acid if the group (*e.g.* amino) is sufficiently basic, but involving partial proton transfer in the case of a weakly basic group like OR or OH. This mechanism (*general acid catalysis*) is involved, though not

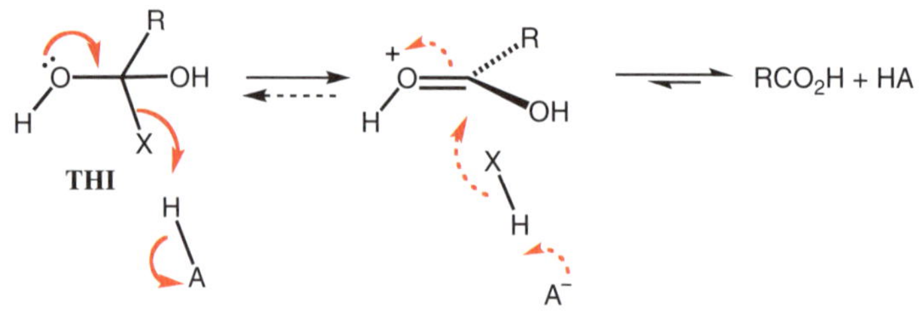

Scheme 1.7   General acid catalysis of the second step in the hydrolysis of carboxylic acid derivatives (Scheme 1.3). Note that general acid catalysis is the microscopic reverse of general base-catalyzed addition (dashed arrows).

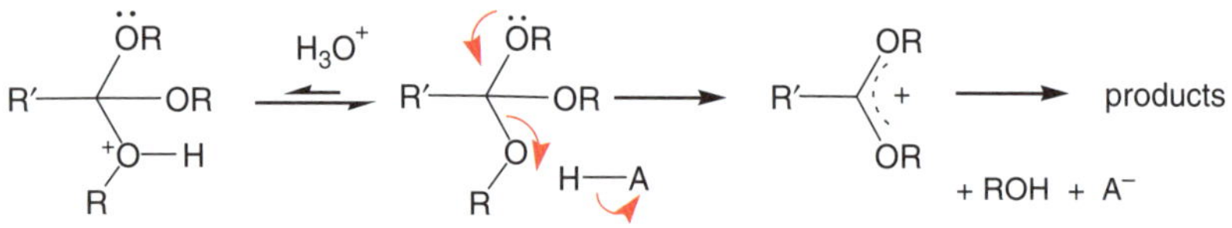

Scheme 1.8   Classical general acid catalysis involves proton transfer becoming concerted with the cleavage of a covalent bond between heavy atoms.

easily observed, in the breakdown of the tetrahedral addition intermediates involved in the acyl transfer reactions of esters and amides (Scheme 1.7).

General acid catalysis is most conveniently observed in the reactions of stable tetrahedral species with two or more O, S or N atoms attached to a central atom (though only in special cases for simple acetals, as discussed in Section 4.3.1). That general acid catalysis is observed in the hydrolysis of *ortho* esters **1.3** (Scheme 1.8) is explained as follows: the C–OR oxygens are very weakly basic and the dioxocarbocation intermediate **1.4** particularly stable, a result of π-donation from the two remaining oxygens. Under these conditions C–OR cleavage can occur before proton transfer is complete, thus becoming concerted with proton transfer (Scheme 1.8).

Given a viable electrofuge (*e.g.* the cation **1.4** in Scheme 1.8) proton transfer can be expected to become concerted with the making or breaking of covalent bonds if the reaction is sufficiently thermodynamically favourable. The practical requirements are summarized in Jencks' so-called "libido rule".[4,5]

"Concerted general acid–base catalysis of complex reactions in aqueous solution can occur only (a) at sites that undergo a large change in pK$_a$ in the course of the reaction, and (b) when this change in pK converts an unfavorable to a favorable proton transfer with respect to the catalyst; i.e. the pK of the catalyst is intermediate between the initial and final pK$_a$ values of the substrate site."

Thus, general acids with pK_as of 7 ± 4, of potential interest in biological systems, are well qualified to assist in the cleavage of bonds to oxygen leaving groups: since the pK_as of ester oxygens are typically negative, while those of their alcohol cleavage products are usually > 14.

1.3.3 Kinetic Equivalence

These simple examples might suggest that the observation of general acid or general base kinetics is *prima facie* evidence for the corresponding general acid or general base catalysis mechanisms. This is not the case, for the usual reasons of (a) kinetic equivalence (the proton is a uniquely mobile species), and (b) the possible involvement of the solvent (*e.g.* the water molecules in Scheme 1.3) in the mechanism. For example, ketone enolization below pH 7 is general acid catalyzed (*i.e.* is first order in the concentration of general acids HA), but is explained not by the general acid-catalysis mechanism (**gac**, Scheme 1.9) but by the kinetically equivalent specific-acid general-base catalysis (**gbc** route in Scheme 1.9): which requires only bimolecular encounters for the rate-determining step of the reaction).

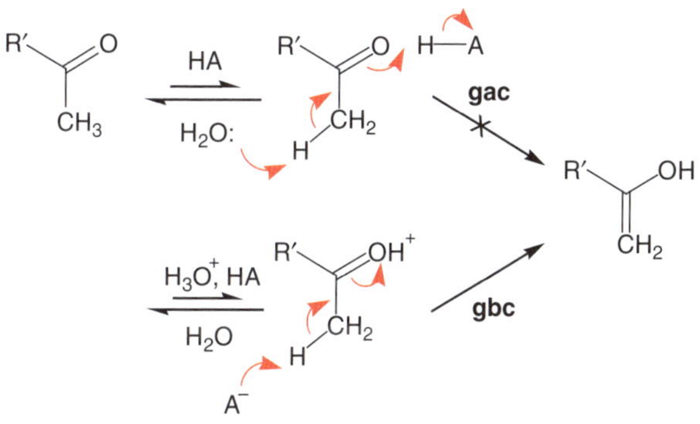

Scheme 1.9 The general-acid catalyzed enolization of ketones involves the kinetically equivalent general-base catalyzed removal of the CH proton from the substrate conjugate acid.

In practice, the entropic cost of bringing four molecules together is far too great for the complete mechanism of Scheme 1.4 to be observable in free solution: even a three-molecule collision is unfavourable enough that the mechanism of Scheme 1.5 is observed only when the enthalpy of activation for the reaction is intrinsically low: and even nucleophilic catalysis, with the more modest entropic requirements of a bimolecular encounter, is observed in water only with activated substrates. Since many enzymes catalyze reactions of highly unreactive substrates, it follows that they have evolved ways of reducing the free energies of activation. The simplest of these involves bringing the reacting groups into close proximity in the substrate-binding step, in such a way that the entropy debt is paid off "in advance", in the binding step. So the simplest models of fundamental enzyme processes are intramolecular reactions, in which reacting groups are brought into close proximity by being positioned close together on the same molecule.

1.4   Intramolecularity

Though the entropic cost of bringing four molecules together is too great for the complete mechanism outlined in Scheme 1.4 to be observable in free solution, this is not a problem for a substrate bound in close proximity to the catalytic groups of an enzyme-active site. A classic example is the mechanism (Scheme 1.10, **A**) used by many serine proteinases (enzymes that hydrolyze the intrinsically unreactive (at pH 7) peptide (*i.e.* amide) bonds of proteins, using the OH group of an active-site serine as a nucleophile). The earliest work on these mechanisms involved studying the component parts in simple bimolecular systems, as described above: and then – more appropriately, and also more conveniently – in systems where the reacting functional groups are held in close proximity on the same molecule. Thus, the central nucleophilic reaction can be modeled by the cyclization (lactonization) of an appropriate hydroxyamide (Scheme 1.10, **B**), which can be much faster than the neutral hydrolysis of the amide group under the same conditions, allowing the use of unactivated substrate groups.

Intramolecular reactions, and particularly intramolecular nucleophilic reactions (*i.e.* cyclizations) are the only simple reactions that begin to rival their enzyme-catalyzed counterparts in rate under physiological conditions. Making the reaction of interest part of a thermodynamically favourable cyclization can produce quite simple systems in which the extraordinarily stable groups of structural biology (amides, glycosides and phosphate esters typically have half-lives of tens to millions of years under physiological conditions near pH 7) can be cleaved in fractions of a second. An early example was **1.5** (Scheme 1.11), where an ordinary aliphatic amide is forced into close proximity with a COOH group in such a way that the half-life of **1.5**, where R is a simple alkyl group, is less than a second.[6]

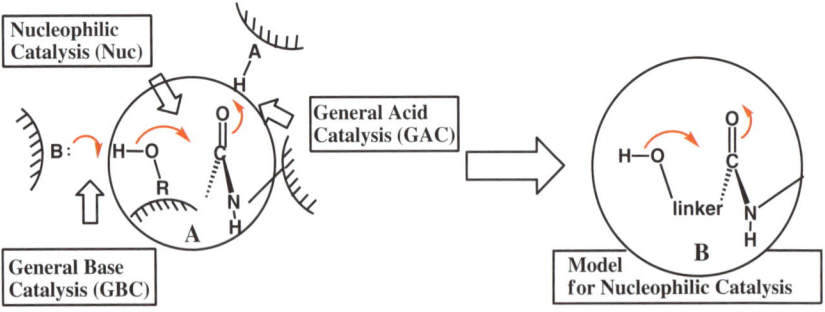

Scheme 1.10   Lactonization model for nucleophilic catalysis by serine proteases.

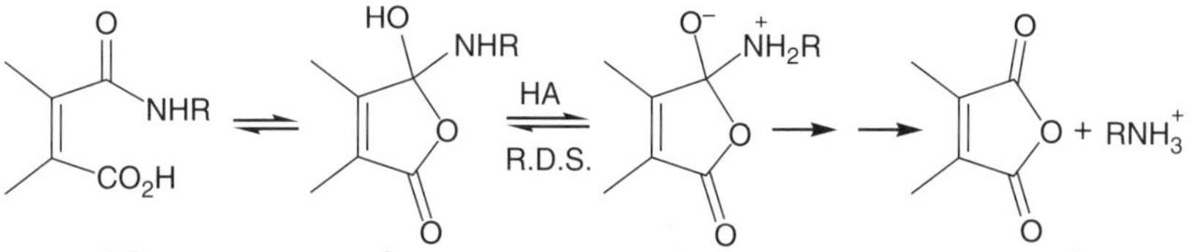

Scheme 1.11 The extraordinarily rapid cyclization of simple, unactivated amides **1.5** of dimethylmaleic acid involves nucleophilic "catalysis" (see the text) by the neighbouring carboxyl group *and* catalysis by external general acids.[6]

Given a suitable cyclization reaction going at a useful rate, it is possible to ask relevant questions about the basic reaction between the two interacting functional groups, specifically chosen from those involved in enzyme reactions as substrate or catalytic group. Varying the length and structure of the linker (Scheme 1.10, **B**) can provide information about preferred geometries of approach. And because amide cleavage can be so fast in this efficient intramolecular reaction, it is possible to observe also external general acid or general base catalysis of the cyclization. This in turn provides the information needed for the rational design of a next generation of models in which the general acid or base is also built in to the reacting system, so that the general acid/base catalysis becomes itself intramolecular. At each stage we interrogate the system about its intrinsic preferences, and then try to improve the efficiency of catalysis in light of the answers.

This is not a trivial exercise, and there can be surprises (positive as well as negative!). For example, the reaction of **1.5** (Scheme 1.11) is general acid catalyzed: the neutral tetrahedral intermediate T^0 breaks down rapidly to the reactant acid amide, and goes on to product only after a double proton transfer has converted it to the zwitterion $T^{\pm}$, with a neutral amine as the leaving group. The double proton transfer is neatly catalyzed by the CO_2H group of a general acid, which can transfer its proton to the N centre of T^0 while simultaneously accepting one from the OH group on its other oxygen (Scheme 1.12).

The observation of general acid catalysis suggests that reaction would be faster if the general acid was a built-in CO_2H group. This would have to be attached via a linker long enough to support the geometry required for the double proton transfer, but not so long that catalysis would be inefficient. A study of a series of amides derived from amino-acids showed that the most reactive were derivatives of β-amino-acids, *e.g.* **1.5a** (Scheme 1.12) from β-alanine: but the catalytic group involved turned out to be not CO_2H but the carboxylate anion, CO_2^{-}.[7] The CO_2H-catalyzed reaction might still be most effective in an enzyme-active site, where the geometry can be optimal (as for T^0 in Scheme 1.12): but this geometry cannot be achieved in the intramolecular system **1.6**. The mechanism of this reaction is discussed in more detail in Section 4.1.3.1.

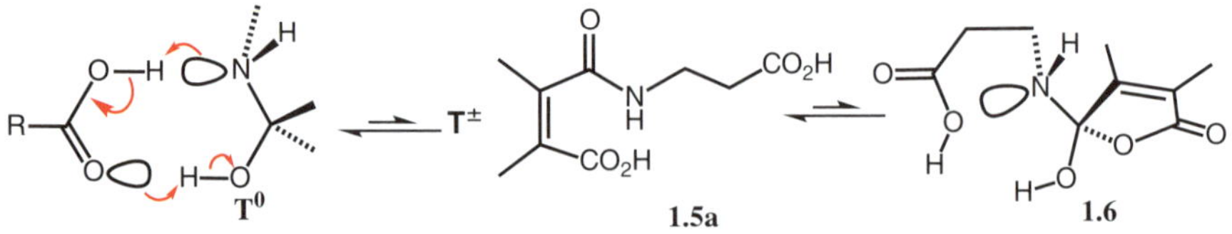

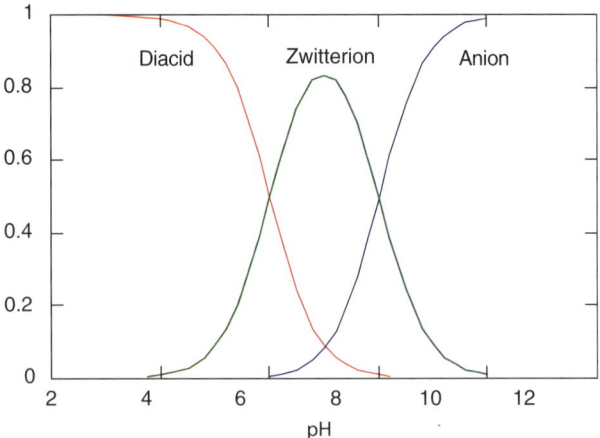

Scheme 1.12   The synchronous mechanism for the general acid-catalyzed conversion
of T^0 to $T^\pm$ is not geometrically possible (**1.6**) for the reaction of the
β-alanine derivative **1.5a**. (**1.6** shows the OH of the CO_2H group in its
preferred z conformation).

This result contains several important lessons.

1. Experimentally the pH–rate profile shows a rate maximum in the region
   between the pK_as of the two carboxyl groups, because one is involved as
   CO_2H and the other as carboxylate. This is the simple, general explanation for the pH–rate optima observed for many enzyme reactions (see
   Figure 1.2). If the mechanism depends on two (or more) functional
   groups, one in the basic and one in the acidic form, the concentration of
   the reactive form (Figure 1.3), and thus the rate of the catalytic reaction, is
   at a maximum at a pH half-way between the two pK_as.

Figure 1.3   Distribution as a function of pH of the three ionic forms of a dibasic acid
(*e.g.* an amino-acid) with pK_as of 6 and 8. (The "diacid" is a cation in the
case of an amino-acid.)

2. Intramolecular reactions are useful if limited models for enzyme reactions.
   The neighbouring groups can be held in close proximity, but the linker
   may place constraints on their geometry of approach.
3. In reactions involving the making or breaking of bonds between heavy
   (nonhydrogen) atoms the intramolecular reaction is a one-off event –
   there is no possibility of turnover. The term intramolecular catalysis can

be used correctly, because the catalytic group may be regenerated: but the neighbouring substrate group is not, so no turnover is possible.

1.4.1 Efficiency of Intramolecular Catalysis

Despite these limitations the study of intramolecular reactions, apart from being an integral part of the broader study of organic reactivity, has informed our thinking on enzyme catalysis by defining, in detail in simple systems, the basic mechanisms involved in most enzyme reactions. (Enzyme evolution has involved a great deal of attention to mechanistic detail!) For example, general acid catalysis of acetal hydrolysis was first observed as an intramolecular reaction, after being suggested as the likely role for one of the two active-site groups of lysozyme: and is now a well-understood process. Furthermore, since intramolecular reactions offer the best prospect of observing reactions of the least reactive naturally occurring functional groups in simple systems, the efficiencies as well as the mechanisms of the reactions are of special interest.

The efficiency of intramolecular catalysis is defined most simply in terms of the effective molarity (EM) of the catalytic group.[8] The EM is defined as the ratio of the first-order rate constant for the intramolecular reaction and the second-order rate constant for the corresponding intermolecular reaction (Scheme 1.13) proceeding by the same mechanism under the same conditions. It has the dimensions of molarity and is simply the concentration of the catalytic group that would be required in the equivalent intermolecular process to match the rate of the intramolecular reaction. (This concentration is purely nominal in most cases, since EMs are often greater than 10 M.)

Two generalizations are of major importance:

1. Intramolecular catalysis can be highly efficient in cyclization reactions, which involve the formation of bonds between heavy (nonhydrogen) atoms. EMs of 10^{8-9} can be observed even in flexible systems,[8] and higher figures are possible in rigid systems where ground-state strain is relieved as part of the reaction: the EM of the carboxyl group in the cyclization of **1.5** (Scheme 1.11), for example, is over 10^{13} M.[8] Enzymes cannot exert the sorts of very strong nonbonding interactions that can be built by synthesis into structures like **1.5**, because protein structures are relatively flexible;

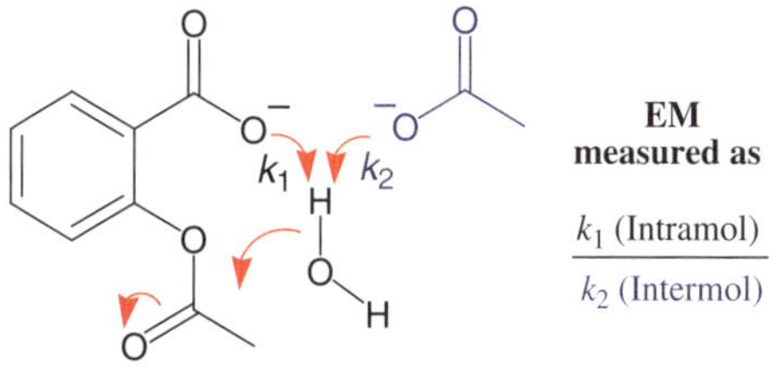

Scheme 1.13 Rate constants defined for intra- *vs.* intermolecular catalyzed reactions going by the same (general base catalysis) mechanism.

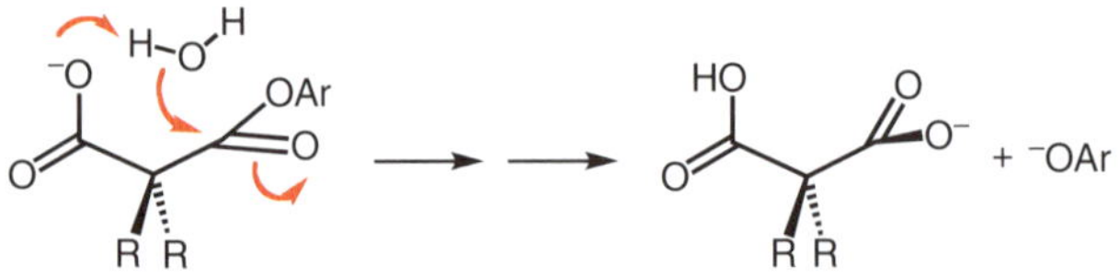

Scheme 1.14   The neighbouring carboxylate group of malonate monoesters **5** with good leaving groups acts as a general base, rather than a nucleophile: because it is *too* close.

but studies of such rigid systems can give us access to valuable mechanistic information about the reactions of relevant functional groups going at rates comparable to those of the corresponding enzyme-catalyzed processes. There has been much discussion of the origins of the rate enhancements in these systems, in terms of at least a dozen theories,[9] from orbital steering to the spatiotemporal hypothesis.[10] Some of these provide useful guidance for systems design, but the fundamental requirement for a high (kinetic) EM is a thermodynamically favourable cyclization with no stereoelectronic impediment to reaction. The entropy term is always more favourable for an intramolecular process (the basis of the description of enzymes as "entropy traps"): but this factor is common to all simple intramolecular processes, and does not vary a great deal for many favourable reactions; whereas the enthalpy term can depend strongly on the way the reacting groups are brought together.[11]

2. Intramolecular reactions involving rate-determining intramolecular proton transfer – *i.e.* general acid and general base-catalyzed reactions – are typically far less efficient, with EMs generally below about 10 M.[8] For example, the EM of the carboxylate group in a series of half-esters **1.7** of dialkylmalonic acids, hydrolyzed by the general base-catalysis mechanism shown in Scheme 1.14 (and observed "by default" – because cyclization, to form a 4-membered ring, is thermodynamically strongly disfavoured) ranges from less than 1 up to no more than 60 M.[12]

These low numbers would appear to present something of a problem, since proton transfer is the commonest reaction catalyzed by enzymes. There is no doubt that enzymes catalyze such reactions efficiently, so the reasonable conclusion is that the models, not the enzymes, are deficient. More recent work suggesting a solution is discussed in Section 4.3.

1.5 Energetics

The efficiency of enzyme catalysis depends on a large number of factors, no single one of which is uniquely responsible for the accelerations observed in any particular case. The most important of these are discussed in the following

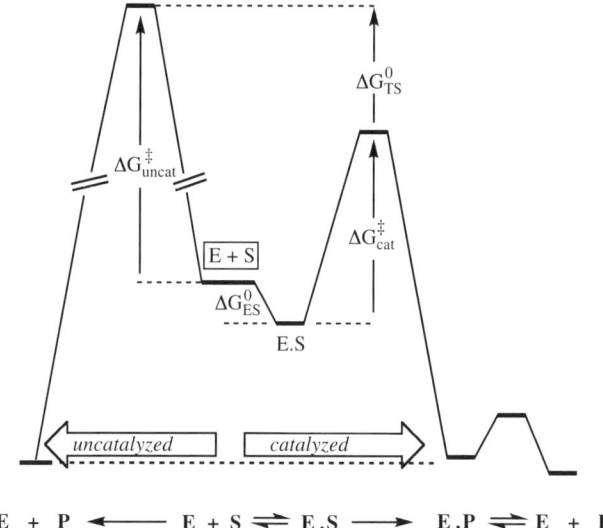

Figure 1.4 Energy-profile diagrams compared for a reaction catalyzed by an enzyme or enzyme model **E** and for the same reaction under the same conditions with no catalysis.

pages, together with attempts to identify and to quantify them in model systems. But we must always bear in mind the overall thermodynamics of the process: which can be summarized in the energy-profile diagram of Figure 1.4.

Figure 1.4 compares simple energy-profile diagrams for a general reaction of a substrate **S** catalyzed by an enzyme **E** (left to right arrow) with that (right to left arrow) for the same, uncatalyzed, reaction, involving the same functional groups, in water. The picture is oversimplified, greatly so for many enzyme-catalyzed processes, to consider only a single (chemical) rate-determining step in each direction: but has the advantages (i) that it is based on the Michaelis–Menten equation, and thus on standard, readily measurable parameters, k_{cat} and K_M: and (ii) that it can be applied equally well to measure the efficiency of an enzyme model or mimic. The key parameter derived from the direct comparison of catalyzed and uncatalyzed reactions is the binding energy of the transition state of the catalyzed reaction, defined in terms of the rate ratio $(k_{cat}/K_M)/k_{uncat}$ (for more detailed discussion see Chapter 2).

$$\Delta G_{TS}^0 = -\ln[(k_{cat}/K_M)/k_{uncat}] = \Delta G_{uncat}^{\ddagger} + \Delta G_{ES}^0 - \Delta G_{cat}^{\ddagger}$$

In the simple case this is a measure of the stabilization of the rate-determining transition state by the catalyst compared with stabilization by solvent water. (This assumes that substrate binding is thermodynamically neutral under the conditions. Stronger substrate binding stabilizes the effective ground state for the reaction (Figure 1.4), and hence reduces catalytic efficiency.)

A complete description of the mechanism of a reaction must include a description of the transition state, but – as a species with no significant lifetime – its structure is rarely well defined, especially for reactions taking place in the active site of a complex protein enzyme. However, we can draw some general conclusions.

An enzyme has evolved to bind – and thus stabilize specifically – the transition state for its particular reaction. For "binding the transition state" to lead to catalysis the interactions of molecular recognition concerned must be strongest in the transition state. But what exactly do we mean by "binding the transition state" – this dynamic entity with no significant lifetime?

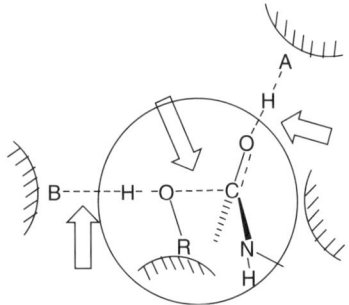

Scheme 1.15   Binding the transition state. The arrows indicate three of the five partial covalent bonds that contribute directly to transition-state binding.

Take the transition state for the serine protease reaction discussed previously (Scheme 1.10, **A**). The transition state is conventionally drawn as shown in Scheme 1.15. Note that there is no well-defined boundary between enzyme and "transition state," because enzyme and substrate are linked by partial covalent bonds. The details will depend on the particular reaction, but this partial covalent bonding gives rise to particularly strong, stabilizing, dynamic interactions specific to the short-lived transition-state structure, which can contribute significantly to the "binding of the transition state" that makes catalysis efficient. This contribution is particularly strong for partial covalent bonding between heavy (nonhydrogen) atom centres. Thus, those enzymes that have to be the most efficient, because they catalyze reactions of substrates of very low intrinsic reactivity, commonly do so by mechanisms in which the enzyme acts as a nucleophile (as in Scheme 1.10, **A**): using one of the groups available on amino-acid side-chains (Table 1.1), perfectly positioned in the active site for the purpose.[13,14]

1.6   Binding and Recognition

Enzymes work by a sophisticated, integrated system of binding and catalysis. The essential first step of any enzyme reaction (Figure 1.4) is substrate binding, by which the specific substrate is selected from many others available after diffusing into contact with the protein, and then into the active-site "reaction vessel", where it comes into contact with the catalytic apparatus. The rest of the reaction coordinate for the catalyzed reaction is accessible only to the bound substrate. Binding and catalysis are thus complementary, and if we accept that an enzyme has evolved to bind specifically the transition state for its particular reaction both can be regarded as binding phenomena. Discrimination between closely similar substrates – enantiomers are obvious examples[i] – depends on the selective binding of both substrate and transition state. Of these, transition-state binding is more important because it affects the rate-determining step directly.

Comparisons with the results of much recent work on molecular recognition show that substrate binding falls quantitatively into the class of typical host–guest interactions involving the binding of small molecules to synthetic hosts, which are characterized by association constants in water in the range 10 to $10^4 \, M^{-1}$.[16] The direct noncovalent intermolecular interactions involved – hydrogen bonding, electrostatic and van der Waals attractions – are fairly well understood, with multiple interactions the main source of stronger binding. They are conveniently studied in nonpolar solvents, because hydrogen-bonding and charge–charge-based associations are generally at their weakest in strongly polar water. Water is a good hydrogen-bonding solvent for individual charged or strongly polar molecules, and the major role in stabilizing host–guest complexes involving neutral systems in aqueous solution is played by the more subtle hydrophobic interactions.

This "hydrophobic effect" refers to the tendency for nonpolar organic compounds – or groups – to avoid water in favour of a nonpolar environment (oil and water don't mix!). It is responsible for the formation from nonpolar compounds of aggregates of all sizes, from short-lived "encounter complexes" and well-defined host–guest complexes – thus stabilized by "hydrophobic binding" – to (relatively long-lived but less well-defined) micelles and vesicles. It is also one of the most significant factors controlling protein folding: nonpolar (hydrophobic) groups on amino-acid side-chains are typically buried in the interior of a globular protein: the way they pack together, together with the preference for charged groups to lie on the surface in contact with solvent water, controls the folding of the peptide chain. Similarly, small molecules with nonpolar groups aggregate in water to form micelles or vesicles, in which hydrophobic groups pack together and charged groups lie on the surface in contact with the solvent. Last, but not least, because charged species are disfavoured in an apolar environment, the pK_a values of ionizable catalytic groups (at least one form is always charged) may be significantly perturbed in a hydrophobic active site (see Section 3.2).

---

[i] Enzymes are well able to discriminate with remarkably high selectivity between closely similar substrates, from large molecules differing by just one carbon atom to enantiomers. In practice, not all enzymes show perfect selectivity, including enantioselectivity.[15]

1.6.1   Hydrophobic Binding[17,18]

When water molecules come into unavoidable contact with a nonpolar (hydrophobic) surface the solvent's unique, dynamic three-dimensional network of hydrogen bonds is disrupted. These hydrogen-bonding interactions are dominant as the system adjusts to the disruption in two important ways. (i) The local network of solvent hydrogen bonds at the interface becomes close to two-dimensional, with O–H bonds making tangential contact with the nonpolar surface. This enforced ordering process involves strongly unfavourable (negative) entropy effects.[ii] (ii) The extent of the interaction is minimized by bringing nonpolar molecules, or parts of molecules, together. This *hydrophobic binding* is reinforced by (much weaker) van der Waals attractions, and depends for small molecules on their nonpolar surface area, though not with great precision. In terms of equilibrium constants at room temperature, from 70 Å^2 to as little as 10 Å^2 of surface area buried can contribute a factor of 10 to the equilibrium constant for association.[16]

A useful practical measure of hydrophobicity is log P, based on the partition coefficient P for a given molecule between water and 1-octanol.[19] Values are more or less additive, so can be broken down into group contributions and used in linear free-energy relationships for simple binding processes; or for structure–activity relationships involving binding equilibria, from enzyme reactions to anaesthesia. Such relationships do not need to be very precise to allow interesting and sometimes useful predictions from simple extrapolation.

We know that enzymes, to be efficient catalysts, must bind transition states much more strongly than their substrates (see Section 1.5): presumably by a combination of mechanisms including those discussed above; plus, perhaps, additional special factors. Uncertainties arise because as we have seen, binding a transition state is not the same as binding a stable molecule; and because we need to estimate how strongly a protein *could* bind its substrate, using only a combination of the familiar intermolecular interactions plus hydrophobic binding. We can get a good idea by considering the properties of antibodies, proteins produced by the immune system and evolved specifically to bind stable antigens of various sorts.

Antibodies bind small organic molecule antigens with association constants of $10^{7 \pm 2} \text{ M}^{-1}$, corresponding to a binding free energy of $-10 \pm 3 \text{ kcal mol}^{-1}$ – about

---

[ii] The hydrophobic effect near room temperature is now generally considered to result from entropy effects, resulting from the increased ordering of water molecules in contact with a nonpolar solute, as discussed. However, at high temperatures the unfavorable enthalpy of interaction of water with a nonpolar solute rather than with other water molecules becomes the dominant effect.

double that observed for cyclodextrins (see Section 4.1.1.3.1) and typical synthetic hosts.[16] It is reasonable to suppose that this approaches the limit for binding small molecules under physiological conditions. As might be expected, bigger natural ligands like proteins and oligosaccharides offering multiple binding interactions can be bound up to two orders of magnitude more strongly *in vivo* ($K_a = 10^{9 \pm 2} \, M^{-1}$).

In general, the strongest protein–small molecule binding interactions, matching these values ($K_a = 10^{9 \pm 4} \, M^{-1}$), are found for enzymes binding transition-state mimics (or analogues): stable compounds that bind specifically to active sites, and of practical interest because they can act as enzyme inhibitors (see Section 3.3). It is not surprising that enzyme-active sites should bind particularly strongly molecules designed to mimic their specific transition states. However, no stable molecule can be a perfect mimic for the dynamic transition state that the enzyme has evolved to recognize. There remains a substantial gap between K_a values for the binding of the best transition-state mimic inhibitors and "equilibrium constants" for transition-state binding to enzymes. These fall typically in the range $10^{16.0 \pm 4.0} \, M^{-1}$, equivalent to free energies of binding of $22 \pm 6 \, kcal \, mol^{-1}$, but can reach as high as $32 \, kcal \, mol^{-1}$ for reactions of common, very unreactive substrates.[16]

In a small number of cases proteins are known that bind relatively small molecules with free energies of binding that fall in this elevated range. The best-known example is the formation of the avidin:biotin complex, with a free energy of binding of $-21 \, kcal \, mol^{-1}$. This can be interpreted in terms of multiple complementary contributions, from hydrogen bonding, van der Waals interactions and hydrophobic binding.[20] The "substantial gap" referred to above, of the order of $10 \, kcal \, mol^{-1}$ remains. But it is clear that the very large free energies of binding by enzymes of transition states for reactions of highly unreactive substrates also represent the summation of multiple, complementary contributions from recognized effects. The main effect specific to transition-state binding that is not involved in the binding of ordinary molecules is likely to be the partial covalent bonding between substrate and nucleophilic groups of the enzyme-active site.

Finally, it is important to stress that very strong binding of ordinary molecules inevitably involves slow release of the bound guest: for example, biotin is bound effectively irreversibly to avidin. Tight binding of either substrate or product to an enzyme would inhibit turnover (and slow product release is a common problem for enzyme mimics). Only a high energy, and thus short-lived species – like a transition state – can be bound so strongly without dissociation becoming unacceptably slow. Binding (or association) constants K_a can be dissected into rates of binding (k_1) and rates of release (k_{-1}), such that $K_a = k_1 / k_{-1}$. k_1 has an upper limit, the rate constant for diffusion through water. This is of the order of $10^8 \, M^{-1} \, s^{-1}$ or less for the binding of small molecules to proteins, whose conformations may have to change and binding sites be desolvated. For example, k_1 is typically $\sim 10^6 \, M^{-1} \, s^{-1}$ for the binding of small molecules to antibodies, which, combined with an association constant of $10^7 \, M^{-1}$ or more gives k_{-1} of the order of $10^{-1} \, s^{-1}$ or less for antigen release.

1.6.2   The Special Environment of the Active Site

The active site of an enzyme is far more than a passive "reaction vessel." It provides a dynamic environment tailored specifically to the whole of the developing reaction coordinate of the reaction it has evolved to catalyze. In terms of molecular recognition it acts as host to a particularly demanding guest: starting with a cautious welcome, followed – if the guest's "face" fits – by the tightest of embraces: after which the guest – changed perhaps permanently by the experience – can hardly get away fast enough. At each stage the micro-environment of the active site is adjusting to match the electronic profile of the guest in a preprogrammed "vectorial" process.[21] The process is only complete when the active site is vacant, and ready for the next guest.

Of course the "vacant" active site in the "resting" enzyme is no less dynamic. Fischer's seminal lock-and-key hypothesis introduced the important concept of enzyme–substrate recognition,[2] but we understand now that there is no such thing as an active-site cavity with a fixed shape precisely complementary to that of the substrate. A "vacant" active site must either collapse (reversibly, of course), or be occupied by (more or less rapidly exchanging) water molecules. A specific substrate will replace some or all of these, as the local geometry adjusts to the requirements of the new guest: opening wider to accept it, closing, per-haps all round it in the transition state; and opening again to release the pro-duct. All these changes involve reciprocal interactions with the surrounding protein structure, which can play a significant part in the overall process.

The active-site microenvironment is of course not homogeneous: it is formed from individual peptide groups and amino-acid side-chains, and may be largely hydrophobic, largely polar, or – significantly – a specifically "designed" com-bination of the two: with the functional groups involved in catalysis positioned to optimize the stereoelectronic interactions involved in bond making and breaking. It is sometimes suggested that a particular organic solvent, or dielectric constant, offers a better model than water for an "active-site envir-onment". There can be an element of truth in this for some simple systems. But no homogeneous medium can model the dynamic three-dimensional arrange-ment of molecular diversity available in a protein microenvironment.

1.7   Cofactors

The functional groups available on the side-chains of the 20 amino-acids that occur naturally in protein enzymes (Table 1.1) may make up an unimpressive selection of reagents, but they perform extraordinarily effectively in a vast number of reactions. Nevertheless, they do have their limitations: for example they include no oxidizing or reducing agent, and no significant electrophilic centre. (All this is for good reasons: proteins contain groups that are relatively sensitive to oxidation, and have to operate, and thus be stable in (relatively nucleophilic) water.)

The missing functionality for these, and some other types of essential chemistry not available from side-chain groups is provided by cofactors, auxiliaries that are bound, usually reversibly, to extend the capabilities of the active sites of large numbers of relevant enzymes. Cofactors include metal ions, which provide precisely positioned electrophilic centres and the possibility of single-electron transfer processes; and the coenzymes, a range of specialized organic systems. Familiar examples are NADH, an organic reducing agent (sometimes called biological borohydride); and pyridoxal **1.8**, which acts as a "portable electron sink" (Scheme 1.16).[22]

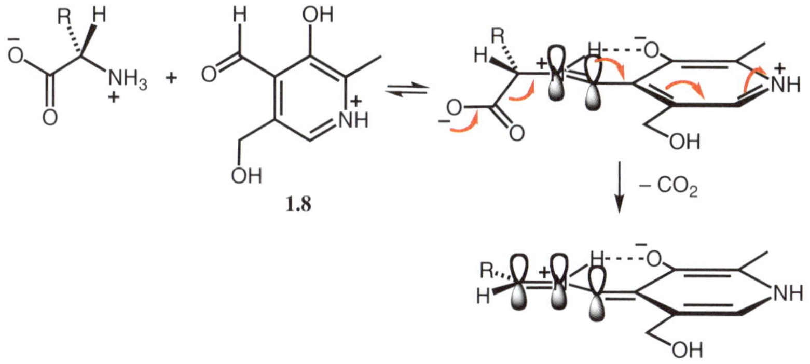

Scheme 1.16 Outline mechanism for the decarboxylation of an α-amino-acid by pyridoxal (**1.8**). The pair of sigma-bonding electrons forming the C–C bond that is broken can be delocalized into the electron sink provided by the pyridinium system most efficiently in the geometry shown, which optimizes overlap with the (π*-orbital of the) π-system. The same process can lead to cleavage of the C–H or C–R bonds, in conformations with these bonds perpendicular to the plane of the π-system.[22]

In both cases the coenzyme acts primarily as an extension of the enzyme. The chemistry can be shown to work perfectly well in the absence of any enzyme, albeit far less efficiently. The enzyme provides the environment for catalysis, just as it does for reactions not involving cofactors. Thus, the principles of catalysis, which enzyme models are designed to test and illustrate, are also no different, and we will not discuss coenzymes (or coenzyme models) as such in this book: beyond a single high-profile example (Section 4.2), chosen to illustrate this powerful way of extending the capabilities of protein enzymes.

1.8 Many Enzymes Use the Same Basic Mechanisms

There is always one (or a handful of) best – meaning most efficient – mechanisms for the catalysis of a particular reaction by a protein enzyme under physiological conditions, and biochemical evolution is efficient enough – and may be assumed to have had long enough – to have found them. So, it should be no surprise that large numbers of the thousands of enzymes use basically the same mechanism to carry out the relatively small number of particular common tasks, like amide hydrolysis, phosphate transfer or reduction using NADH. This means that the active sites of such enzyme "families" have in common the same set of basic features: including for example recognizable nucleotide binding domains for cofactors like ATP and NADH; and in particular the same side-chain groups involved in catalysis. The need for large numbers of different enzymes to catalyze the same basic reaction comes from various extra-kinetic requirements, such as mechanisms for feedback control of activity, and especially the need for substrate specificity. This economy of mechanism is a major advantage in the design of enzyme models. It is also the basis of fascinating insights into biochemical evolution, because homologies in primary sequence can be used to construct evolutionary "family trees" of groups of such enzymes.

The first such family to be identified was the serine proteases, enzymes like chymotrypsin with a common tertiary structure (and thus thought to be derived from a common ancestor proteinase). These enzymes use an active-site serine (part of a "catalytic triad" that includes a histidine general base) as a nucleophilic catalyst for amide hydrolysis (the first step of the mechanism is outlined in Scheme 1.10). This mechanism has probably inspired more enzyme models and mimics, of various sorts, than any other. Nature has also been involved: the same catalytic triad, in the same 3-dimensional arrangment, is also found in a second, smaller group of serine proteinases (e.g. subtilisin) that have no significant sequence homology with the first group. Evidently, evolution has arrived at the same solution to this particular mechanistic problem by a quite different route: a clear case of convergent evolution. (Divergent evolution has also been identified, for example in the very different activities of the enzymes tyrosinase, hemocyanin, and catechol oxidase, which have almost identical active sites.[23])

1.9   Enzyme Models, Mimics and Pretenders

This multiplicity of different enzymes all using basically the same chemical mechanism naturally makes the reactions involved attractive targets for enzyme modelling. Apart from the serine proteases one of the biggest enzyme classes is the glycosidases, which have been classified on the basis of similarities of primary sequence into over 100 families (for the current score go to *http://www.cazy.org/fam/acc_fam.html*). All of these, with the exception of a few special cases, use one of just two different mechanisms. In both cases, the hydrolysis of the glycosidic bond is accomplished by two catalytic residues of the enzyme: a nucleophile/base and a general acid.

The observation that glycosidases "choose" to use a general acid, rather than a general base, clearly makes good chemical sense. We know that acetals (a glycoside is a particular sort of acetal) are stable in base, and hydrolyzed in acid. But when the first glycoside active site (of lysozyme) was shown to contain just two carboxyl groups general acid catalysis of acetal hydrolysis was not a known reaction.[24] Given the hint, it did not take long for the first enzyme model to appear, when it was shown that the glucoside **1.9** containing salicylic

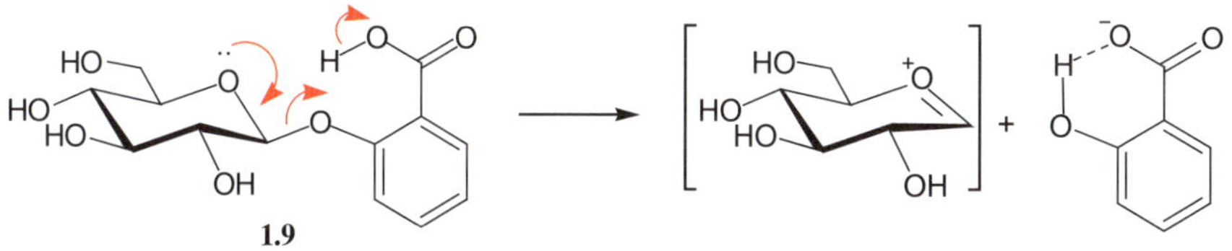

Scheme 1.17 Intramolecular general acid catalysis can account for the rapid hydrolysis of 2-carboxyphenyl β-D-glucoside **1.9**. See the text.

acid was hydrolyzed some 10^5 times faster than its isomer with the COOH group in the *para* position.[25] The mechanism, written as classical intramolecular general acid catalysis, is shown in Scheme 1.17. Since **1.9** has a half-life in water of 8 min even at 91.3 °C, and a phenol is a much better leaving group than a sugar OH, this result is only a first indication of what might be happening in a fraction of a second in an enzyme-active site under physiological conditions, and **1.9** is only the first – albeit the key first example – of a series of models designed to address specific questions about the mechanism of the enzyme reaction (see Section 4.3.3).

It is, however, a start. Given a convincing working model – in this case efficient catalysis near pH 4 of the hydrolysis of an aryl β-D-glucoside – a type of substrate normally showing no observable reaction under the conditions in the absence of an enzyme – it becomes possible to ask questions about the detailed mechanism of catalysis. Questions relevant to catalytic efficiency include the identification of significant electronic and stereoelectronic effects on the general acid, the leaving group, and their geometry of approach; and the influence of the medium on the reaction. The deeper understanding that emerges can then be applied – with appropriate reservations – to the same part-reaction going on in the enzyme-active site. It can also inform the design of improved enzyme models, as discussed further in Section 4.3.3.

We reserve the term enzyme mimics for more highly developed enzyme models, which combine (at least) substrate binding with catalysis: and have the evidence to prove it. It is not unusual to come across work that claims to mimic enzyme reactions but provides no evidence for the formation of a catalyst–substrate complex (either directly or in the form of saturation kinetics: see Section 2.3). In the absence of such evidence we file such systems (even if only provisionally) as pretenders. Always, the ultimate target is a properly qualified enzyme mimic that can also achieve turnover at a reasonable rate. This could in many cases be called, not unreasonably, an artificial enzyme. We now prefer (see the discussion in the Preface) the more general term model enzyme.

Recommended Further Reading

W. Aehle, ed., *Enzymes in Industry*; 3rd edn., Wiley-VCH, Weinheim, **2007**.
J.-P. Behr, ed., *The Lock and Key Principle*, Wiley, Chichester, **1994**.

R. Breslow, ed., *Artificial Enzymes*, Wiley-VCH, Weinheim, **2005**.

A. R. Fersht, *Structure and Mechanism in Protein Science*, Freeman, New York, **1999**.

P. A. Frey, A. D. Hegeman, *Enzymatic Reaction Mechanisms,* OUP, Oxford and New York, **2006**.

H. Maskill, *The Physical Basis of Organic Chemistry*, OUP, Oxford and New York, **1985**.

X. Y. Zhang and K. N. Houk, Why enzymes are proficient catalysts. *Accts. Chem. Res.*, **2005**, *38*, 379–385.

CHAPTER 2

Evaluation of Catalytic Efficiency in Enzymes and Enzyme Models

2.1 Introduction

It is clear that enzymes achieve extraordinary effects – but just how extraordinary are they? What sort of rate accelerations and free-energy benefits can enzymes confer? A meaningful assessment of the catalytic proficiency of an enzyme – or enzyme model – requires a comparison of the catalyzed and uncatalyzed rates of the reaction concerned. This provides an evaluation of the transition-state stabilization (Chapter 1) it achieves for its native (or any promiscuous[i]) reaction. The result benchmarks the proficiency of the enzyme, allows informed comparisons with other systems and provides a starting point for a quantitative discussion of the effects that lead to catalysis, and in particular the selective recognition of transition states.

The challenge for the experimentalist is twofold. First, a common problem is that many biological substrates are extraordinarily unreactive. DNA, for example, has evolved in Nature as a viable long-term data-storage polymer because the phosphate diester bond that connects the nucleotide building blocks is very stable: the half-life of a phosphate diester bond in aqueous solution under physiological conditions is of the order of 30 million years (see below). Second – compounding the problem – to allow the fast reaction sequences that are catalyzed by nucleases, splicing enzymes and other phosphotransferases these same phosphate diester bonds have to be processed – activated, cleaved, rearranged and reorganized – on subsecond timescales. This often enormous

---

[i] Some enzymes catalyze reactions other than those of (what we consider to be) their natural substrates. See Section 3.6.3.

From Enzyme Models to Model Enzymes
By Anthony J. Kirby and Florian Hollfelder
© Anthony J. Kirby and Florian Hollfelder 2009
Published by the Royal Society of Chemistry, www.rsc.org

difference in timescale means that the catalyzed and uncatalyzed reactions cannot simply be measured and compared under the same conditions. However, the necessary practical procedures are now well established.

2.2 Measurement of Uncatalyzed Rate Constants, k_{uncat}

The starting point of any kinetic investigation is the measurement of a time course of product formation. This requires a way to detect the product, either continuously (*e.g.* by monitoring spectroscopically the disappearance of a reactant or the appearance of a product) or – less conveniently – discontinuously by withdrawing aliquots, in which reaction progress can be analyzed separately.

A rate constant k_{obs} is then derived from the time course. In the best case the appearance of product can be described by and thus fit to the exponential equation $[P] = [P]_0 \times e^{-kt}$. Historically a large number linearization methods have been used, but nonlinear curve fitting is now a straightforward operation (*e.g.* using readily available software packages).

There are situations where the kinetics are less straightforward:

  (i) The product inhibits the reaction and an exponential fit is not possible.
  (ii) It may not be practically possible to record a full exponential time course, because the reaction is too slow, because taking a large number of time points is too cumbersome, or because the dynamic range of the assay makes it impossible to follow the full variation of product concentrations.
  (iii) Two-phase kinetics or more complicated scenarios are observed, or the product reacts spontaneously (*e.g.* by oxidation or bleaching) on the same timescale.

In all these cases recording an initial, linear period of the reaction is an alternative. Of course it must be possible to measure a relatively small increase in product concentration in the presence of a large excess of starting material. The linear slope (concentration units/time) is converted into a k_{obs} (units of time^{-1}) by dividing it by the final product concentration.[ii] Accurate determination of the exact concentration of product is therefore important, and this is easily done by measuring the endpoint of the reaction, ideally after 10 (or at least 7) half-lives.

The observed k_{obs} is formally a first-order rate constant (usually in units of s^{-1}) which is a summary measure containing all reactions that might be occurring simultaneously under a given set of conditions. Formally, it is a

---

[ii] This should of course be equal to the initial substrate concentration – but actually measuring it makes for a useful – and sometimes vital – independent check.

pseudo-first-order rate constant: more than one reagent may be involved in the reaction, but – under the conditions usually preferred – all concentrations relative to that of the observed substrate are high and so stay effectively constant. Separation of the variables affecting the reaction rate is possible by measuring k_{obs} as a function of the concentration $[C_i]$ of known or suspected reactants. Plotting k_{obs} against $[C_i]$ typically gives a straight line, with its slope defining the second-order rate constant (in $M^{-1} s^{-1}$). A linear plot indicates that the reactant whose concentration has been changed contributes a first-order term to the reaction rate: curvature that things are more complicated.

Many biological substrates, and especially the biopolymers (proteins, nucleic acids and glycosides) are very stable. The functional groups that connect their monomers react only exceedingly slowly at physiological (near neutral) pH, and break down in the absence of enzymes only at extremes of pH, where catalysis by OH^- and H_3O^+ becomes significant. The original approach to this problem was to measure the rates (k_{uncat}) of the uncatalyzed reactions concerned using activated substrates. The immediate advantage was that a sufficiently activated substrate would react at a measurable rate. The various disadvantages emerged more slowly: in particular, any derived estimate of k_{uncat} for a natural, unactivated substrate requires a long extrapolation, based on structure–reactivity correlations not designed for long extrapolations: while the mechanisms of reactions of activated substrates are not necessarily those appropriate for natural, unactivated substrates.

The enormous differences in timescale mean that long extrapolations of some kind are unavoidable. The single most reliable technique for estimating k_{uncat} values for very slow reactions under physiological conditions is now considered to be to measure the rate of the slow reaction itself, using representative unactivated substrates, at a series of temperatures sufficiently high for reaction to be followed.[26] The derived Arrhenius plot (log k_{obs} is normally a linear function of $1/T$) allows a simple extrapolation to 298 K, or any other desired temperature. The experimental difficulties are considerable: reactions followed above the boiling point of water have to be followed under pressure (conveniently in sealed tubes, made of quartz or Teflon to avoid slow reactions with glass surfaces; themselves often sealed for safety inside steel containers), and the reactions of interest are still likely to be very slow even at these "convenient" temperatures. The good news is that very slow reactions by definition have high enthalpies of activation, so that rates are more sensitive to temperature than those of familiar synthetic reactions.

For example, only recently has it been possible to make an accurate estimate of k_{uncat} for the P–O cleavage of an unactivated phosphate diester, as a guide to the intrinsic reactivity of the phosphodiester linkage in DNA.[27] The model substrate used was dineopentyl phosphate **2.1**, chosen to preclude the alternative C–O cleavage observed with less sterically hindered alkyl groups. Measurements in buffered solutions, in pressure tubes, at up to 250 °C gave thermodynamic parameters of $\Delta H^{\ddagger} = 29.5\,kcal/mol$ and $T\Delta S^{\ddagger} = -8.5\,kcal/mol$. for a pH-independent reaction that could thus be estimated to be over 10^{12}

times slower at 25 °C. This is necessarily a long extrapolation, and any such result is strengthened if other, independent estimates are consistent.

2.1

2.3 Characterizing the Catalytic Reactivity of an Enzyme or an Enzyme Model

The rates of reactions catalyzed by enzymes are generally, and those catalyzed by enzyme models almost invariably, measured under steady-state conditions: with the thermodynamic equilibrium involving catalyst and substrate already established. Presteady-state measurements, of the setting up of this equilibrium, can give important insights into enzyme mechanisms,[28] but are rarely relevant for reactions catalyzed by enzyme models. On the other hand it is convenient in both cases to measure initial rates (of the first few per cent of reaction): in the case of enzymes, because changes in substrate concentrations during the experiment are minimized, but in the case of enzyme models because even the catalyzed reactions under observation are often inconveniently slow.

Figure 2.1 shows the characteristic effect on the rate of an enzyme-catalyzed reaction of increasing substrate concentration, at a constant, catalytic concentration (E_0) of enzyme.

Fitting the curve (Figure 2.1) to the basic Michaelis–Menten equation (inset) gives the standard parameters k_{cat} and K_M, which define the kinetic behaviour of a particular enzyme – or enzyme model – towards a particular substrate. The Michaelis constant K_M can be read off as indicated, as the substrate concentration at which the rate is half the maximum value V_{max} reached at high substrate concentrations: and k_{cat} is obtained from this limiting rate, as V_{max}/E_0, where E_0 is the concentration of enzyme used.[iii] The initial rate at low substrate concentration ($[S] \to 0$) is linear in $[S]$, with slope $= V_{max}/K_M$.

The parent equation was defined in Section 1.1.2

$$E + S \underset{}{\overset{1/K_M}{\rightleftharpoons}} E \cdot S \overset{k_{cat}}{\rightleftharpoons} E \cdot P \longrightarrow E + products$$

Chemists will expect the diffusion away of product from catalyst to be fast, and not kinetically significant, and will recognize this as the equation of a standard two-step reaction, either of which might be rate determining (depending on the rate constants k_2 and k_{-1} for the forward and back reactions

---

[iii] These are useful visual checks. For accurate values with proper error estimates a curve-fitting programme is preferred.

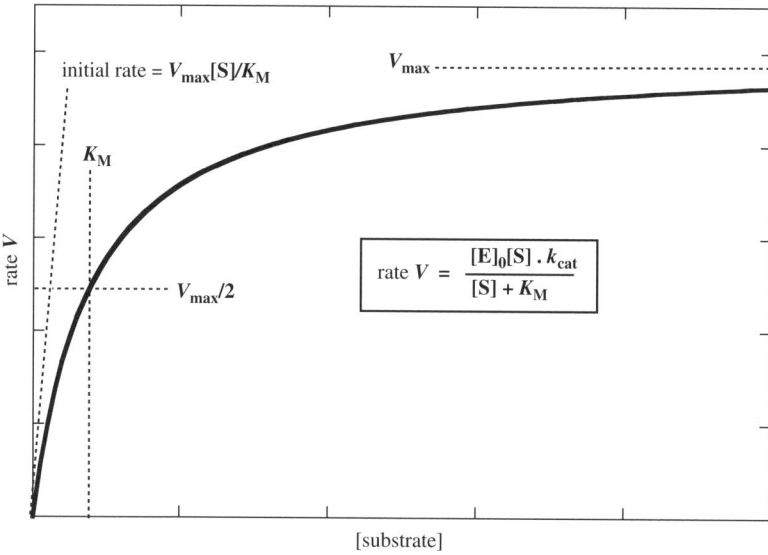

Figure 2.1 Characteristic Michaelis–Menten curve for an enzyme-catalyzed reaction. The rate V of the catalyzed reaction reaches a maximum V_{max} at high, "saturating" concentrations of substrate, S. Inset: the Michaelis–Menten equation (see Section 1.1.2).

of the intermediate E.S).

$$E + S \underset{k_{-1}}{\overset{k_1}{\rightleftharpoons}} E \cdot S \xrightarrow{k_2} E + products$$

Note that the equilibrium constant k_1/k_{-1} for the binding equilibrium is equivalent in the simple case, involving the rapid pre-equilibrium formation of $E \cdot S$, to $1/K_M$: *i.e.* K_M is a dissociation constant (identified, when this has been properly established, as K_S).

We use the parameters k_{cat} and K_M to characterize catalysis by enzyme models as well as enzymes (see Section 1.1.2). These familiar, readily available parameters allow direct comparisons with enzyme reactions. So it is important to understand what they mean. k_{cat} is simply the (first-order) rate constant for the forward reaction of the enzyme–substrate complex $E \cdot S$: it is also known as the *turnover number* (literally the number of times per second the enzyme "turns over" substrate molecules). K_M in the simple Michaelis–Menten mechanism represents the dissociation constant of the Michaelis complex $E \cdot S$: so the initial rate constant k_{cat}/K_M gives the apparent second-order rate constant, for the reaction of the substrate with free enzyme. It can therefore be compared with second-order rate constants for simple second-order reactions. (More generally, K_M corresponds to the dissociation constant for all enzyme bound species.)

To summarize:

k_{cat} is the first-order rate constant for the reaction of the bound substrate (thus $V_{max} = k_{cat}.[E_0]$). It can be compared directly with other first-order reactions, *e.g.* intramolecular reactions of model systems.
It is easy to interpret – bigger is better.

K_M is a measure of binding – of all enzyme-bound species, but simply of the substrate in the simplest case, when the binding step is a rapid pre-equilibrium (*i.e.* $k_{-1} \gg k_2$). K_M then becomes equal to K_S, the dissociation constant of the enzyme–substrate complex **E·S**.
Smaller values correspond to tighter binding.

k_{cat}/K_M is the *apparent second-order rate constant* at very low substrate concentrations (below K_M); under these conditions it is a useful measure of the overall efficiency of the catalyst, *including the binding step*. It can be compared directly with other second-order rate constants.

2.3.1 Comparing Catalyzed and Uncatalyzed Rates

Comparisons are conveniently made using outline schemes of the sort illustrated in Figure 2.2. The free-energy profile (**B**, based on Figure 1.4) applies strictly only to the most basic one-step reaction in which the mechanisms of the catalyzed and uncatalyzed reactions are identical. Enzyme reactions are generally complicated multistep processes and steps other than the chemical transformation can be rate determining, so whether the following generalizations apply has to be checked on a case by case basis.

The thermodynamic box (Figure 2.2, **A**) and the accompanying energy-profile diagram **B** define the relationship between the dissociation constants K_S and K_{TS} of the substrate and the transition state from the enzyme, and the (pseudo)equilibrium constants $K_{uncat}^{\ddagger}$ and $K_{cat}^{\ddagger}$ for the uncatalyzed and the enzyme-catalyzed reactions. The following equation (2.1) relates two opposite corners of the thermodynamic box, representing the lowest and the highest energy species **E·S** and **E + S**‡, in terms of the two energetically equivalent pathways:

$$K_S.K_{uncat}^{\ddagger} = K_{cat}^{\ddagger}.K_{TS} \qquad (2.1)$$

This leads to the important relationship

$$k_{cat}/k_{uncat} = K_{cat}^{\ddagger}/K_{uncat}^{\ddagger} = K_S/K_{TS} \qquad (2.2)$$

which tells us that the ratio of first-order rate constants for the enzyme-catalyzed and uncatalyzed reactions is given by the ratio of dissociation constants

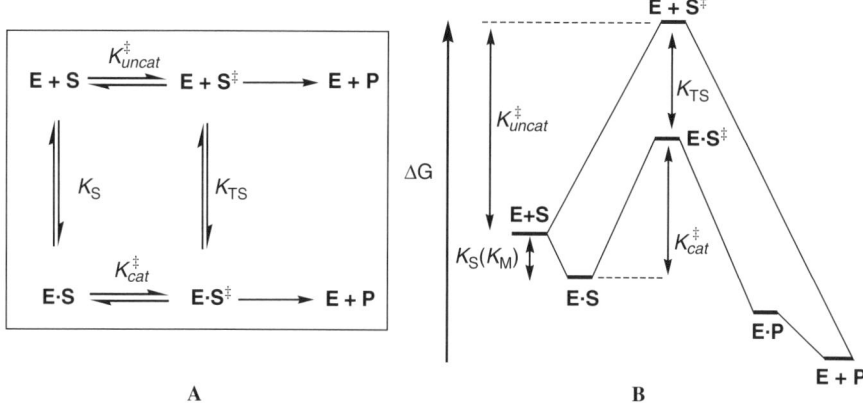

Figure 2.2　Simplified thermodynamic schemes for comparison of an idealized enzymatic one-step process with the corresponding background reaction. **A**. Thermodynamic box illustrating ground-state (K_S) and transition-state (K_{TS}) binding. The bottom line represents the reaction in the enzyme-active site, the upper line is the uncatalyzed or background reaction, with the enzyme present but not involved. K_{TS} is formally the dissociation constant of the transition state from the enzyme. The free-energy scheme (**B**, derived from Figure 1.4) shows the relationships between the kinetic parameters and the corresponding free-energy differences.

from the enzyme for the transition state and the substrate. This simple equality (2.2) is the basis of the generalization that the rate acceleration by the enzyme is attributable to the selective binding of the transition state relative to the substrate (Figure 2.2, **B**), and of the convenient definition of *catalytic proficiency* in terms of this selective binding, $K_{TS} = (k_{cat}/K_S)/k_{uncat}$, or in the general case, $K_{TS} = (k_{cat}/K_M)/k_{uncat}$.

2.3.2　Calculating Rate Accelerations

Rate accelerations can be expressed in several different ways, in terms of different useful comparisons, and it is important to be aware of the basis of estimates in the literature. For example, V_{max}, the rate of the enzyme reaction at saturating substrate concentrations, can be compared with the rate of the background reaction in water under the same conditions, in terms of rate constants as ($V_{max}/[E_0] = k_{cat}/k_{uncat}$): this is a direct measure of the difference in energies between the transition states of the enzyme catalyzed and the competing uncatalyzed reaction(s). However, the details require careful attention, because a particular k_{obs} is simply a first-order rate constant measured under a given set of reaction conditions. Observed rates may vary with pH (Figure 2.3, **B**), and may depend on other components

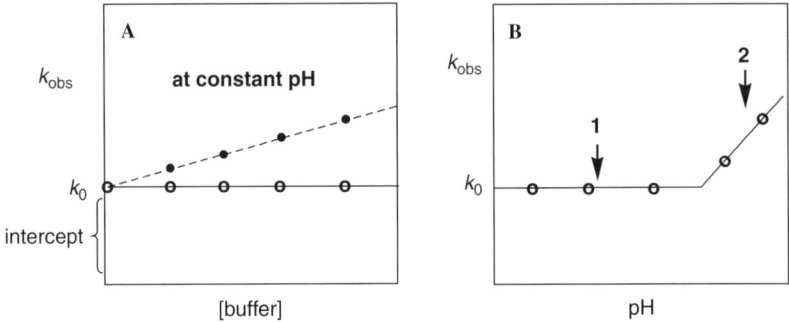

Figure 2.3 Derivation of background rate constants k_{uncat}. Plot **A** contrasts two different reactions that are buffer catalyzed (dashed line) or independent of buffer concentration (solid line). Plot **B** shows a pH–rate profile that can help to interpret the observed rates. The intercept at zero buffer concentration (k_0 in **A**) represents the spontaneous reaction with water (k_0, case 1 in **B**) or the sum of the hydroxide- and water-catalyzed rates ($k_0 + k_{\text{OH}}[\text{OH}]$, case 2 in **B**).

of the reaction mixture, such as the buffer (Figure 2.3, **A**) used to maintain a constant pH. In which case the ratio $k_{\text{cat}}/k_{\text{obs}}$ would compare the enzyme reaction with the sum total of several chemically distinct processes occurring in water.

Figure 2.3, plot **A** shows how the reactions in water can be more clearly defined, and independent processes isolated, by extrapolating observed rate constants k_{obs} back to zero buffer concentration. If this extrapolation is made at a pH where k_{obs} is pH independent (case 1 in Figure 2.3, plot **B**), then the y-axis intercept (k_0 in **A**) will be $k_{\text{H}_2\text{O}}$, representing the pH-independent reaction with H_2O. In cases where this extrapolation yields an intercept k_0 on the y-axis that depends on pH (*e.g.* case 2 in Figure 2.3, **B**), there is an additional reaction catalyzed by OH^- (or H_3O^+). An extrapolated rate constant in this situation is not $k_{\text{H}_2\text{O}}$, but yet another k_{obs}, this time the sum of the hydroxide (or hydronium) and water-catalyzed reactions. (k_{OH} is a second-order rate constant that can be determined by plotting these values for k_{obs} against the hydroxide ion concentration: see below).

However, $k_{\text{cat}}/k_{\text{uncat}}$ defined in this way does not necessarily compare reactions of the same functional groups. For example, if the reaction in the enzyme-active site involves a nucleophile other than water a direct comparison with a reaction of this other nucleophile will be more appropriate. This comparison can be made by using the relevant second-order rate constants k_2. The comparison between the second-order rates in solution and in the enzyme-active site is given by $(k_{\text{cat}}/K_{\text{M}})/k_2$: this ratio provides a more informed comparison of transition states, because the same functional group (*e.g.* a nucleophile) is involved in both cases. In this comparison a further correction may be made to account for the different pK_{a} values of the nucleophile in solution and in the enzyme-active site (by multiplying with $10^{\Delta pKa}$). It is also possible to make a correction for nucleophiles of different strengths, which requires knowledge of the dependence of the reaction rate on a change of pK_a of the nucleophile. A formal treatment of this correction uses the parameter, $\beta_{nucleophile}$, derived from a linear free-energy relationship describing the change of the reaction rate with change of pK_a of the nucleophile.[29] (The correction factor then becomes $10^{\Delta pKa \times \beta nuc}$.)

For hydroxide-catalyzed reactions, a second-order rate constant k_{OH} can be obtained from a pH–rate profile as the slope of k_0 against [HO⁻] at high pH (after initial extrapolation of the rates to zero buffer concentration, if necessary). A similar procedure at low pH will give k_{H_3O+} for acid-catalyzed reactions. Under conditions where the water reaction is dominant the first-order rate constant (k_{H_2O}) can be converted into a second-order rate constant [$k_2(H_2O)$] by dividing by the concentration of water in water (55 M), thus formally accounting for the concentration of water acting as a reagent.

Comparisons of second-order rate constants also take account of the net entropic advantage that an enzyme confers when it brings all reaction partners together in the active site,[iv] by including the ground-state binding contribution (contained in K_M). Though this is not directly catalytic, the unrivalled economy of active-site design often uses residues in both capacities, thus involving the same functionality in both ground- and transition-state interactions. As a result, a clear separation of active-site features contributing to binding and catalysis functions is often impossible. These additional energetic contributions explain why the values for the second-order rate comparisons calculated in Table 2.1 are larger than the corresponding first-order comparisons.

According to the definition of catalysis as the selective recognition of the transition state with respect to the ground state, the selective transition-state binding can be evaluated in terms of binding constants. This so-called *catalytic proficiency* is simply calculated as the ratio $(k_{cat}/K_M)/k_{uncat}$ (Section 2.3.1). (The inverse relation $k_{uncat} \cdot K_M/k_{cat}$ gives a formal dissociation constant K_{TS} of the transition state from the enzyme.)

[iv] For reactions that involve reagents other than water, first-order comparisons can only be made when an appropriate standard state is assumed for all reagents. While 55 M, the "standard state" for water, has physical reality, a standard state for the nonenzymatic reactions of any other reagent is arbitrary. The use of second-order comparisons circumvents this problem, defining the standard state as 1 M.

Table 2.1 Turnover numbers (k_{cat}), catalytic efficiencies (k_{cat}/K_M), rate enhancements (k_{cat}/k_{uncat}), and values of K_{TS} (($k_{cat}/k_{uncat})/K_M$) for water-consuming enzyme reactions at 25 °C. Obtained by comparison of enzyme reaction rate constants with rate constants for the corresponding uncatalyzed reactions (k_{uncat}).

Enzyme	k_{uncat}, s^{-1}	k_{cat}, s^{-1}	k_{cat}/K_M, s^{-1} M^{-1}	k_{cat}/k_{uncat}	K_{TS}, M
Fructose-1,6-bisphosphatase	2.0×10^{-20}	21	1.5×10^7	1×10^{21}	7×10^{-26}
Staphylococcal nuclease	7.0×10^{-16}	95	1.0×10^7	1.4×10^{17}	7×10^{-23}
β-Amylase	1.9×10^{-15}	1.4×10^3	1.9×10^7	7.2×10^{17}	1.0×10^{-22}
Fumarase (37 °C)	3.5×10^{-14}	880	2.4×10^8	3.5×10^{15}	1.0×10^{-21}
Jack bean urease	1.2×10^{-11}	3.6×10^4	9×10^6	3×10^{15}	1.3×10^{-18}
Chloroacrylate dehalogenase	2.2×10^{-12}	3.8	1.2×10^5	1.8×10^{12}	1.8×10^{-17}
Carboxypeptidase	4.4×10^{-11}	240	6×10^6	1.3×10^{13}	3.3×10^{-17}
E. Coli cytidine deaminase	2.7×10^{-10}	300	2.7×10^6	1.1×10^{12}	1.0×10^{-16}
Phosphotriesterase	2.0×10^{-8}	2.1×10^3	4.0×10^7	1.8×10^{11}	1.9×10^{-16}
Hamster dihydroorotase	3.2×10^{-11}	1.2	1.1×10^5	3.7×10^{10}	2.9×10^{-16}
Carbonic anhydrase	0.13	1.0×10^6	1.2×10^6	7.7×10^6	1.1×10^{-9}

Note. Data from Wolfenden.[26]

The Enzyme Model Scorecard

The following tests and parameters can be used to assess the quality of an enzyme model.

1. Does the reaction catalyzed by the enzyme model show saturation kinetics?
2. What are the steady-state parameters K_M and k_{cat}?
3. How many product molecules are generated (*i.e.* is the enzyme model a stoichiometric reagent or a multiple turnover catalyst)?
4. To what extent do the quantitative comparisons match comparable data for enzymes (see Figure 2.4)?
   - the *catalytic proficiency*, i.e. $(k_{cat}/K_M)/k_{uncat}$: or its inverse, namely
   - the formal dissociation constant K_{TS} for the transition state from the enzyme (calculated as $k_{uncat} \cdot K_M/k_{cat}$).
   - the rate acceleration over the background reaction in water (k_{cat}/k_{uncat})
   - the second-order rate comparison $(k_{cat}/K_M)/k_2$
   - the effective molarity, EM (Section 1.4.1) for reactions between functional groups.

Not all of these data will be available for all enzyme- and enzyme-model-catalyzed reactions. But the more comparisons are available, the better we can understand the specific strengths and weaknesses of artificial catalytic systems.

While the spontaneous or uncatalyzed rates of the great majority of reactions of biological interest vary over an enormous range, from 10^{-20} to $10^{-1}\,s^{-1}$, the rates of the same reactions catalyzed by enzymes fall within the much narrower range of 10 to $10^6\,s^{-1}$, and most of them between 10 and $10^3\,s^{-1}$.[26] Depending on the difficulty of the chemical background reaction, estimated catalytic proficiencies $(k_{cat}/K_m)/k_{uncat}$ range from $10^8\,M^{-1}$ (for the thermodynamically undemanding hydration of CO_2 by carbonic anhydrase) up to $10^{26}\,M^{-1}$ (for the much more difficult phosphate monoester hydrolysis by fructose-1,6-bisphosphatase).[26] Values for K_{TS} can be as large as $10^{-26}M$, corresponding to a binding energy of some $19\,kcal\,mol^{-1}$. Transition-state binding is thus formally much stronger than ordinary noncovalent binding processes, though such figures include the thermodynamic benefits of catalytic effects: including those that cannot be classed as straightforward binding events, *e.g.* the partial bond formation involved in the TS for nucleophilic catalysis (Section 1.5). For comparison, K_M serves as an *indicator* for the strength of ground-state binding, although it should be borne in mind that K_M represents a simple dissociation constant only for a one-step reaction in which association of substrate and enzyme is fast. In more complex scenarios it is a function of binding and a number of rate constants that contribute to a multistep reaction.

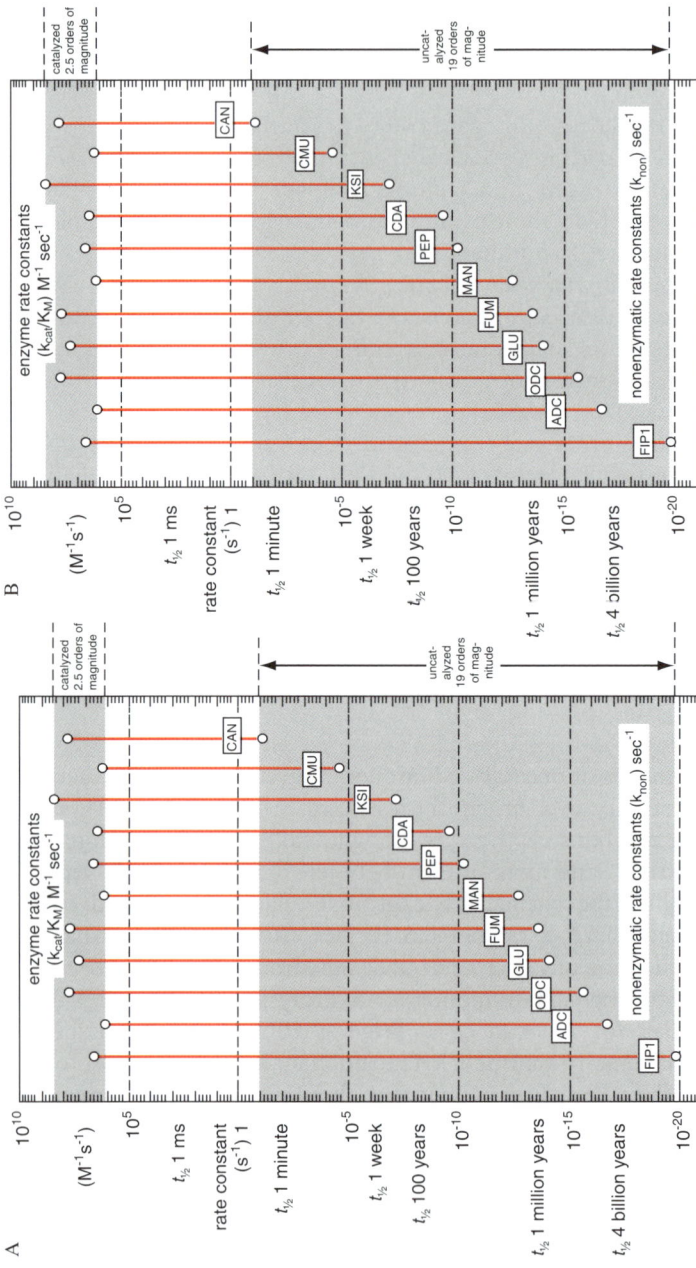

Figure 2.4 Comparisons of k_{cat}/k_{uncat} and $(k_{cat}/K_M)/k_{uncat}$ for representative enzymes. Notes. CAN = carbonic anhydrase; CMU = chorismate mutase; KSI = 17-ketosteroid isomerase; CDA = cytidine deaminase; PEP = carboxypetidase; MAN = mandelate racemase; FUM = fumarase; GLU = α-glycosidase; ODC = OMP decarboxylase; ADC = arginine decarboxylase; FIP1 = FBPase, inositol phosphatase. We thank Professor Wolfenden for this updated version.

2.4 Catalytic Efficiencies of Representative Enzymes: The Size of the Challenge

Typical values of turnover numbers k_{cat} and k_{uncat} for a number of representative enzymes are compared logarithmically in plot **A** of Figure 2.4, where the lengths of the vertical lines indicate the accelerations involved. The individual values of k_{cat} and k_{uncat} are also translated, revealingly, into half-lives for the reactions concerned. Data for the same reactions expressed in terms of catalytic proficiencies $(k_{cat}/K_M)/k_{uncat}$ are shown in Plot **B**. The numerical data on which these plots are based appear in Table 2.1. The table and Figure 2.4 together provide a systematic indication of the magnitude of the challenge facing the enzyme modeler. They also provide a sobering reality check when assessing the quality of the enzyme models discussed in Chapter 4.

Recommended Further Reading

A. R. Fersht, *Structure and Mechanism in Protein Science*, Freeman, New York, 1999.

P. A. Frey and A. D. Hegeman, *Enzymatic Reaction Mechanisms,* Oxford, New York, 2006.

A. Williams, A. *Free-energy Relationships in Organic and Bio-organic Chemistry*, Royal Society of Chemistry, Cambridge, 2003.

R. Wolfenden, Degrees of Difficulty of Water-Consuming Reactions in the Absence of Enzymes, *Chem. Rev.* **2006**, *106*, 3379–3396.

Constructing Enzyme Models – Building up Complexity

Chapter 1 defined the individual chemical tasks that constitute the catalytic package of an enzyme, some or all of which must be integrated into an enzyme model for it to be efficient. Then, Chapter 2 spelled out the daunting challenge presented by the observed efficiency of enzyme catalysis, and introduced the parameters used to quantify it. The defining characteristic of natural enzymes is that they manage this integration of chemical tasks extremely well, while current enzyme models generally fail this basic test. Consequently, the most significant advances in the enzyme-model field involve bringing together more of the general features of enzyme catalysis, even in some cases going as far as using parts of natural enzymes.

This chapter attempts to give an overview of the approaches and strategies on which the construction of a model are based. Models educate our expectation of the contribution that each effect can make, and of how delicate a task is its combination with other effects. Table 2.1 summarized typical targets that enzyme models must match if they are to rival what enzymes can do. The table makes a sobering read, but it is essential to identify the problems that seem so far to preclude the development of an ideal, fully functioning enzyme model. Current models should be seen as staging posts on a long road to enzyme-like efficiency, and the following paragraphs identify the significant conceptual steps involved. In this context we expand our definition of enzyme models to illustrate these steps. On the one hand, we use "top-down" examples for enzyme models based on modified protein catalysts, *i.e.* show how a protein scaffold can be adapted to new tasks – and what the challenges are when natural enzymes are reused in this way. On the other hand, we introduce "bottom-up" examples of how functional complexity is organized in synthetic catalysts. Their practical application is illustrated by the detailed examples of

From Enzyme Models to Model Enzymes
By Anthony J. Kirby and Florian Hollfelder
© Anthony J. Kirby and Florian Hollfelder 2009
Published by the Royal Society of Chemistry, www.rsc.org

design-based systems made by the tools of organic synthesis to be found in Chapter 4. Chapter 5 details the more recent advances based on iterative approaches, using either synthetic chemistry or the tools of modern molecular biology.

3.1 Solvents – Catalysis Without Functional Groups?

Max Perutz noted in describing the crystal structure of lysozyme – shortly after this first enzyme structure had been solved: "*Organic solvents have the advantage over water of providing a medium of low dielectric constant, in which strong electrical interactions between the reactants can take place. The nonpolar interior of enzymes provides the living cell with the equivalent of the organic solvents used by chemists.*"[30] Putting a substrate molecule into an organic solvent is thus considered comparable to its binding to an active site in the "hydrophobic core" of an enzyme: where reactivity may be enhanced by the hydrophobic effect, especially specific solvation or desolvation, and the stronger electrostatic interactions between dipoles in a medium of low polarity (see Section 1.6).

Despite the frequent observation of hydrophobic pockets at or around enzyme-active sites, it is not a simple matter to measure and quantify the effect of the local microenvironment. Measurement of a "local" dielectric constant, for example, is experimentally impossible. We have to rely on indirect evidence to define the solvent properties of an active site: assessing the character of local amino-acid side-chains and the potential for preferential binding of hydrophobic substrates, paying particular attention to the presence and arrangement of hydrophobic and functional groups: and estimating resultant pK_a shifts of ionizable groups.

For an active-site functional group to carry out its designated role it must be in the correct protonation state. A hydrophobic environment destabilizes charges, so that the familiar pK_a values of ionizable active-site groups (Table 1.1) may be significantly perturbed in a hydrophobic active site (Table 3.1). The pK_as of carboxylic acids typically go up and the pK_a values of amine bases go down, in each case because the charged equilibrium form of a functional group is destabilized compared with the fully solvated ion in aqueous solution (Scheme 3.1).

pK_a values in enzyme-active sites are often so strongly perturbed by local environmental effects, which can include hydrogen bonding as well as hydrophobic effects, that it becomes seriously risky to attempt to assign a role to a particular catalytic group simply on the basis of an observed kinetic pK_a. This ability of enzymes to change the pK_as of potential catalytic groups, making an active ionic form available at physiological pH, is an important part of their catalytic armoury.

We have seen in Section 1.3.2 how even bulk solvent can be catalytic. Solvents can accelerate reactions by interacting with transition states, ground states, or both, adjusting their relative energies to decrease differences in energy between ground and transition states (Scheme 3.2). In practice, ground-state

Table 3.1 Ranges of observed pK_a values in enzyme-active sites.

Amino acid	Side-chain functional group	pK_a (Normal)[a]	Range observed in active sites[b]
Asp/Glu		4–4.5	<2–6.7
His		6.4	2.3–9.2
Cys		9.1	2.9–10.5
Lys		10.4	As low as 6.0
Tyr		9.7	As low as 6.1

Notes. [a] pK_as measured in small peptides. (Fersht, p. 170.[28])
[b] From Fersht, p. 189, Frey, p. 116,[31] and Li.[32]

destabilization and transition-state stabilization are not easily distinguished, because kinetics report only on differences between the two states. In qualitative terms we know that rates of reaction can be affected by the presence or absence of solvent molecules (resulting in the effective solvation and desolvation of reaction centres) as well as the changes in the interactions between point charges in the presence and absence of bulk solvent.

The effects can be very large: simply changing the solvent can result in rate accelerations as large as 10^9 in quite simple reactions. A classical example is the Kemp decarboxylation (Scheme 3.3), in which the cleavage of the $C–CO_2^-$ bond is concerted with the opening of the benzisoxazole ring. The negative charge on the carboxylate group is further delocalized in the transition state **3.2‡**. This reduces the strong solvation by hydrogen-bonding solvents, which stabilizes the reactant in water, and contributes to the up to 10^7-fold rate acceleration observed on going to a dipolar aprotic solvent.[33]

Computational studies have attempted to quantify the contribution of putting a substrate into a hydrophobic protein interior and exposing it to the effects of point charges arising from amino-acid side-chains and reinforced by the hydrophobic environment.[34] Early studies treated the enzyme-active site effectively like the gas phase, with all solvent molecules squeezed out.[35] The unsurprising result (electrostatic interactions between point charges in the gas phase can amount to hundreds of kilocalories) was that the calculated effects could explain in principle *all* of enzyme catalysis, which tells us more about

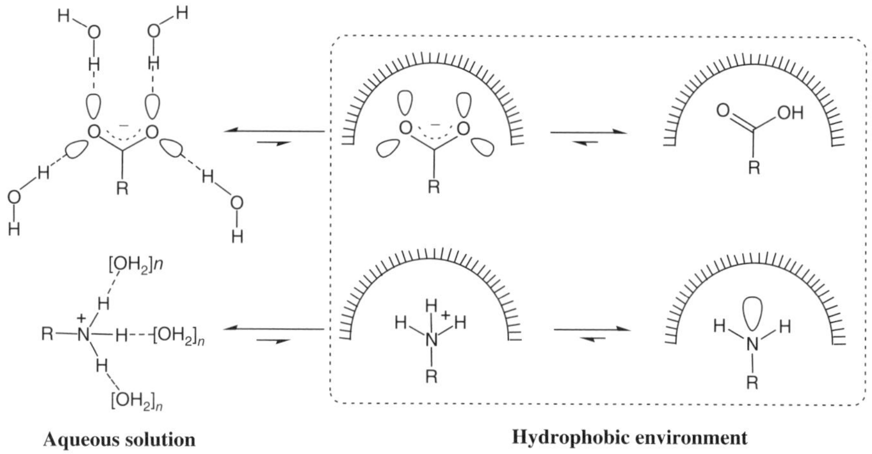

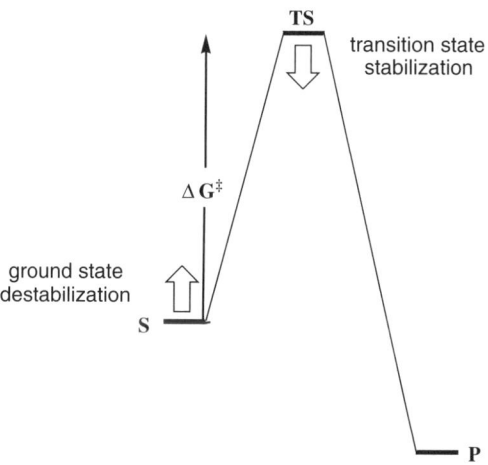

Scheme 3.1 The charged form of an ionizable group present in aqueous solution is disfavoured in a hydrophobic environment, so that the pK_a values of active-site groups may differ significantly from those in aqueous solution. The pK_a values of active-site groups are thus shifted towards neutrality (see Table 3.1).

Scheme 3.2 A catalytic effect of changes in solvation can result from stabilization of the transition state or destabilization of the reactant ground state: or of course both. Selective stabilization of the bound reactant will evidently be anticatalytic.

the computations than about reality. The approach, and the tools available, have gradually evolved, to take account of the complex, dynamic, structured environment, quite different from water – or indeed any homogeneous solvent – presented to a substrate by the enzyme-active site.[36–38]

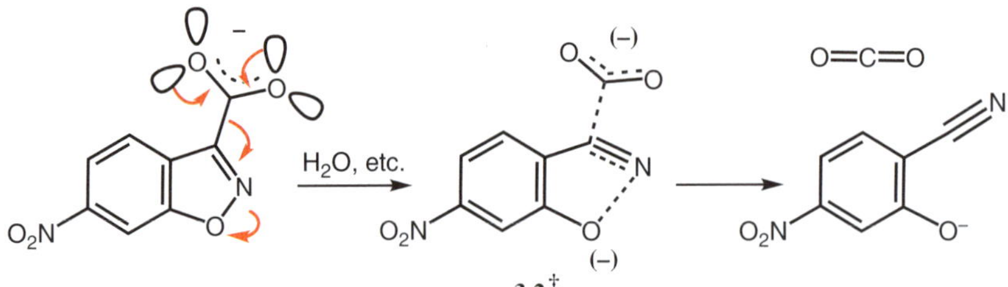

Scheme 3.3 The reactant **3.1** in the Kemp decarboxylation is stabilized by H-bonding solvation of the carboxylate group (see Scheme 3.1); accounting in part for the enormous rate accelerations observed on going from water to dipolar aprotic solvents.

A crucial difference between reactions in solution and in an active site is that the enzymic environment is preorganized. Functional groups are in place to bind substrate, and especially to bind more strongly to the TS (transition-state complementarity), with the result that the reorganization penalty as the reaction proceeds is relatively small. The medium effects exerted by bulk solvent and the structured microenvironments of enzyme-active sites can evidently be quite different. So, the magnitudes of solvent effects on model reactions (compared to water) give us an approximation of possible effects, without making clear whether they represent lower or upper limits: the observation of a bulk solvent effect on a reaction gives no more than an indication of what is possible. Interactions involving specific groups are addressed in Section 3.2.

Thus, solvent "catalysis" provides an important reference point, though it cannot be convincingly regarded as an enzyme model. The substrate has been actively introduced into the more or less unwelcoming solvent by the experimentalist. Enzymes, by contrast, recognize their cosolute substrates in solution and attract them into the active site spontaneously. These deficiencies are addressed in the improved models introduced in Sections 3.3ff.

3.2 Introducing Catalytic Groups – Without Positioning Them

Transition states are characterized by the localized charge rearrangements involved in the making and breaking of bonds. The reorganization of electrons leads to increases in charge density at reaction centres, and such accumulation of charge is energetically costly. Intermolecular interactions with functional groups can help to offset the development of charge and lower the energetic price for making and breaking bonds: by direct electrostatic interaction, or by the partial making or breaking of bonds involved in nucleophilic and general acid/base catalysis. The mechanisms of such catalysis can be studied by measuring the effects on the rates of reaction of varying the concentrations of buffer constituents, as described in Chapters 1 and 2, or by adding appropriate acids, bases or nucleophilic reagents to the solution: but this intermolecular catalysis is enormously inefficient compared with the enzyme reactions of interest; and typically only observed by using artificially activated substrate systems.

3.3 Positioning of Substrate and Catalytic Groups by Covalent Design

Systems where the reacting substrate and catalytic functional groups are held in close proximity on the same molecule have been described in Section 1.4, in the context of studying the mechanisms of the reactions involved. The design of relatively simple systems can provide information about the preferred geometries as well as the mechanisms of reactions of interest, but the single most important point for the enzyme modeler is that intramolecular reactions are faster, sometimes very much faster, than their intermolecular equivalents: so that reactions can be observed, and their mechanisms studied without recourse to activated substrate groups. However, in terms of fully developed enzyme models intramolecular reactivity is only indirectly relevant, because the primary interaction that brings the reacting groups together is covalent and permanent, and established by synthesis rather than by a binding step. Nevertheless, the clear implication is that a binding mechanism that brings catalytic and substrate groups into close proximity in the Michaelis complex is likely to favour also stabilization of the transition state for the reaction between them; and can provide a basis for very substantial enhancements of reactivity.

3.4 Binding the Ground State by Noncovalent Interactions

Enzymes work by a sophisticated, integrated system of binding and catalysis. The essential first step of every enzyme reaction is substrate binding; by which a specific substrate is selected from the many others available after diffusing into contact with the protein, and enters the active-site "reaction vessel", where it comes into contact with the catalytic apparatus. So, a primary task for the model builder is creating a binding site for a substrate, because the rest of the reaction coordinate for an enzyme-catalyzed reaction is accessible only to bound substrate. Consequently, the first and basic measure of success of a model is the observation of saturation or Michaelis–Menten kinetics (illustrated in Figure 2.2). Significantly, this hallmark of enzyme catalysis is not observed in the three most basic approaches to models – solvolysis, buffer catalysis and intramolecular catalysis – discussed in this chapter so far.

Nature provides a number of relatively small molecules, nothing directly to do with enzymes, that have been found to act as hosts for small-molecule guests: some of the most successful examples are discussed in the context of design-based enzyme models in Chapter 4. One of the simplest of these is β-cyclodextrin (Scheme 3.4). Cyclodextrins are readily available water-soluble hosts, known to bind hydrophobic guests in their well-defined central cavity. Even in this primitive mode of attachment, the substrate in the cyclodextrin cleft is typically protected from attack by outside reagents, but open to what is effectively intramolecular attack by the OH groups of the host; or even in rare cases by an appropriately reactive second guest.

The battery of hydroxyl groups on each rim of the cyclodextrin cavity is well positioned to react with electrophilic functional groups of bound substrates, suggesting cyclodextrins early on[39] as possible mimics for enzymes like the chymotrypsins and alkaline phosphatase, which rely on an active-site serine OH group as the primary nucleophile. A great deal of work designed to take advantage of these properties has been summarized in expert reviews (for earlier references see Breslow, in a special cyclodextrin issue of *Chemical Reviews*[40]).

The neutral hydroxyl group is not particularly reactive, and simple cyclodextrins are not significant catalysts near pH 7. But at high pH in water-activated esters with hydrophobic components acylate cyclodextrins more rapidly than they are hydrolyzed. These reactions are not catalytic, because the initial products are unactivated esters of the cyclodextrin (Scheme 3.4), but provide useful models for the initial, acylation steps of the enzyme reactions. For example, *m*-nitrophenyl acetate acylates β-cyclodextrin more rapidly than does the (intrinsically more reactive) *p*-nitrophenyl acetate. The well-established

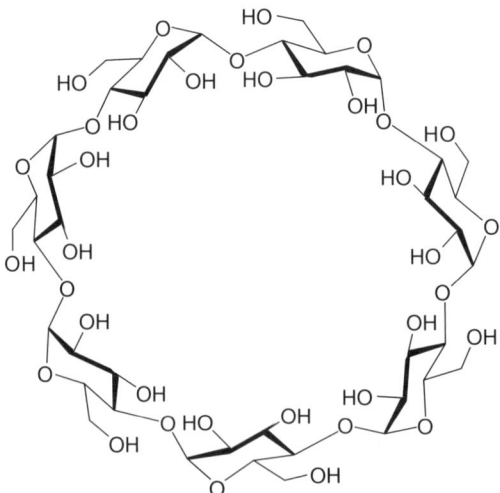

Scheme 3.4   β-cyclodextrin has a hydrophobic cavity (C–H bonds and oxygen lone pairs line the central cavity) with two rings of 7 and 14 OH groups encircling the two entrances.

mechanism (**3.4 → 3.5** in Scheme 3.5) shows how this can be simply explained by the different geometries imposed by binding.

Several points of general importance are well illustrated by this work:

(i) The initial step involves substrate binding. Reactions follow Michaelis–Menten kinetics and show competitive inhibition in the presence of inert hydrophobic solutes.[41] These compete with substrate for hydrophobic contacts in the cyclodextrin cleft.

(ii) The nitrophenyl ring is bound relatively weakly in the hydrophobic cavity of the cyclodextrin: the observed K_M is typically several milli-molar. This is a low affinity compared to the K_M values of many – but not all – enzymes.

(iii) Reaction involves nucleophilic attack by the anion derived from a secondary 2- or 3-OH group (Scheme 3.4), showing that productive binding occurs at the wider end of the cavity. This could be the result of stronger or more geometrically favourable binding at this end, or because the 2- or 3-OH groups are more acidic (the apparent pK_a of 12.3 results from intramolecular hydrogen bonding between the OH groups): or of course result from a combination of these factors. (We will see in Chapter 4 that other reactions may involve productive binding at the narrow end of the cavity.) More generally, the positioning of the reactive groups with respect to substrate is not planned or designed.

(iv) The C=O group of *m*-nitrophenyl acetate is held in closer proximity (**3.4**) to the ring of OH groups on this face than that of the *p*-nitrophenyl ester (**3.3**), accounting for its faster reaction. Evidently the geometry and energetics of approach of the nucleophilic OH group to the substrate C=O group are critical for reactivity (and remember that β-cyclodextrin has 14 such potential nucleophiles available).

(v) "Catalysis" of nitrophenyl acetate hydrolysis is not particularly efficient in either case (Scheme 3.5). The rate accelerations (k_{cat}/k_{uncat}) over

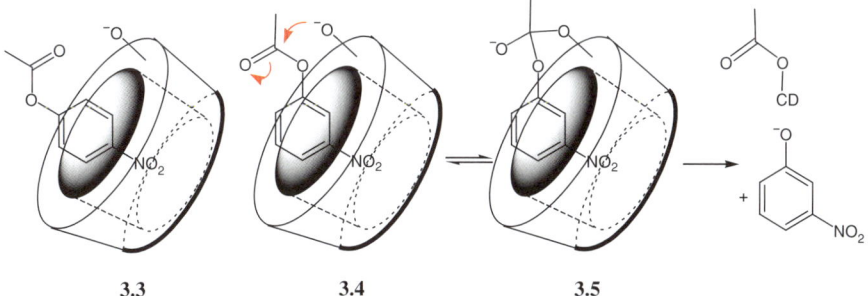

3.3   **3.4**   **3.5**

Scheme 3.5 The intrinsically less reactive *m*-nitrophenyl acetate is cleaved more rapidly (**3.4**) than the *p*-nitrophenyl ester (**3.3**), in a reaction involving nucleophilic attack by a secondary alkoxide anion. However, the *transition state* for the reaction (close in geometry to the tetrahedral intermediate (**3.5**)) is still not optimally bound (see the text).

background at pH 10.6 amount to factors of 9 and 96, respectively, for the reactions of *p*- and *m*-nitrophenyl acetates with β-cyclodextrin.[42]

Similar generalizations hold for a large number of enzyme models where the design rationale is based solely on substrate binding. In particular the observed rate accelerations are small compared to the corresponding values for enzymes. For catalysis to be effective it is not enough – it is necessary but not sufficient – that the model binds the substrate. The accurate positioning of substrate with regard to catalytic groups, which can be so efficient in the best intramolecular models discussed in Section 3.3, is missing. Furthermore, efficient binding of substrate is in principle *anti*catalytic: if the bound substrate is stabilized and the energy level of the TS unaffected, the barrier between free substrate and TS is increased. The problem of selective transition-state binding is addressed in the following section.

3.5   Binding the TS More Strongly than the GS by Noncovalent Interactions

Enzyme efficiency is characterized by a high rate of *turnover* of substrate (k_{cat} is also called the turnover number), and the process is correspondingly sensitive to inhibitors. Any nonsubstrate molecule that binds in the active site will reduce turnover. The more tightly it binds the more effective is the inhibition: so some of the most effective enzyme inhibitors are transition-state analogues (TSAs) – stable compounds designed to be as close as possible in structure to the putative TS of the natural reaction. Of course, a stable analogue of the *substrate* will also compete with it for binding in the active site, but since TS binding is stronger a transition-state analogue should be bound more strongly.

A simple example is the use of phosphonic acid derivatives to model the transition states for the hydrolysis of carboxylic acid amides and esters: tight binding of these analogues is observed in suitable cases. For example, phosphonamidate analogues of the tetrahedral intermediates involved in the hydrolysis of ZGly Leu-X substrates bind strongly to – and so inhibit – the proteolytic enzyme thermolysin. The binding of a series of phosphonamidates (Scheme 3.6(A): X is another amino-acid residue) shows a strong correlation with K_M/k_{cat} for the reactions of the corresponding natural peptide substrates, as expected for true transition-state analogues (Scheme 3.6(B)).[43] The linearity of this double-logarithmic plot indicates that any effect that is relevant for catalysis (measured by K_M/k_{cat}) is reflected to the same thermodynamic extent in the inhibitory interactions of TSA with the enzyme (measured by the relevant dissociation constant K_i), pointing to closely similar molecular recognition in the two cases. (Note that there is no correlation with K_M – as would be expected for the binding of substrate analogues.)

Catalytic antibodies (Section 5.2) provide instructive examples of enzyme models in which the selective recognition of transition states is key. The idea is that the binding energy of antibodies could be used to promote chemical

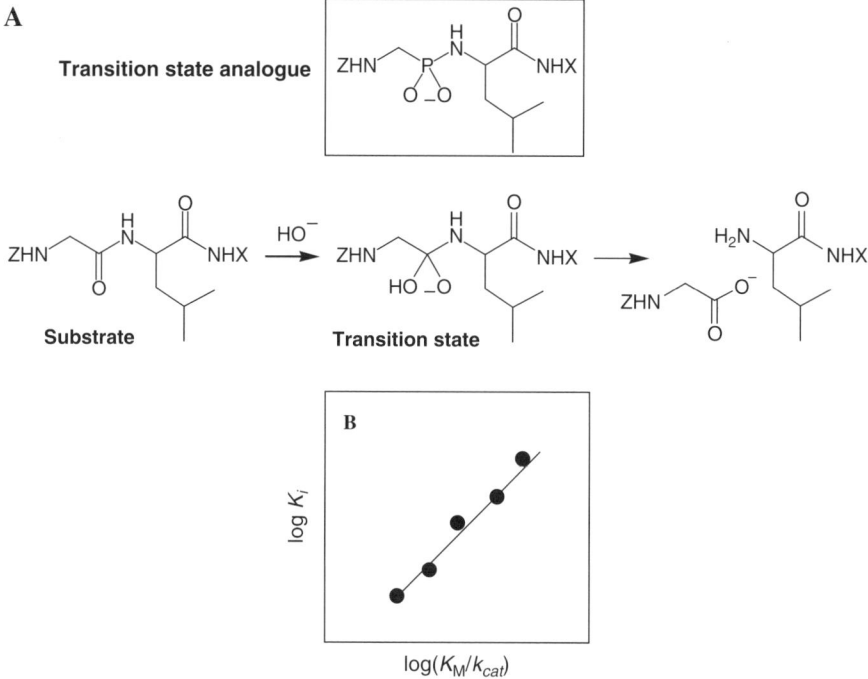

Scheme 3.6 (**A**) The stable tetrahedral phosphonamidate (boxed) mimics the high-energy tetrahedral addition intermediate involved in the hydrolysis of a peptide substrate, and thus also the transition states involved in its formation and breakdown.[43,44] (**B**) The observed linear correlation of K_M/k_{cat} and K_i (see the text).

reactions. Antibodies generated (in large numbers, using the immune system and library approaches explained in Chapter 5) against a TSA as antigen will be complementary to, and thus bind transition-state analogues. Insofar as they also bind the transition state the TSA is designed to mimic more strongly than the substrate, they should catalyze the reaction concerned, and successful examples of antibody catalysis are discussed in Chapters 4 and 5.

In principle, this approach can lead to selective binders, and thus catalysis, if the dissociation constant of the enzyme-TSA complex (K_i) is equal to the actual transition-state dissociation constant K_{TS}. However, eqn (2.2) derived in Section 2.3.1, comparing the Michaelis constant K_M for the antibody-catalyzed reaction and the dissociation constant for the enzyme-TSA complex (K_i), sets limits for the efficiency of antibody catalysis:

$$(k_{cat}/k_{uncat})_{max} = K_S/K_{TS} = K_M/K_i \qquad (3.1)$$

An antibody raised against a transition-state analogue could provide sufficient binding, if the binding affinity for the substrate ground state (K_M) were sufficiently low. However, the strengths of binding interactions of small molecules (like TSAs) to proteins have limits. Dissociation constants K_i are typically greater than nanomolar, while K_M values need to be at least millimolar to be practically useful. Consequently, k_{cat}/k_{uncat} ($= K_M/K_i$) is limited to about 10^6 for this particular class of enzyme model. (For further discussion of efficiency in antibody catalysis see Section 5.2.3.) This result makes very clear why the differential stabilization of TS *vs.* GS routinely achieved by enzymes – giving k_{cat}/k_{uncat} values up to 10^{21} – is not easily achievable even in the best models.

The use of TSAs as templates is a general approach, and is also the basis of a substantial body of work involving synthetic polymers. Molecular imprinting[45] (see Section 5.1.3) typically involves the attachment, covalent or noncovalent, of polymerizable functional groups to a suitable TSA template. The monomer thus produced is then copolymerized by radical initiation under carefully controlled conditions. Removal of the template should in principle produce an imprinted polymer, with cavities complementary to the original TSA and thus potentially active sites for the reactions concerned, embedded in what can be a robust polymer framework.

Similar principles are involved in the dynamic combinatorial chemistry approach,[46] which produces libraries based on a mixed set of small molecules that bind covalently but reversibly with each other. The introduction of a template can lead to the amplification of a "library member" that is particularly stable, and binds – because it is complementary to – the TSA. This "library member" is thus in principle a potential catalyst for the reaction the researcher first thought of.

3.6   Existing Enzymes as Catalytic Scaffolds to Accommodate New Functions

At this point we depart for a moment from the consideration of synthetic model systems and turn to natural proteins for comparisons, to inform us about the more complex strategies they employ to bring about catalysis. We start not with natural, highly efficient, fully evolved enzymes, but – to keep it simple – with catalysts that are less than perfect – mutants, or enzymes catalyzing reactions other than those they have evolved for.

Imperfect as they may be these systems can provide an appropriate starting point for thinking about the design of protein enzyme models with new functions, extending the range of reactions that can be catalyzed beyond known natural processes. If, for example, a pharmaceutical chemist needs a catalyst for a particular reaction, should he or she go to the "chemical drawing board" to design a suitable system? Or to an enzymes database to look for a known enzyme that catalyzes the same or a similar reaction, or handles a similar substrate; and might be modified to catalyze the new reaction of interest, using the methods of directed evolution (involving iterative screening/selection cycles from biological libraries, as discussed in Chapter 5). As protein chemistry develops in countless new directions, existing protein structures can provide appropriate starting points because they are already folded, thus bypassing the definitely nontrivial challenge of creating a stable protein fold.

Can modified enzymes be considered enzyme models? Strictly "enzyme model" may seem a misnomer in this context, but relaxing our definition slightly – to include enzyme proteins catalyzing reactions of non-natural or nonspecific substrates – allows us to draw useful comparisons that suggest significant potential advantages over synthetic enzyme models.

An enzyme model based on an existing protein can offer several potential advantages:

(i) *Recognition of substrate features can be provided.* Such interactions are not necessarily catalytic (*i.e.* may not become stronger as the TS is approached), but can localize the substrate in the active site, increasing its effective concentration with respect to potential catalytic groups. A logical extension of this idea is to combine a suitable binding protein in combination with a second enzyme or enzyme model that can provide the catalytic functionality.

(ii) *Reactive generic functionality in the active site.* Enzyme active sites are highly functionalized protein regions comprising binding sites, useful catalytic functionality and sometimes cofactors. These features can be combined with new functionality by site-directed mutagenesis and chemical modification, to catalyze new reactions. The well-defined arrangements evolved for catalysis of the original enzymatic reaction will not of course remain equally effective, but the close approximation of the bound alternative substrate and potential catalytic groups may provide useful rate accelerations in favourable cases.

(iii) *Provision of a proactive binding pocket.* When proteins sequester their substrates from bulk solvent the bound substrate can be subject to specific effects of the microenvironment that may alter its reactivity. For example, changes in pK_a values can increase the amount of a catalytic group in its reactive form (see Section 3.1). In addition, there may be effects that increase the strengths of local polar, ionic and dispersion interactions when water is excluded from an active site. This is a catalytic contribution that most synthetic enzyme models cannot emulate due to their small size.

Working with existing proteins also has potential disadvantages. In practice, proteins are often less robust (*e.g.* may be denatured at extremes of pH and temperature) than synthetic catalysts. Building an enzyme from first principles would undeniably be a stronger demonstration of our understanding of enzyme catalysis, but since this understanding is still imperfect these disadvantages could be outweighed by the higher levels of activity that existing proteins can provide. Enzyme-active sites are purpose-made for catalysis and should be adaptable to fulfil new functions: by virtue of preorganized reactive functionality that can potentially be redeployed to serve a similar catalytic function in processes different from an enzyme's native reaction. The methods of directed evolution (see Chapter 5) provide the means to improve enzyme function in this way.

3.6.1 Enzymes Modified by Addition of Functionality

Adding catalytic functionality to enzymes can in principle provide active sites with functionality that Nature does not – or cannot – make available. Coenzymes represent Nature's solution to the problem, and specialist artificial cofactors can also be attached to enzymes, to explore chemistry that is not normally accessible using the chemical repertoire of the 20 natural amino-acids.

The first practical question is how the additional group is to be introduced. An active-site serine can be modified, *e.g.* the conversion to selenocysteine brings in new reactivity,[47] including new redox chemistry that cannot be realized with the conventional 20 natural amino-acids.[47–49] More convenient is the use of an intrinsically reactive cysteine thiol(ate) nucleophile to attach a reactive artificial group or cofactor to a protein. Carbohydrates,[50,51] hydrophobic and negatively[52] or positively[53] charged groups can be introduced in this way. An intrinsic limitation of the approach is that the usually nonspecific attachment chemistry can lead to the modification of every functional group of a particular kind: so that only one such potential target group should be present in the protein.

Alternatively, non-natural amino-acids can be introduced, by appropriate modifications to the protein biosynthesis machinery. For example, tRNA can be charged with non-natural amino-acids by evolved aminoacyl-tRNA synthetases or some ribozymes.[54–56] Codons have been extended to include four-base codons for non-natural base pairs, to allow the site-specific incorporation of new, synthetic amino-acids into proteins. In this way, fluorescent, glycosylated, metal-ion-binding, and redox-active amino-acids, as well as amino-acids with unique chemical and photochemical reactivity can be incorporated into specific positions of proteins.[55,56] Alternatively, the *residue*-specific incorporation of new amino-acids is possible, because aminoacyl-tRNA synthetases can activate *e.g.* fluorinated amino-acid analogues and incorporate them globally into proteins instead of the original amino-acid (which must be missing from the growth medium). The protein thus generated carries the replacement amino-acid in *all* positions where the original amino-acid was located, allowing engineering of the overall physical and chemical behavior of proteins. Proof of concept is now established for all these techniques, though low translational efficiencies of amino-acid incorporation, typically producing only small amounts of proteins, have so far precluded their widespread use.

3.6.1.1 *Exploiting Modular Build-Up of Binding and Catalytic Features: Chimeras of Binding Proteins and Reactive Chemical Functionality*

Enzymes generally combine binding and catalytic features in a single protein, and in some enzymes structural elements corresponding to binding and

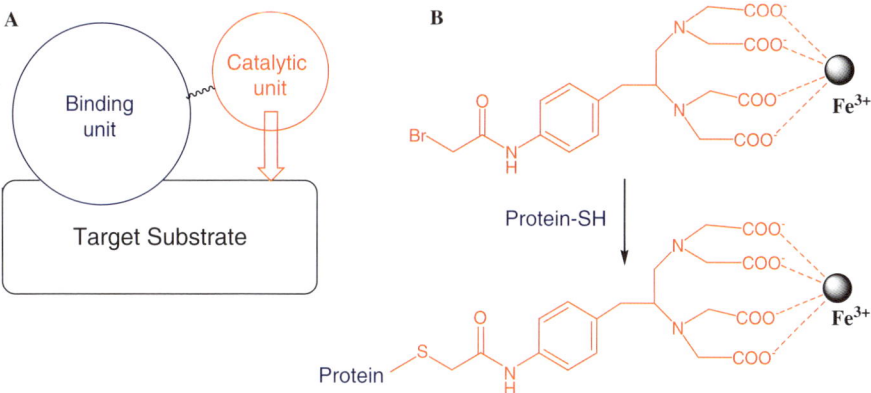

Scheme 3.7 (**A**) Schematic modular design of a catalyst in which binding and catalytic subunits are structurally separate. (**B**) shows a reactive iron(III)-complex (Fe-BABE)[61] that can be attached to a nucleic-acid-binding protein to generate a site-specific nuclease.

catalysis can be separately identified.[57] In natural enzymes the coordination between the binding and catalytic modules is of course highly efficient. However – as we have seen – *de novo* integration of the two is difficult. Simply attaching a binding protein or a binding domain to a catalytic unit would at least increase the local concentration of the catalyst, and could provide a nonspecific, reactive catalyst with a sense of direction.

A successful application of this principle is illustrated by the combination of a staphylococcal nuclease with a small piece of DNA, which was attached to the enzyme via a cysteine residue (actually mutated for the purpose from an original lysine residue). The DNA directs the nuclease to a specific site on a single-stranded DNA (by simple Watson–Crick base pairing), and the resulting conjugate was found to cleave DNA predominantly at the unique site targeted by the attached DNA, creating a semisynthetic nuclease.[58–60] Noller has used a similar reactive probe (Fe-BABE, Scheme 3.7, an iron complex generating hydroxyl radicals) in combination with RNA-binding proteins to define interactions between the ribosomal RNA and ribosomal proteins.[61]

3.6.1.2 Noncovalent Introduction of Reactive Cofactors: Transition-Metal Catalysts in a Protein

Streptavidin is a protein that binds the small molecule biotin exceptionally tightly, and the streptavidin/biotin pair is widely used in biotechnology as a convenient all-purpose affinity tag. The streptavidin-binding pocket is relatively large, and so can accommodate other molecules in addition to biotin. This gave rise to the idea of attaching a "cofactor" catalyst to biotin that would accompany it into the microenvironment of the streptavidin pocket. A series of transition-metal based hydrogenation catalysts were coupled to streptavidin and gave useful stereoinduction (up to 96% ee) in the catalytic reduction of acetamidoacrylic acid to AcNH.Ala.[62–64] As catalysts, these chimeras are less than impressive – the rate accelerations (measured in terms of k_{obs}) are well below 100-fold. Medium effects are small and no other catalytic functionality of the streptavidin pocket is used. But the provision of a chiral environment – enzymes are of course intrinsically chiral reagents – achieved something that is still difficult in many other enzyme models.

3.6.1.3   *Covalent Derivatization of Active-Site Residues to Introduce Reactive Cofactors*

Similar to the hydrogenation catalysts described above in (ii), Section 3.6, covalent chemistry can be used to attach a catalyst. A simple protease, papain, with a reactive cysteine (Cys-25) that normally acts as a nucleophile in peptide cleavage can be derivatized with a monodentate phosphite ligand. The modified side-chain can now complex a rhodium ligand, turning the modified enzyme into a hydrogenation catalyst: though in this case one lacking enantioselectivity.[58] In contrast to (i) the entire active site is refunctionalized here, without a separation of binding and catalytic features.

3.6.2   Site-Directed Mutants of Enzymes: Minimalist Protein Redesign[65]

Recombinant DNA technology makes it very straightforward to generate mutants of existing enzymes in which one or more residues have been changed to another naturally occurring residue. A mutation can be introduced into a particular position by well-established procedures that involve synthesis of a mutagenic primer, one or more PCR reactions and cloning of the altered gene into an expression vector, which is then used to produce protein.

Occasionally, it is possible to turn a noncatalytic into a catalytic protein by changing just one amino-acid. For example, the catalytically inactive phospho(serine/threonine/tyrosine) binding protein STYX was converted to a phosphatase by change of a Gly to Cys, suggested by structural comparison of STYX with homologous catalytic proteins.[66]

Scheme 3.8 shows another successful single amino-acid conversion, of 4-oxalocrotonate tautomerase (4-OT) into an oxaloacetate decarboxylase. The N-terminal residue Pro-1 is converted into an alanine that, with a free amino rather than an imino group, can form a covalent adduct for subsequent decarboxylation.[67] The mutant enzyme catalyzes the decarboxylation reaction relatively well, with a k_{cat} of 0.08 s^{-1} and k_{cat}/K_M of 114 M^{-1} s^{-1}. Interestingly, there are also synthetic model catalysts available for oxaloacetate decarboxylation: a designed peptide ($k_{cat} = 0.0067$ s^{-1}, $k_{cat}/K_M = 28$ M^{-1} s^{-1})[68] and a synzyme (a synthetic polymer: see Section 5.1.1), with $k_{cat} = 0.035$ s^{-1} and $k_{cat}/K_M \sim 35$ M^{-1} s^{-1}. In each case, the decarboxylation is believed to involve Schiff base formation.[69,70] All these systems catalyze the reaction comparably

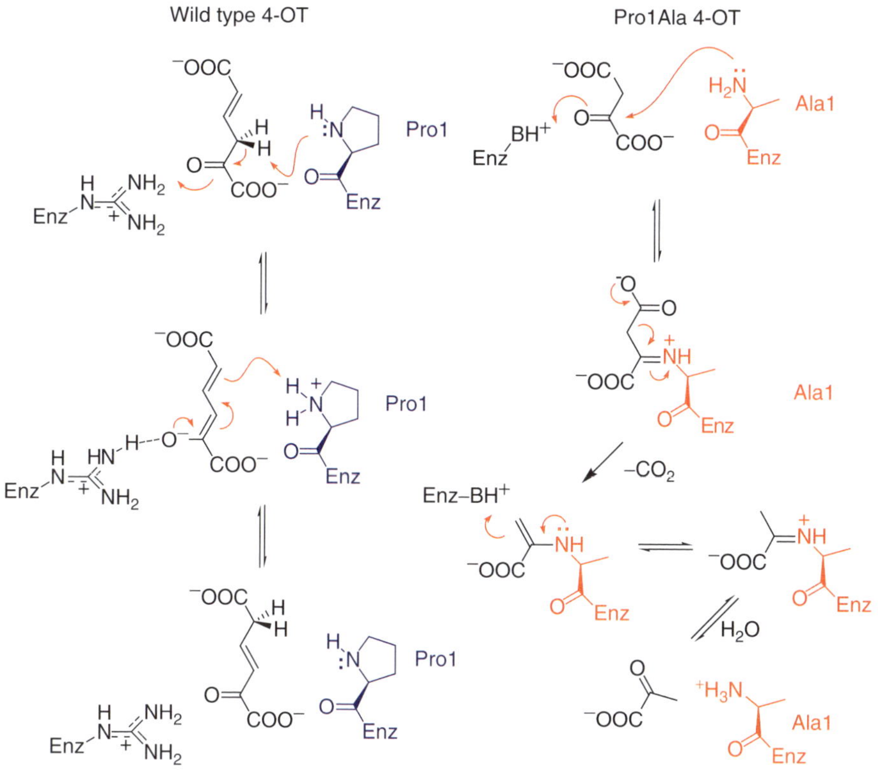

Scheme 3.8   A new activity can be introduced into 4-oxaloacetate tautomerase (4-OT) by switching one residue (Pro-1 to Ala-1) enabling 4-OT to decarboxylate oxaloacetate.[71]

well (with $k_{cat}/k_{uncat} \sim 10^3$–$10^4$), demonstrating that the different approaches have converged on a common, moderately efficient solution.

3.6.3   Exploring Enzyme Promiscuity

Predicting the single key amino-acid residue that might trigger such a switch in activity is naturally difficult. Alternatively, one can ask whether access to new activities is intrinsic in the evolution of enzyme structure. Numerous enzymes show rather broad substrate specificity, performing the same reaction on a variety of structurally related substrates. This is a commercially highly important phenomenon known as *substrate* promiscuity. Familiar examples are lipases and esterases used in bioprocessing and fine chemical synthesis.[72,73] This level of substrate promiscuity is perhaps not very surprising: it means simply that the enzyme pocket can tolerate slight mismatches in the substrate structure, while still doing the same specialized chemistry it has evolved for.

An extension of this concept, to the catalysis of chemically distinct reactions, is more remarkable. Catalytic promiscuity, also called cross- or polyreactivity

Table 3.2 Selected examples of catalytically promiscuous enzymes: with their native and promiscuous activities and the central catalytic feature involved in the native reaction (but not necessarily all other reactions).

Family/Superfamily	Enzyme	Catalytic feature	Native reaction	Promiscuous reaction	Ref.
Chymotrypsin-like serine protease superfamily	α-Chymotrypsin	Asp-His-Ser triad	peptide hydrolysis	ester hydrolysis, phosphate triester hydrolysis	83
Alkaline phosphatase superfamily	Alkaline phosphatase	Metal-activated serine	phosphate monoester hydrolysis	phosphate diester hydrolysis, phosphonate monoester hydrolysis, sulfate ester hydrolysis, phosphite oxidation	84–87
	Arylsulfatase A	Metal-activated formylglycine	sulfate ester hydrolysis	phosphate diester hydrolysis and phosphate diester hydrolysis	79,80
	Phosphonate monoester hydrolase	Metal-activated formylglycine	Phosphonate monoester hydrolysis	phosphate monoester, sulfate monoester, phosphate diester and sulfonate hydrolysis.	Table 5.7
	Exonuclease III	metal-activated water	DNA phosphate diester hydrolysis	DNA phosphate monoester hydrolysis	88
Amidohydrolase superfamily	Phosphotriesterase	metal-activated water	phosphate triester hydrolysis	phosphate diester hydrolysis, ester hydrolysis, lactone hydrolysis	89,90
	Urease	metal-activated water	urea hydrolysis	phosphoramidate hydrolysis	75
	Dihydroorotase	metal-activated water	reversible dihydroorotase hydrolysis	phosphate triester hydrolysis	90

(or even "moonlighting"), is defined as the ability of some enzymes to catalyze the making and breaking of completely different bonds in the same active site (involving different transition states): albeit – not surprisingly – with lower efficiency than the native reaction. Textbooks usually describe enzymes as highly specific in their substrate-recognition properties and in the reaction they perform, with one enzyme carrying out only one function. However, there are now many known exceptions to the "one enzyme – one activity" rule. And the increasing number of enzymes identified as catalytically promiscuous suggests that promiscuity is not a rare but a rather widespread feature of enzymes: currently about two dozen examples of catalytically promiscuous enzymes are known, some with surprisingly high efficiency.[74–78] In some cases, the catalytic proficiencies for the promiscuous reaction (measured by $(k_{cat}/K_M)/k_2$) can be as high as 10^{13} to 10^{18}.[79,80] These figures match – or even exceed – the typical rate accelerations exhibited by some enzymes. Apparently, the nucleus of reactivity located in an enzyme-active site, equipped as it is with activated functional groups in a functional microenvironment, can go at least some of the way towards providing more "open-access" catalysis. These examples challenge the idea that efficient catalysis necessarily requires specialization.

The idea that promiscuity might play an important role in the evolution of new enzyme functions was first proposed by Jensen in 1976[81] and later by O'Brien and Herschlag.[76] It is widely accepted that a gene-duplication event is needed in order to evolve a new enzyme function, thereby relieving one gene copy of the selective pressure and allowing it to mutate randomly.[82] Providing a promiscuous "head-start" activity that might be further improved under selective pressure provides an attractive scenario for the development of new functions – in the laboratory as well as for natural or directed evolution. Table 3.2 gives an indication of the sorts of promiscuity discovered to date, and the development of novel catalytic functions using this route is discussed in more detail in Chapter 5.

Conceptually the observation of catalytic promiscuity shows us that enzymes too can be considered models for other enzymes: Nature also starts slowly, using existing protein architectures that are further improved by iterative evolution cycles. Similarly cyclodextrins, for example, show a number of different activities that teach us about catalytic starting points. The difference between Nature's method and model design is that the cycles of improvement can now be carried out much faster, by randomization/selection (see Chapter 5), than by rational redesign followed by painstaking synthesis. Fatal to both, approaches, however, is to observe no catalysis at all, because "negative hits" are of no help in either: the model builder cannot learn from them and Nature cannot sustain evolutionary cycles if all clones are lost.

Recommended Further Reading

Houk, K. N., Leach, A. G., Kim, S. P., Zhang, X. Y., Binding affinities of host-guest, protein-ligand, and protein- transition-state complexes. *Angew. Chem., Int. Ed. Engl.* **2003**, *42*, 4872–4897.

Kraut, D. A.; Carroll, K. S.; Herschlag, D., Challenges in enzyme mechanism and energetics. *Ann. Rev Biochem.* **2003,** 72, 517–71.

Mader, M. M.; Bartlett, P. A., Binding Energy and Catalysis: Implications for TS Analogs and Catalytic Antibodies. *Chem. Rev.* **1997**, *97*, 1281–1301.

Toscano, M. D.; Woycechowsky, K. J.; Hilvert, D., Minimalist active-site redesign: teaching old enzymes new tricks. *Angew. Chem., Int. Ed. Engl.* **2007**, *46*, 3212–36.

CHAPTER 4

Enzyme Models Classified by Reaction

Enzymes achieve their extraordinary catalytic efficiencies by the application of ordinary chemistry: using reaction mechanisms that are normal – under the special conditions of the active site. In this chapter we introduce the major classes of reaction they catalyze, outline the active-site arrangements found in typical enzymes of each class, then summarize progress in attempts to model particular features of the mechanisms thought to be involved. Model systems are discussed in each case in increasing order of complexity, emphasizing the design principles involved. Design becomes more difficult, and eventually even less relevant, for systems catalyzing pericyclic reactions (Section 4.6), which may use no functional groups. At this point combinatorial or iterative methods, often based on biological technology, become important.

Most of the enzymes discussed in this chapter are hydrolases of various sorts. The *biosynthesis* of proteins, polynucleotides, *etc.* involves multienzyme systems working in a production line, with complex arrangements in place for substrate and product handling and quality control. But the individual chemical steps of the reactions concerned use the same basic principles in all cases. To identify those basic principles we base our model systems on the less complicated reactions of enzymes that can work individually. (Of course, most if not all enzyme reactions are subject to some sort of regulatory control.)

4.1 Acyl Transfer

The first enzymes to be studied and (broadly) understood mechanistically were hydrolases such as chymotrypsin, which catalyze the hydrolysis of amides or esters (or in some cases, both). Acyl transfer is the most common group- (as opposed to proton-) transfer reaction, and is catalyzed by thousands of different enzymes.

From Enzyme Models to Model Enzymes
By Anthony J. Kirby and Florian Hollfelder
© Anthony J. Kirby and Florian Hollfelder 2009
Published by the Royal Society of Chemistry, www.rsc.org

Proteolytic enzymes cleaving specific peptide bonds regulate various aspects of the life cycle in all organisms, and their action is so central for biochemical processes that they are encoded by at least 2% of every known genome.[91]

The synthesis and degradation of peptides, proteins and glycerides all involve transfers of carboxylic acyl groups. Phosphoric acyl (phosphate, or phosphoryl) transfers are different – and important – enough to deserve separate treatment, but these two main groups of acyl transfer reactions have many features in common. In particular, high specificity is essential in the synthesis of both proteins and nucleic acids. Once formed, both biopolymers are extremely stable compounds, so that their degradation reactions are extraordinarily slow under physiological conditions in the absence of the relevant enzymes. Much of the earliest work on enzyme mechanisms and models involved reactions in which carboxylic acyl groups are transferred. This work had a substantial influence on the way the area developed, and makes a logical and convenient starting point for this Chapter.

4.1.1 The Serine Proteases. Typical Active Sites

(Carboxylic) acyl transfer involves the displacement of a leaving group by a (usually better) nucleophile, typically by way of a high-energy, and thus short-lived, tetrahedral intermediate (Scheme 4.1 (**A**)).

A complete hydrolase reaction involves transfer of the acyl group to water, but early evidence for acyl enzymes as intermediates in the reactions of activated substrates indicated nucleophilic catalysis by the enzyme, acting through an amino-acid side-chain nucleophile present in the active site (Scheme 4.1 (**A + B**)). Most of the nucleophilic groups available on amino-acid side-chains (Table 1.1) can be shown to act as nucleophiles towards the amide group in simple intramolecular reactions near pH 7 (see Section 1.4), and most of them do in fact act as nucleophiles in the various classes of peptidase.

The active-site nucleophile was identified in a handful of landmark cases, by classical sequencing of the acyl enzymes, as a specific serine. Subsequent protein crystallography revealed the 3-dimensional arrangement of functional groups in

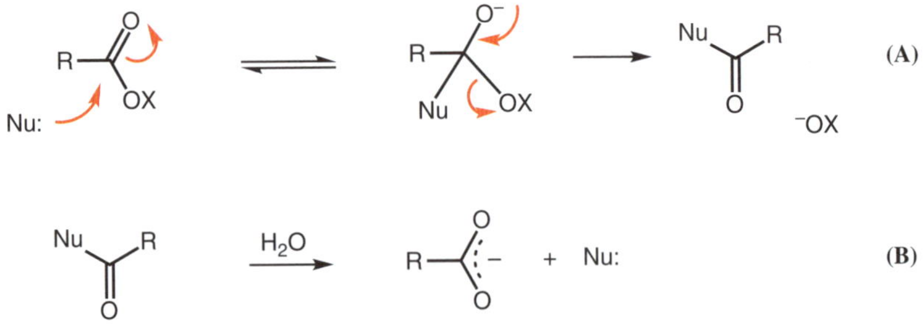

Scheme 4.1   Acyl transfer involving nucleophilic catalysis of the transfer of the acyl group to water.

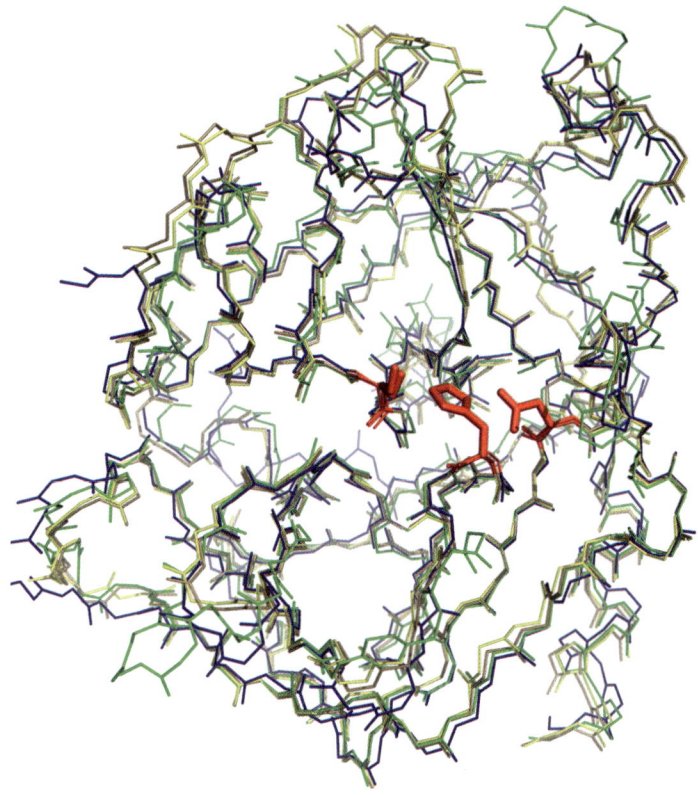

Figure 4.1   Structural alignments of the backbones of α-chymotrypsin (blue; pdb code: 5cha), porcine pancreatic elastase (green, 1est) and two bovine trypsin structures (5ptp, 1tld; yellow and sand) show closely similar active-site arrangements of Ser, His and Asp (red) for these (as for many other) related serine proteases. (Produced using PyMol.)

the vicinity of this serine in the active site (Figure 4.1). Such detailed evidence cannot by itself define the mechanism of a reaction catalyzed by an enzyme, but it does at least define the structure of the starting material: in the sort of detail one would expect to have when starting a mechanistic investigation involving reactions of small molecules: and may allow possible mechanisms to be ruled out.

The same basic three-dimensional arrangement of the same three functional groups (the "catalytic triad") is found in hundreds of serine proteases. These are distinguished by their different substrate specificities, and most are presumed to have evolved from the same parent enzyme, as evidenced by extensive homologies in their primary amino-acid sequences. (Thus accounting for the obvious similarities in tertiary structure shown in Figure 4.1.) A fascinating exception is a second, separate group of serine proteases, with the same active-site arrangements but quite different primary sequences (the classical example is subtilisin from *Bacillus amyloliquefaciens*). Evidently, enzyme evolution arrived at the same solution to the mechanistic problem of how to hydrolyze the amide bond efficiently by two independent routes (convergent evolution): and this solution is so successful that it is has come to be used by a large number of different enzymes (divergent evolution).

4.1.1.1   The Active-Site Environment . . . and Mechanism

We know (Section 1.1.2) that the active site must provide a "highly sophisticated reaction vessel", tailored to the needs of the specific reaction and the specific substrate. Substrate recognition is taken care of by an appropriate binding site, in exactly the right place for the interactions with the catalytic apparatus. Chymotrypsin, for example, is specific for peptide sequences with an aromatic amino-acid at the reaction site by virtue of a "hydrophobic pocket", 10–12 Å deep, of just the right dimensions to accommodate an aromatic ring (Ar in Figure 4.2), which is typically about 3.5 Å thick. But the catalytic apparatus and the binding interactions have to operate in three dimensions, and the "catalytic triad" identified in Figure 4.1 needs further mechanistic support, specifically stabilization of the developing negative charge on what was the substrate $C=O$ oxygen atom. This is taken care of, to a perfectly tailored extent, by hydrogen bonding to the N–H protons of two backbone amide groups, belonging to Ser-195 and Gly-193 (Figure 4.2), the key components of the "oxyanion hole". The tetrahedral intermediate **E.THI** produced is stabilized sufficiently only for it to exist briefly as a high-energy species. This will be close in structure as well as energy to the transition state for the acylation step (and thus for the rate-determining TS for the complete protease reaction under most conditions). However, its geometry *as it forms* is inevitably optimal for reversion to substrate, and for reaction to proceed further the local conformation has to change, in such a way that the departing amino-group of the leaving peptide $R'NH_2$ can acquire the necessary proton, most likely from the imidazolium group created in the first step of the reaction.

This sequence of events produces a more stable intermediate, the **acyl enzyme** (Figure 4.2). This is an ester, and thus more reactive than the amide (though still of relatively low *intrinsic* reactivity), which must itself be rapidly hydrolyzed: most economically by a rerun of the mechanism by which it was formed, using the catalytic triad but with water in place of the serine OH as the nucleophile. The acyl enzyme does not accumulate during reactions of natural substrates: intermediates with significant lifetimes would tie up the enzyme and reduce turnover.

It is worth noting that enzymes (esterases and lipases) that have evolved specifically to hydrolyze ester linkages (the same basic reaction as the hydrolysis of the acyl enzyme) nevertheless mostly use the familiar catalytic triad mechanism of Figure 4.2; with serine as the primary nucleophile in conjunction with a recognizable anion hole. However, the phospholipase A2 group of enzymes do use the direct mechanism, with a His-Asp dyad catalyzing the attack of water as the nucleophile.[92] These enzymes are of special interest because they often handle substrates that are little or not at all soluble in water, and must act at the lipid/water interface of membranes.

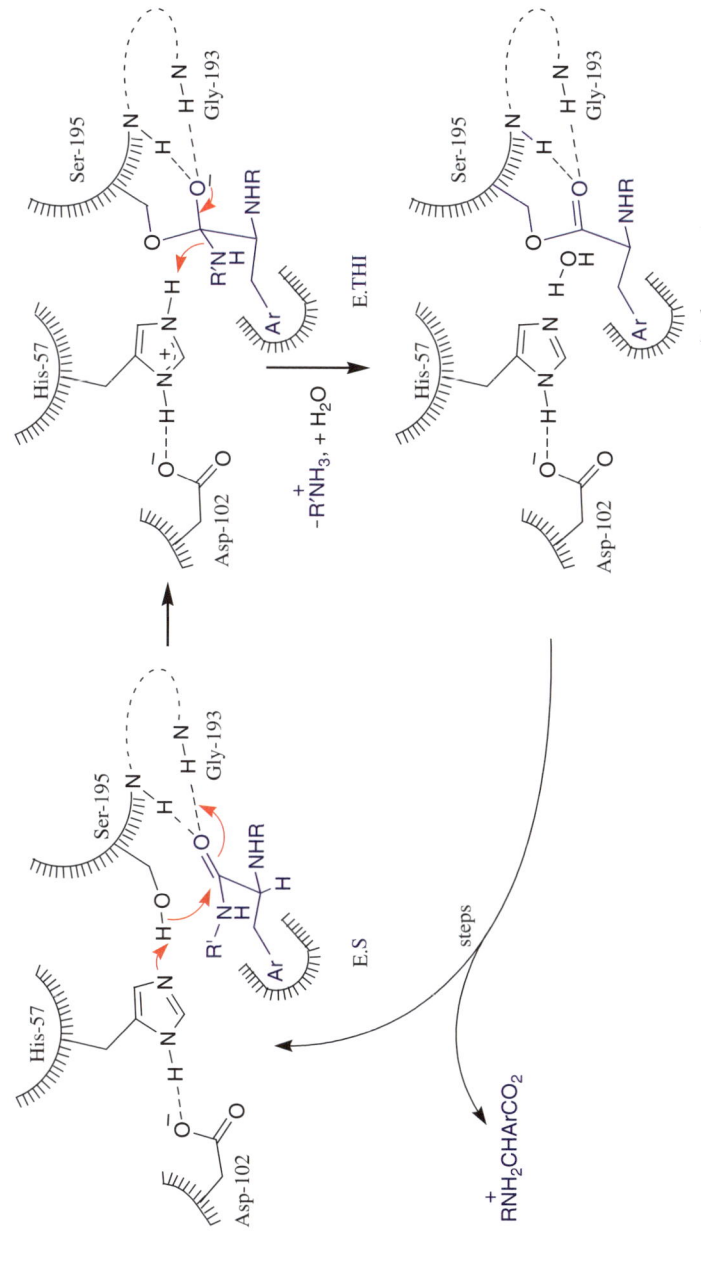

Figure 4.2 Key steps in the cleavage of a peptide bond by α-chymotrypsin. The substrate RNH.CH(CH₂Ar).CO–NHR′ is orientated in the Michaelis complex **E.S** by binding interactions involving its functional groups and especially the aromatic side-chain CH₂Ar in the hydrophobic binding site. For further discussion see the text.

This short summary has identified four key steps in which bonds between heavy (nonhydrogen) atoms are formed or broken, but glosses over a great deal of mechanistic detail. An enzyme uses properly placed functionality in the tailor-made microenvironment of its active site to reduce the free-energy barriers to what would be slow, rate-determining steps: to a point where processes that are normally too fast to be rate limiting can themselves become significant. These include the diffusion of reactants in and out of the active site, conformational changes that are an essential part of the reaction, and proton transfers. These processes were first characterized, and are most easily studied in detail in simple systems, including many designed as models for specific facets of enzyme reactions.

4.1.1.2   Intramolecular Models

The serine proteases catalyze the hydrolysis of peptide amide bonds with accelerations of the order of 10^{10}, compared with the nonenzymatic hydrolysis of amide bonds under the same conditions, and the availability of increasingly detailed structural information for the enzymes has inspired a very large number of mechanistic investigations using models. The three functional groups brought together in the active site are far apart in the protein primary sequence, so – not surprisingly – denaturation of the protein destroys catalytic activity. The conserved 3-dimensional arrangement of Asp, Ser and His is thus clearly essential, and synthetic oligopeptides like Asp-Ser-His, which simply bring them closer, are no more active than the sum of the constituent parts.[93] These parts – namely carboxylic acids, alcohols and imidazoles – have individually no measurable effect on the rates of hydrolysis of ordinary amides in aqueous solution near neutrality. The mechanistic work described in Section 1.3 that defined the detailed mechanisms of general acid, general base and nucleophilic catalysis by these groups used substrates activated in the acyl or the leaving group.

The key to significant reactivity in the hydrolysis of an unactivated amide, or in attack by OH more generally, is (as would be expected – see Section 1.5) the formation of the new covalent bond, between the OH group and the amide $C = O$. This is typically the rate-determining step for the enzyme-catalyzed reactions of natural substrates, and thus the point at which catalysis is most clearly expressed. Making the process intramolecular – and thus a cyclization reaction – can make available reactions going at measurable rates, thus offering the possibility of developing simple systems with reactivities comparable to similar reactions catalyzed by enzymes. An important additional benefit is the availability of accurate, detailed information about the preferred geometry of approach of the nucleophile.

A successful early example placed the OH group of a benzylic alcohol in the *ortho* position of simple benzamides **4.1** (Scheme 4.2).[94,95]

The effective molarity (EM, see Section 1.4.1) of the hydroxyl group in the intramolecular cyclization of **4.1** ($R_1 = R_2 = H$) is $\sim 10^5$. It is some 1000 times higher for **4.1** ($R_1 = Me$, $R_2 = Ph$):[8] this is the well-known gem-dialkyl effect,

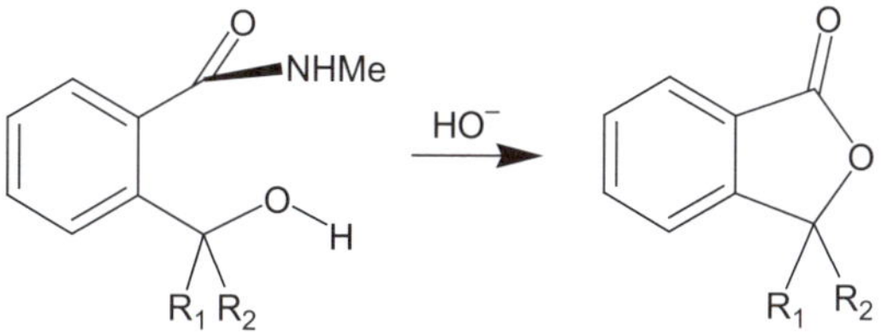

Scheme 4.2 Efficient neighbouring group participation, resulting in the cleavage of the amide group of **4.1**, is enhanced by the gem-dialkyl effect (see the text).

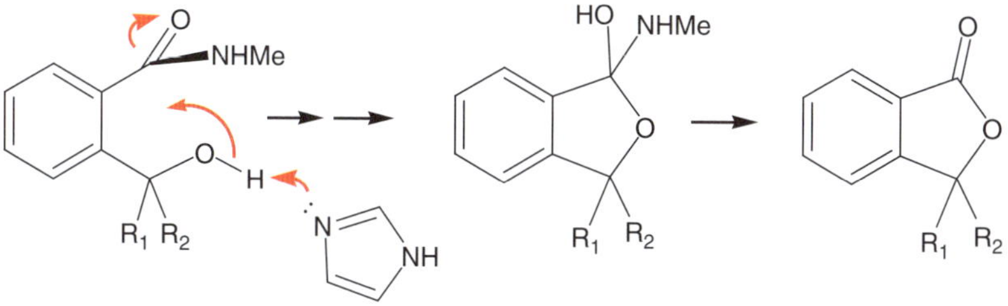

Scheme 4.3 General base catalysis of the intramolecular cyclization of Scheme 4.2.

which in this system favours the equilibrium constant for lactone formation by a similar amount. The cyclizations are fast enough in both cases for buffer catalysis to be easily measurable. Most significant in the present context is the (kinetically) general base catalysis observed with imidazole buffers. The overall reaction (Scheme 4.3) is thus a good model for the acylation step of the reaction taking place at the active site of a serine protease. The OH group models the nucleophilic serine and (external) imidazole the active-site histidine. Imidazole can be shown – with confidence in the simple intramolecular system – to act (mechanistically) as a general base (Scheme 4.3), and the characteristic "kinetic profile" of the reaction provides a model for the assignment of mechanism for the same reaction in an enzyme. This is all the more convincing because the reactions proceed at comparable rates. To make the imidazole-catalyzed lactonization of **4.1** as fast as the chymotrypsin-catalyzed hydrolysis of a natural dipeptide substrate would require a catalyst concentration of about 20 M. Such a concentration is not physically possible for a bimolecular reaction, but an EM of 20 M should be attainable in an intramolecular reaction – and certainly in an enzyme-active site.

It is interesting to compare the development of this model with a similar investigation starting from the general base-catalyzed reaction. An *ortho* imidazole acts as a general base to catalyze the hydrolysis of a neighbouring aryl acetate group (**4.5**, Scheme 4.4), as might be expected from the closely similar

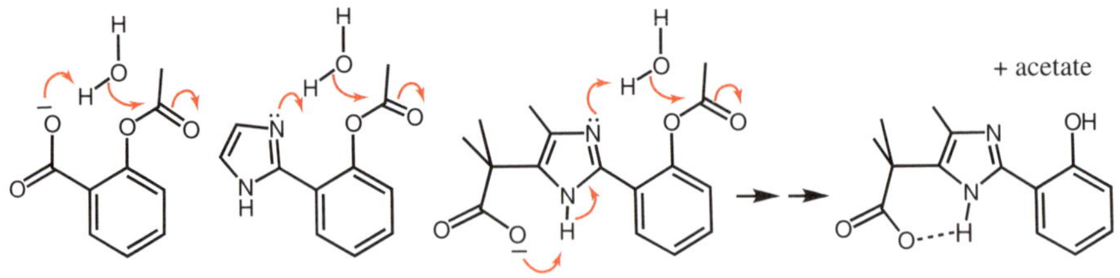

Scheme 4.4   Evolution of a model for the catalytic triad.

reaction of the carboxylate group of aspirin **4.4** (see Section 1.4.1).[96] The reaction is a few times faster for **4.5**, as expected for the more strongly basic imidazole. The system was then extended to model the full catalytic triad by adding a carboxy group, placed to allow efficient proton transfer from the imidazole NH through a 6-membered ring.

The effect was minimal: **4.6** is not hydrolyzed significantly faster than **4.5**, and the effect of the extra carboxyl group amounted to no more than a factor of three.[97] Negative evidence of this sort is by no means conclusive, and general base catalysis of nucleophilic attack by imidazole has been identified even in intermolecular processes. However, the relevant hydrogen bond drawn in **4.7** has not been identified as a strong one in structural studies, and the CO_2^- group appears to be particularly weakly basic for a carboxylate anion, so that the proton transfer never becomes thermodynamically favourable. Thus, though it is clear that Asp-102 plays a significant part in the mechanism of action of the serine proteases (for example, replacing it by Ala reduces k_{cat} for the subtilisin-catalyzed hydrolysis of a synthetic tetrapeptide substrate 3×10^4-fold[98]), its role may be primarily structural, helping – amongst other things – to position the operational (imidazole) general base.

Purpose-built models have yielded little positive information about the likely quantitative contribution towards catalysis of the oxyanion hole,[99] and the most relevant evidence comes from a site-directed mutagenesis experiment on the enzyme. This is possible with subtilisin because one of the two amide groups making up the oxyanion hole is a side-chain asparagine. Replacement of Asn-155 by alanine has the effect of reducing k_{cat} by 2–300-fold for the reaction of a mutant enzyme with a synthetic peptide substrate.[100]

Modest stabilization of a phenolate oxyanion by hydrogen bonding to a neighbouring amide is measurable in salicylamides **4.8** (Scheme 4.5), when pK_as are compared with those of the corresponding tertiary amides **4.9**,[99] but there is still no kinetic demonstration of a significant kinetic effect in a model system. The usual problems of low effective molarities in any relevant system are compounded by competition for any but the strongest intramolecular hydrogen bonding with H-bonding to solvent water.

The most convincing models for the stabilization of oxyanions by oxyanion holes are supramolecular hosts designed to bind anions. The tetralactam **4.10** (Scheme 4.6), for example, binds oxyanions XO^- strongly in CD_2Cl_2, by virtue

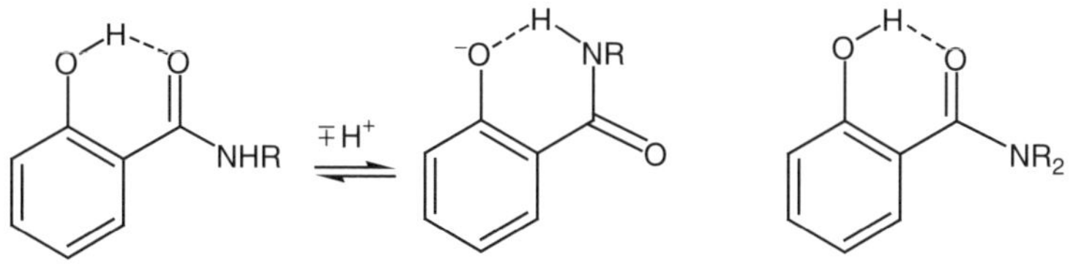

Scheme 4.5 Modest stabilization of the phenolate oxyanion **4.8⁻** by hydrogen bonding to the neighbouring amide NH lowers the pK_a of the phenol OH by 1.2 units compared with the corresponding tertiary amide.[99]

of hydrogen bonding to the two amide groups of one of the isophthalamide centres:[101] consistent with the idea that the microenvironment of the active site is carefully tuned to favour precisely this sort of interaction. However, this property has not been combined with a catalytic system in a complete amide hydrolase model, and the effects are inevitably weaker in solvent water.

4.1.1.3 Supramolecular Models

The low kinetic reactivity of the amide group and the inefficiency of potential artificial enzymes as catalysts have combined to make the attempted systematic development of artificial amide hydrolases necessarily based on the reactions of activated substrates. Esters of *p*-nitrophenol, though not ideal mechanistically, have been a popular choice because they are particularly convenient experimentally: the *p*-nitrophenolate anion has a strong visible absorption at 400 nm, making its release easy to followed at neutral pH and above; and is a good leaving group. (The pK_a of *p*-nitrophenol is 7.14, making its esters relatively reactive: the half-life of *p*-nitrophenyl acetate in water at 25 °C and pH 7 is 2 weeks, compared with some 85 years for ethyl acetate.)

The choice of model systems also depends on practical considerations. The earliest work on nonprotein enzyme models used systems that were readily available, equipped with functional groups, and known to bind compounds – and thus potential substrates – in aqueous solution. Most prominent by far are the cyclodextrins.

4.1.1.3.1 Cyclodextrins. Cyclodextrins were introduced in the previous chapter, Section 3.4, as readily available water-soluble hosts, known to bind hydrophobic guests or parts of guests with distinct shape and size selectivities in their hydrophobic central cavities: these vary conveniently in size from β to γ-cyclodextrin (Figure 4.3). The bound guest is often protected from attack by outside reagents, but normally open to what is effectively intramolecular attack by the OH groups of the host (or even in rare cases by an appropriately reactive second guest).

Aromatic rings in particular are bound in the hydrophobic cavity, with any polar substituents projecting into the solvent. Binding is not tight: dissociation

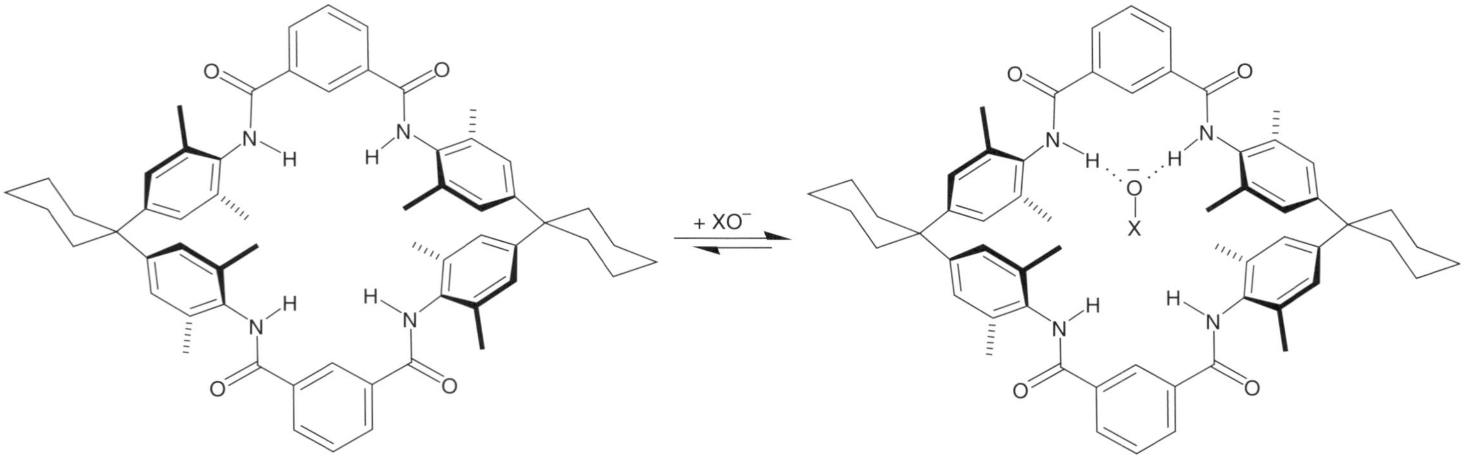

Scheme 4.6 Tetralactam **4.10** binds oxyanions XO⁻ strongly in CD₂Cl₂, by virtue of hydrogen bonding to the amide groups of one of the isophthalamide centres.

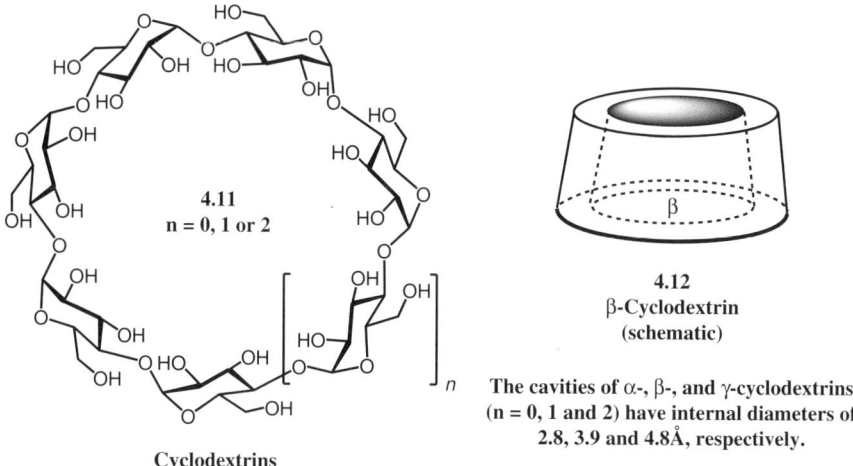

4.11
n = 0, 1 or 2

4.12
β-Cyclodextrin
(schematic)

The cavities of α-, β-, and γ-cyclodextrins
(n = 0, 1 and 2) have internal diameters of
2.8, 3.9 and 4.8Å, respectively.

Cyclodextrins

Figure 4.3 Structures of β-cyclodextrin (**4.11**, $n = 1$) and its readily available homologues α- and γ-cyclodextrin ($n = 0$ and 2, respectively). A simplified representation similar to **4.12** is often used: X-ray structures show that the torus is slightly tapered, giving β-cyclodextrin a ring of seven primary CH_2OH groups at the narrower end and fourteen secondary 2- and 3-OH groups encircling the wider entrance to the hydrophobic cavity.

constants fall in the range $K_d = 10^{-2.5 \pm 1.1}$ M, corresponding to free energies of binding in water ($-\Delta G_a$) of 3–4 kcal mol^{-1}.[16] K_M values for many enzyme reactions similarly fall in the millimolar region, thus allowing rapid exchange with guest (substrate) molecules in solution.

The two rings of hydroxyl groups on the rims of the cyclodextrin cavity are in position to react with electrophilic functional groups of many bound substrates, suggesting cyclodextrins early on[39] as possible mimics for the serine proteases.[40] The neutral hydroxyl group is not particularly reactive, and simple cyclodextrins are not significant catalysts near pH 7. But at high pH in water-activated esters with hydrophobic components acylate cyclodextrins more rapidly than they are hydrolyzed. As discussed in Section 3.4, these reactions are not particularly efficient, with rate accelerations (k_{cat}/k_{uncat}) over background at pH 10.6 of 9 and 96, respectively, for the reactions of *p*- and *m*-nitrophenyl acetates with β-cyclodextrin (Scheme 3.2 and Table 4.1).[42] Nor are they catalytic, because the initial products are unactivated esters of the cyclodextrin (Scheme 3.2), but their reactions provided useful early models for the initial, acylation steps of the enzyme reactions.

Much more efficient substrates for the reaction with β-cyclodextrin have been designed specifically to optimize binding of the transition state. Results for two of the most effective of these compounds (Scheme 4.7) are compared in Table 4.1 with data for the reactions of β-cyclodextrin with the two nitrophenyl acetates. Note first that the conditions for the two sets of measurements are different: the reactions with **4.13** and **4.14** had to be measured in a mixed

Table 4.1 Catalysis by β-cyclodextrin of the hydrolysis of *p*-nitrophenyl ester substrates.

Substrate	$k_{cat}(min^{-1})$	$K_M(mM)$	k_{cat}/K_M	k_{uncat}	k_{cat}/k_{uncat}	Proficiency $(k_{cat}/k_{uncat}/K_M)$	$\Delta G_{TS} kcal\ (kJ)\ mol^{-1}$
pNPA[a]	0.064	6.1	10.5	6.9×10^{-3}	9.1	1520	4.3 (18.2)
mNPA[a]	0.44	8	5.5	4.6×10^{-3}	96	12 000	5.6 (23.3)
4.13[b]	0.094	7.5	12.5	2.83×10^{-7}	3.3×10^5	4.43×10^7	10.4 (43.6)
4.14[b]	0.092	5.7	16.1	1.58×10^{-8}	5.8×10^6	1.02×10^9	12.3 (51.4)

Notes. [a]*p*- and *m*-nitrophenyl acetates: measured at pH 10.6 in water.[42]
[b]At "pH 10" in 60% DMSO – 40% water ("pH 10" corresponds to phosphate buffer with pH 6.39 in water).[102]

4.13 **4.14** **4.15**

Scheme 4.7 Improved substrates for β-cyclodextrin "catalysis" (see the text).

aqueous-organic solvent for solubility reasons: as a result the most striking differences are in k_{uncat}.

A second important difference is that the part of the substrate designed to be bound in the cyclodextrin cavity is the ester acyl (ferrocenyl) group rather than the nitrophenolate leaving group. This makes for increased flexibility for the changes in conformation necessary for rapid cleavage of the bond to the leaving group (its actual departure involves diffusion out of the cavity in the case of the simple aryl esters). The result is the exceptionally high catalytic proficiency for this "optimized" substrate.

A major limitation of these systems is that the host–guest reactions are efficient only for activated esters. Thus, the ethyl ester **4.15** of the same ferrocenyl acrylic acid is bound well enough by the cyclodextrin, but the half-life of the bound **4.15** is more than two years, compared with less than 10 seconds for the *p*-nitrophenyl ester **4.13** under the same conditions (Table 4.1).[103] The difference in intrinsic reactivity of the two esters cannot account for this, because k_{OH} is only some 50 times slower for the ethyl ester. Evidently, the cyclodextrin OH ($pK_a \sim 12.3$) is simply not basic enough to displace ethoxide efficiently (the pK_a of ethanol is 15.9) without further assistance. (Reactions with amides, which are after all the natural substrates of chymotrypsin, must be even less promising.)

An obvious next step in the improvement of a cyclodextrin as a serine proteinase mimic is to add an imidazole group to act as a general base, which should allow catalysis to be followed near pH 7. Synthetic difficulties have prevented extensive exploration of this approach. The only clear conclusions appear to be that an imidazole linked (through N) to C(2) of a sugar (and thus at the wider end of the cavity of β-cyclodextrin) is significantly more effective than the same group at a primary C(6) centre:[104] and that a specific OH group is acylated in the latter case.[105]

4.1.1.3.2 Synthetic Models. The major developments in supramolecular chemistry that began in the 1980s included substantial contributions from leading synthetic groups to the design and construction of state-of-the-art enzyme models, many of them inspired by the familiar active-site arrangements of the serine proteases. The intrinsic unreactivity of the amide group, combined with the general inefficiency of artificial enzymes as catalysts, meant that the systematic development of artificial amide hydrolases was necessarily

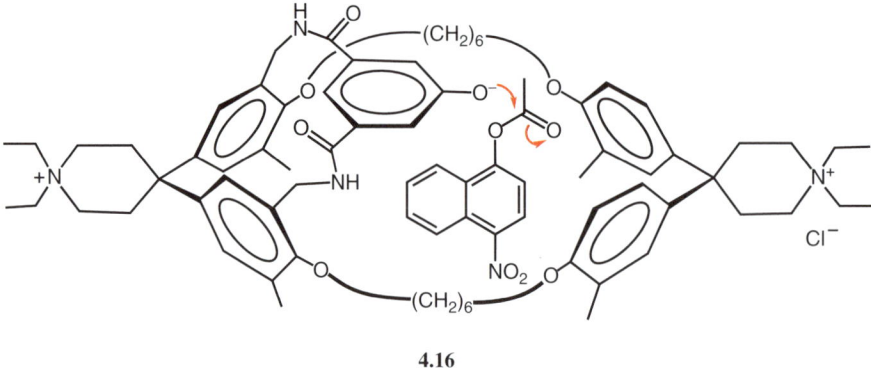

4.16

Figure 4.4   The tetraoxa[6.1.6.1]paracyclophane **4.16** has a hydrophobic binding cavity large enough to accommodate a naphthalene ring, with a phenolic nucleophile (pK_a ~ 8.4) on one face.

based on the reactions of activated substrates. Esters of *p*-nitrophenol are particularly convenient because the release of the *p*-nitrophenolate anion is readily followed at neutral pH and above (see above, Section 4.1.1.3).

The second general problem, that large organic molecules are typically usefully soluble only in organic solvents, whereas most enzymes operate in water, can be approached in two ways. Either by using organic solvents: which has the advantage of allowing hydrogen bonding to play a major role in host–guest binding. Or by building sufficient polar groups into the model system to make it usefully water soluble, as in the tetraoxaparacyclophane host shown in Figure 4.4.[106] The design of the water-soluble host allowed variation of the cavity size, and of the bridging group carrying the nucleophile.

In the most effective system (Figure 4.4) the acylation of **4.16** by 4-nitro-1-naphthyl acetate followed Michaelis–Menten kinetics, with $k_{cat} = 8.52 \times 10^{-3}$ s^{-1} and a K_M of 0.76 mM at pH 8. Acylation by the bound substrate was 472 times faster than its hydrolysis reaction in solution. A system equipped with an aliphatic OH group as the nucleophile showed no significant catalysis.

The most substantial approach to a fully synthetic mimic for the serine proteases was reported in a series of papers from Cram's group.[107] The final target system (**Target**, Figure 4.5: not water soluble) was designed to deliver a nucleophilic primary OH group, hydrogen bonded to the basic N of an imidazole that is itself H-bonded to carboxylate, close to an appropriately oriented binding site: which was expected, from results using the simpler model **4.17**, to bind primary ammonium cations with free energies $-\Delta G°$ of over 10 kcal/mol in solution in CDCl$_3$.

System **4.17** was rapidly acylated by the *p*-nitrophenyl ester **4.18** of L-alanine, faster by a factor of the order of 10^{11} than the acylation under the same conditions of the model nucleophile **4.19**, which lacks the binding site. The active nucleophile was the anion of the benzylic OH group, as was to be expected under the conditions (CDCl$_3$ solvent "buffered" with i-Pr$_2$EtN/i-Pr$_2$EtNH$^+$ ClO$_4$$^-$).

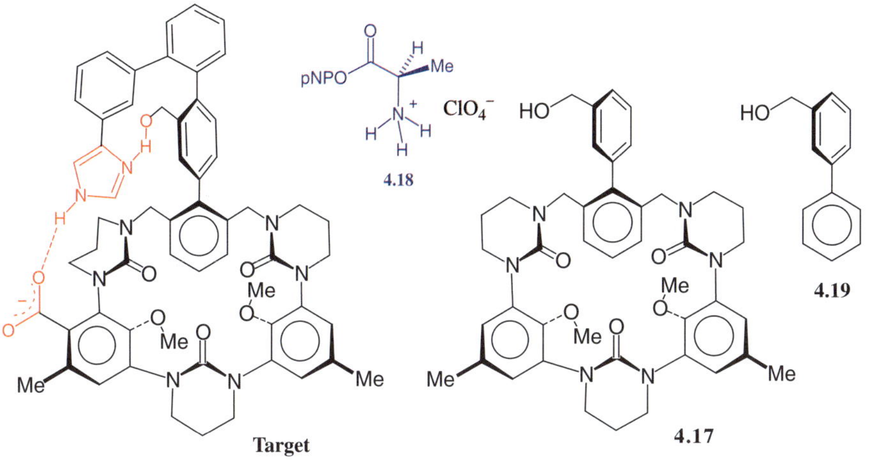

Figure 4.5   Cram's fully featured target system (**Target**) designed to model the serine proteases. The ring of nucleophilic oxygens binds $CH_3NH_3^+$ (as the picrate) in the stripped-down system **4.17** with a free energy of complexation $\Delta G°$ of -12.7 kcal/mol., in solution in $CDCl_3$. Such large organic molecules are not soluble in aqueous media: which would in any case minimize the hydrogen-bonding interactions involved in the binding of the NH_3^+ group.*

The next stage was the incorporation of the imidazole group of the **Target**, to act as a general base and thus support reaction at a pH corresponding more closely to neutrality in water. A 30-step synthesis produced the designed host molecule shown in Figure 4.5, which was indeed rapidly acylated by the *p*-nitrophenyl ester **4.18** of L-alanine.* However, the initial product proved to be the acyl imidazole **4.20** (Figure 4.6). In wet $CDCl_3$ this was subsequently converted to the looked for acyl model-enzyme **4.21**, in a reaction a few times faster than its hydrolysis.[107]

In this carefully designed model the OH and imidazole groups are exhibiting their normal reactivity, rather than the special mechanism supported by the 3-dimensional structure of the serine proteases.

These were important results, which – combined with the emerging evidence that catalysis by even the most efficient model systems could not readily be extended to substrates with unactivated groups[103] – had the significant long-term effect of discouraging the large set-piece synthetic approach to highly designed enzyme models.

4.1.2   SH Hydrolases

Thiolate anions are well known to be powerful nucleophiles for both sp^3 and sp^2-carbon, so it comes as no surprise that enzymes with active-site cysteines are

---

* Following the original authors[107] alanine is shown "for convenience," here and in Figure 4.6, in its D-configuration.

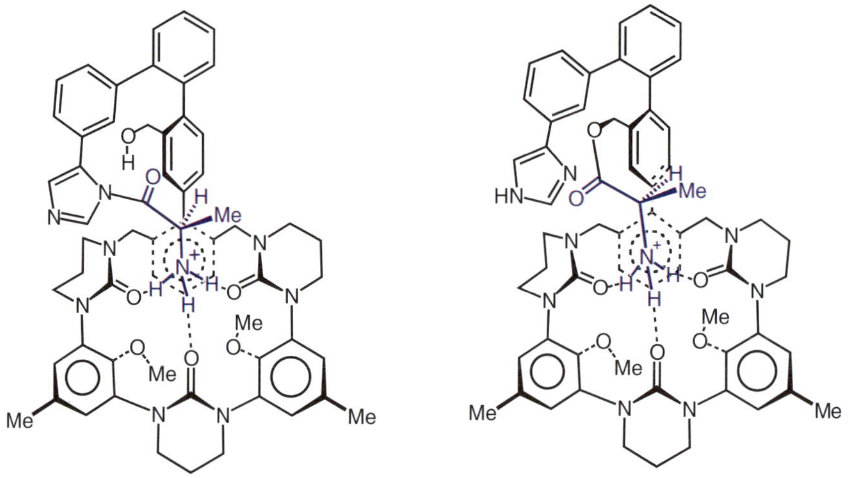

4.20   4.21

Figure 4.6   The improved Target host from Figure 4.5 (see the text) is rapidly acylated on the imidazole nitrogen (**4.20**). In the presence of small amounts of water acyl transfer to the benzylic OH, to give **4.21**, competes with hydrolysis. The acylating alanyl group is picked out in blue.*

involved in many protease reactions. Cysteine proteases resemble the serine proteases in having a histidine imidazole associated with the nucleophilic group, with the papain group using a Cys-His-Asn catalytic triad; but differ from them in that the active form involves the thiolate anion, in an ion-pair with the imidazolium group of the histidine. The generally accepted mechanism is outlined in Figure 4.7; which is drawn to emphasize the marked similarities to the mechanism of action of the serine proteases (Figure 4.2). An oxyanion hole is again a key feature, with hydrogen-bonding stabilization supplied by the amide backbone NH of the active-site cysteine and the side-chain amide group of glutamine-19 [108] (*cf.* the similar arrangement found in subtilisin, described above). With minor variations of detail the oxyanion hole can be seen as the default acceptor for the developing C–O⁻ generated by the addition of a nucleophile to an amide C=O group in an enzyme-active site.[109]

4.1.2.1   Models

There is no measurable reaction between simple thiolates and unactivated amides, so to study the basic process it is necessary to use an activated system. The simplest model reaction uses the activated (because distorted) amide **4.22** (Scheme 4.8).[110] Simple thiolates do not react with **4.22** but specific aminothiols react through sulfur to give thiolesters. Like the cysteine proteases simple aminothiols, including even the imidazole derivative **4.23**, exist in solution

---

*Following the original authors[107] alanine is shown "for convenience," here and in Figure 4.6, in its D-configuration.

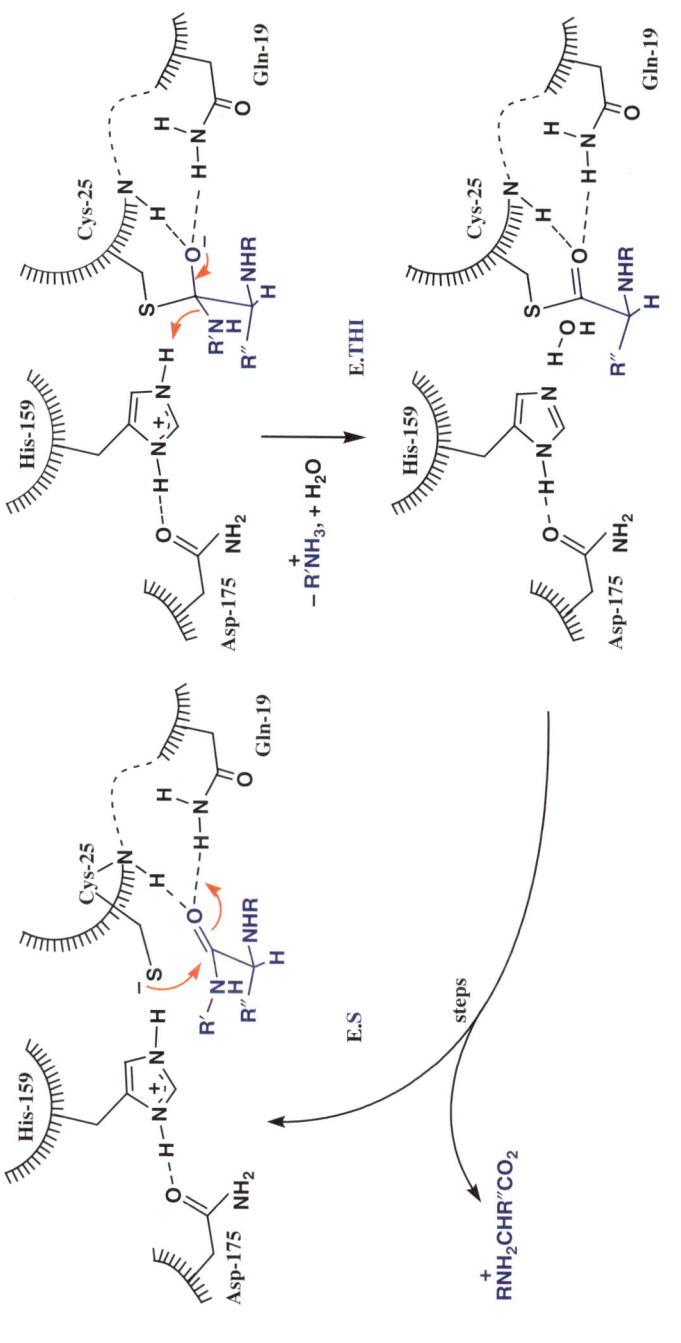

Figure 4.7 Key steps in the cleavage of a peptide bond by papain, a typical cysteine protease. The substrate R′NH.CHR″.CO–NHR is positioned and orientated by binding interactions in the Michaelis complex **E.S** to bring the target peptide bond into close proximity to the nucleophilic thiolate group.

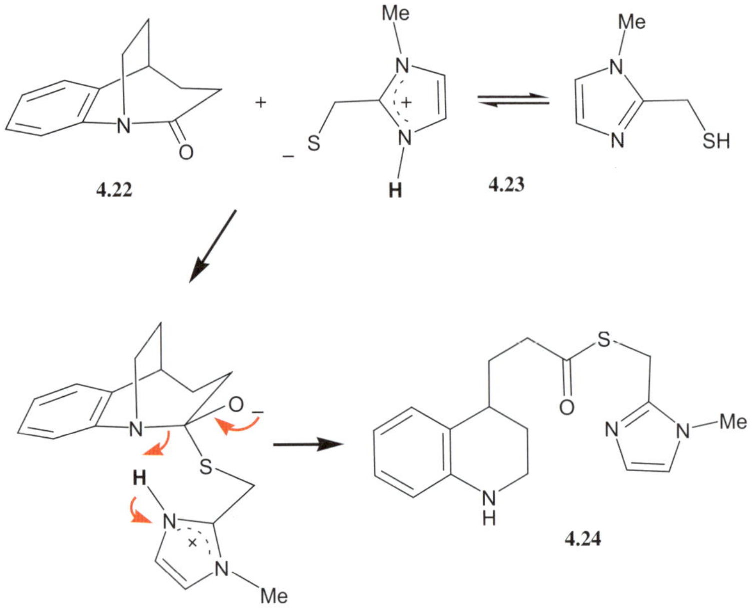

Scheme 4.8 Suggested mechanism for the thiolysis of the twisted (and thus activated) lactam **4.22** (see the text).

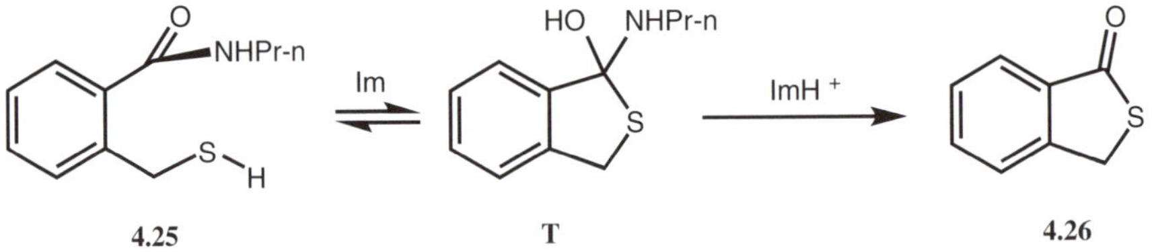

Scheme 4.9 Suggested mechanism for the intramolecular thiolysis of the amide group.[111]

near neutral pH as the reactive ammonium thiolate tautomer. The reaction with **4.22** shows a bell-shaped pH–rate profile because at low pH the nucleophilic thiolate is protonated, while at high pH the imidazole is not. It is essential for amide cleavage that the leaving group amine nitrogen be protonated, and it is likely that the imidazolium group is involved as a general acid, either directly (Scheme 4.8) or after initially "trapping" the high-energy tetrahedral inter-mediate by protonating the developing oxyanion.

4.1.2.2 Intramolecular Models

Alternatively, amide thiolysis can be observed by making the reaction intra-molecular. The cyclization of the amide **4.25** to 2-thiophthalide **4.26** (Scheme 4.9: see the corresponding serine protease model in Scheme 4.2 above) is some 10^7 times faster than the hydrolysis of N-n-propylbenzamide, and can be

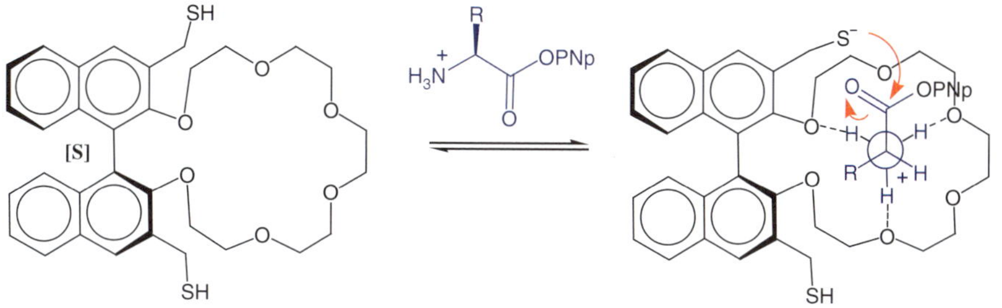

Scheme 4.10 Cram's model **4.27**. One of the two thiolate groups acts as the nucleophile towards the bound amino-acid ester cation (in nonaqueous media).

followed conveniently at 40 °C.[111] The reaction is catalyzed by general acids, most interestingly by imidazole (Im) buffers, which (uniquely) work most efficiently near pH 7, where both imidazole and its conjugate acid are both present in comparable concentrations. This is thought to be because the mechanism is complicated (even for such a simple system!) by an apparent change of rate-determining step. This points to the presence of an intermediate, which can only be the tetrahedral addition intermediate **T** (Scheme 4.9). The general acid-catalyzed breakdown of **T** is the likely rate-determining step above pH 7.

4.1.2.3 Supramolecular Models

Models for the SH hydrolases have been little studied; even the bimolecular ("background") reaction between a thiolate anion and an activated ester is fairly rapid, so there is limited scope for impressive rate increases. A thorough early study by the Cram group covered the ground nicely. The (relatively!) simple dithiol **4.27** (Scheme 4.10) was acylated by *p*-nitrophenyl esters of amino-acids bound to the attached crown ether binding site through their NH_3^+ groups (**4.28**). The nucleophile was the thiolate anion, and rates depended on the solvent: the largest accelerations (of the order of 1000 compared with the ring-opened form of the polyether) were obtained in organic solvents containing a minimal amount of EtOH. As usual, H-bond donors or acceptors competed with the substrate-binding process.[112]

4.1.3 Aspartic Proteinases

Many of the aspartic (or acid) proteinases are digestive enzymes, working in environments of low pH, like the stomach. They are characterized by active sites containing two aspartic acid residues, and it should come as no surprise (see Scheme 1.12, above) that catalysis involves one of these in the COOH form and the other as the carboxylate anion. The combination supports reactions with pH optima typically at moderately acidic pHs between 3 and 5. Early work with intramolecular models strongly suggested a nucleophilic mechanism for the enzyme reactions. Although this turned out to be wrong, the story is instructive, illustrating the limitations of models and, especially, the resource-fulness of enzymes in activating water as a potent nucleophile.

4.1.3.1 *Intramolecular Models*

The amide group of phthalamic acid **4.29** is hydrolyzed below pH 4 almost a million times faster than a simple benzamide (Scheme 4.11), leaving no doubt that the reaction is catalyzed by the neighboring COOH group.[113] Catalysis is nucleophilic,[8] with phthalic anhydride as the first product.

Remarkably, the same reaction of a dialkyl maleamic acid **4.30** is over 10 000 times faster still, with the effective molarity of the COOH group estimated as over 10^{13} M.[8] The rate constant for the cyclization of $>1\,s^{-1}$ is comparable with k_{cat} for some proteinase reactions with good synthetic substrates, sug-gesting that the mechanism is a kinetically viable candidate for the enzyme-catalyzed reactions. So it is particularly interesting that these rapid reactions are general acid catalyzed (Scheme 4.12).[114]

This general acid catalysis has the characteristics of a diffusion-controlled process: in particular the second-order rate constant is independent of the pK_a of the general acid when this is less than 4, but falls off rapidly for weaker general acids.[114] The step concerned was identified as the conversion of the tetrahedral addition intermediate T^0 (which would otherwise revert rapidly to

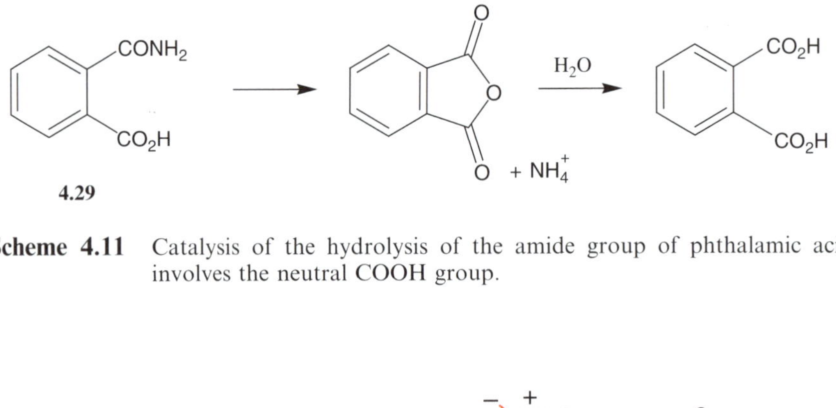

Scheme 4.11   Catalysis of the hydrolysis of the amide group of phthalamic acid involves the neutral COOH group.

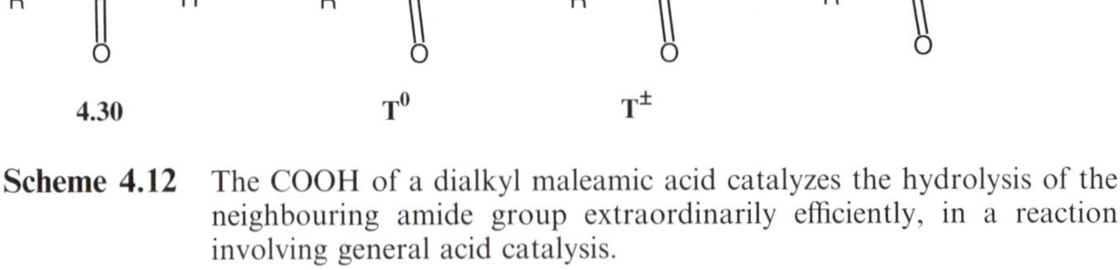

Scheme 4.12   The COOH of a dialkyl maleamic acid catalyzes the hydrolysis of the neighbouring amide group extraordinarily efficiently, in a reaction involving general acid catalysis.

reactant) to the zwitterion $T^{\pm}$. For general acids with $pK_a > 4$ the protonation of the N atom of T^0 becomes thermodynamically unfavourable, because the NH^+ group of the conjugate acid of T^0 is about 4. This tells us that only a carboxyl group, of the general acids available on amino-acid side-chains, is a general acid strong enough to transfer a proton to this sort of key intermediate.

Of course the diffusion together of a high-energy intermediate and a general acid would not be necessary in an enzyme-active site, because the general acid required would already be in place. This extra dimension was added to the model by building carboxyl groups into the reactant.[7] In the resulting system **4.31** derived from aspartic acid the requirement for an external general acid disappeared and the pH–rate profile showed a rate maximum at pH 4.4, indicating that amide cleavage is catalyzed by one COOH and one carboxylate group, just as in the aspartic proteases. It was suggested that the proton is transferred by way of a bifurcate hydrogen bond, as shown in **4.32** (Scheme 4.13).

Though the mechanisms of Schemes 4.12 and 4.13 remain valid as kinetically viable routes for amide hydrolysis catalyzed efficiently by two carboxyl groups under physiological conditions, evidence from work on various aspartic proteases has accumulated that the carboxylate group involved in the enzymes acts not as a nucleophile but as a general base. So what are the limitations of intramolecular models exposed by this result?

A significant limitation of simple models for hydrolysis reactions is that enzymes have evolved ways of presenting a water molecule as a powerful nucleophile, which models cannot. The most efficient intramolecular catalysis – by many orders of magnitude – is nucleophilic catalysis, by which the intramolecular nucleophile completes a cyclization reaction by making a fifth or sixth covalent bond. High reactivity is attained when the cyclization is thermodynamically strongly favourable (hence the basic requirement that the ring formed should be 5 or 6-membered). In the absence of a metal (see Section 4.1.4) a water molecule can be incorporated into a cyclic transition state only by hydrogen bonding, which does not normally support high thermodynamic stability.

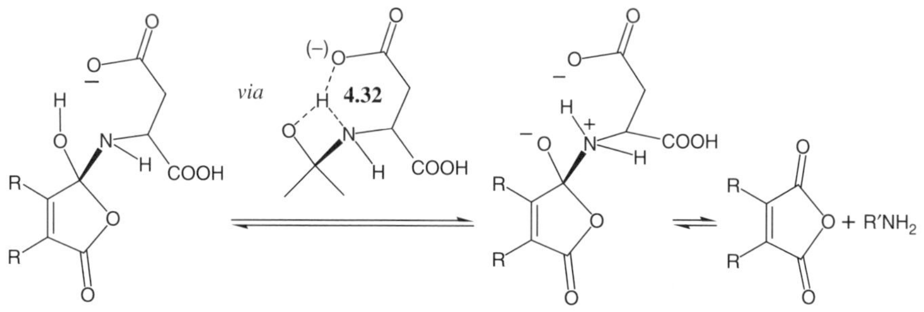

T^0 (**4.31**) $T^{\pm}$ (**4.33**)

Scheme 4.13   Suggested mechanism for intramolecular general base catalysis of the breakdown of the tetrahedral intermediate T^0 involved in amide cleavage catalyzed by a neighbouring carboxy group.

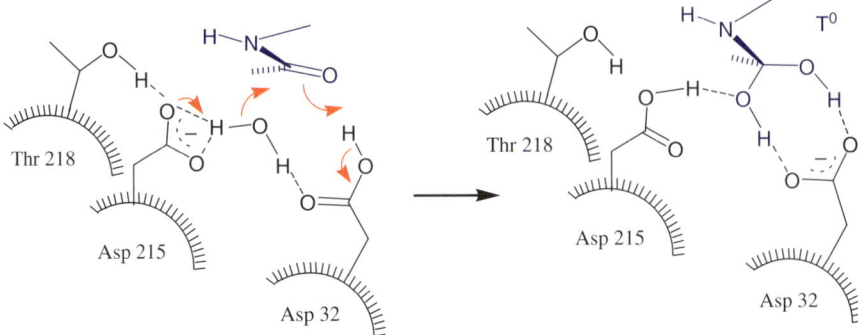

Scheme 4.14   Active-site arrangements for a typical aspartate protease (the number-
ing is based on the pepsin structure).[115]

How can a hydrolase enzyme turn H_2O into a powerful nucleophile? The
short answer is, not easily. Enzymes operating near pH 7 on unreactive substrates typically work by nucleophilic catalysis, for the same reasons that this is
the most efficient mechanism available to simple intramolecular models. Thus,
the initial cleavage of a specifically bound amide (or phosphate ester) substrate
generally involves nucleophilic attack by an active site OH, SH or imidazole
group to give a more reactive acyl-enzyme intermediate, which then reacts with
water to complete the hydrolysis reaction. What is different about the aspartate
proteases is that they operate at low pH, where stronger general acids are
available. In the typical aspartate protease reaction the active H_2O appears to be
hydrogen bonded to a carboxylate anion, which can act as a general base, and
also to the C=O oxygen of the other, COOH group. General acid catalysis of
addition to C=O groups is generally not observed even in intramolecular reactions, but it evidently becomes a simple process when the substrate amide group
is bound in the active site of a typical aspartate protease (Scheme 4.14).[115,116]
The general base is the carboxylate group of Asp-215, already hydrogen
bonded to, and thus helping to position the nucleophilic water molecule.
Formation of the new C–O bond can be further assisted by protonation of the
developing C–O⁻ by the COOH group of Asp-32, actively involved as a general
acid. (Compare the passive roles of the amide protons in the oxyanion holes of
serine and cysteine protease active sites in Schemes 4.2 and 4.7.) The result of
this first step is a tetrahedral addition intermediate perfectly set up for the
second (Scheme 4.14): Asp 32 is now a carboxylate anion, in position to act as a
general base for the breakdown of T^0: which is further assisted by protonation
of the leaving group N by the newly created COOH group of Asp-215.
No effective simple model is available for this mechanism (at least in part
because at the pHs at which COOH and *p*-nitrophenolate coexist there is little (a
few per cent) of either). An intriguing claim that the monoanion of the bis-m-
aminobenzoyl acid diamide of fumaric acid **4.34** (Scheme 4.15) catalyzes the
hydrolysis of a model peptide with Michaelis–Menten kinetics and "the same
biological activity as pepsin"[117] was (unsurprisingly) not confirmed by a later

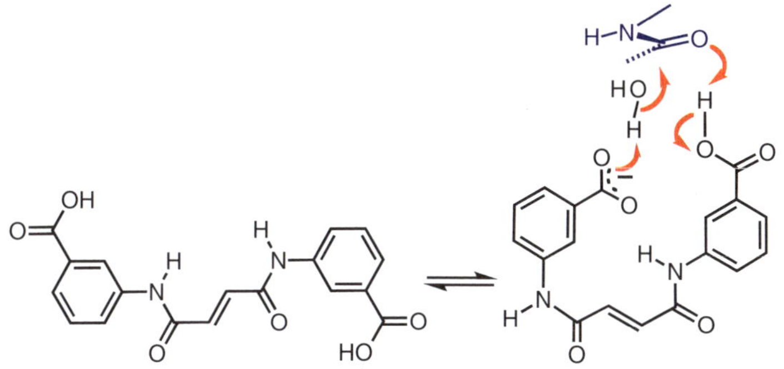

Scheme 4.15 Possible mechanism for the surprisingly efficient catalysis of peptide cleavage by the dicarboxylic acid **4.34**.

investigation, which found its catalytic activity "much lower".[118] However, the difference in activity claimed (of "only a few thousand-fold" in the apparent k_{cat}) would itself be impressive for a simple model if confirmed under conditions of controlled pH. Structure **4.34** could in principle adopt the conformation shown in Scheme 4.15, which brings the two carboxy groups into close but flexible proximity. (Though how this could bind the model peptide (Pro–Thr–Glu–Phe–(4-NO$_2$)Phe–Arg–Leu) with a K_M of 60 μM (at 293 K) remains to be explained.)

4.1.4 Metallopeptidases/Amide Hydrolases

To present a water molecule as a powerful nucleophile near pH 7 requires both correct positioning and chemical activation: a tall order for an intrinsically unreactive molecule capable of only hydrogen-bonding interactions with the functional groups available on amino-acid side-chains. As we have seen, the basis of high reactivity in intramolecular cyclizations is a strong thermodynamic driving force, typically dependent on an uninterrupted connecting sequence of covalent bonds to support the formation of the new one: ordinary hydrogen bonds are not strong enough to complete the efficient cycle.

Coordination to a metal can provide a flexible solution to many of these difficulties, including an uninterrupted sequence of covalent or near-covalent bonds in the transition structure. Enzymes take advantage of the special properties of metal-bound water in various ways, and these have been analyzed in detail in the light of work on model systems. Amide hydrolases in particular typically use zinc as the catalytic metal, for good chemical reasons:[119]

(i) Zn^{2+} has no significant redox chemistry and
(ii) is "borderline hard", so that N, O and S ligands all interact favourably, offering maximum scope for binding to proteins.
(iii) Its low ionic radius is the basis of a strong preference in enzymes for tetrahedral (rather than octahedral) coordination. This promotes the

Lewis acidity of the Zn centre and thus the Brønsted acidity of a coordinated H_2O.

(iv) Ligand exchange is rapid enough (for monodentate ligands like water) to support the catalytic rates typical of enzyme chemistry.

The specific examples discussed below involve Zn^{2+} as the metal cation.

H_2O bound to a metal cation can be positioned correctly if the metal cation is itself in the right place. In an enzyme a zinc cation involved in catalysis is typically bound in the active site by three ligands from amino-acid side-chains. Such binding is generally much stronger than the sum of three separate metal–ligand interactions (the chelate effect[120]), so that the metal – and any attached fourth ligand – becomes effectively part of the enzyme. In this way, bound water can be bound more strongly than through hydrogen bonding alone, and positioned for its role in hydrolase reactions as effectively as the active-site hydroxyl group of a serine protease. Furthermore, binding to a metal cation increases the acidity of H_2O, enough in the right system for its pK_a to be much reduced – to below 7 for systems $[XYZ]$-Zn-OH$_2$ $\rightleftharpoons$ $[XYZ]$-Zn-OH with the right ligand profile. This makes available "bound hydroxide," a much stronger nucleophile than water, in substrate concentrations at or near neutrality. The HOMO involved in the nucleophilic reaction of the Zn–OH group (Scheme 4.16) is the *p*-orbital lone pair on the hydroxide oxygen.[121]

Alternatively, a nucleophilic (Lewis base) centre on the substrate (*e.g.* the $C=O$ oxygen of an amide group) can bind to M^{n+} (Scheme 4.17, step 1), thus enhancing the electrophilicity of the substrate group. If the same metal cation acts in both roles, by accepting an additional ligand, the reaction with bound hydroxide becomes in effect an intramolecular cyclization (Scheme 4.17, step 2). Hydroxide is bound to the metal much more strongly than water, thus reinforcing the thermodynamic driving force for the formation of the new bond.

This mechanism is observed in some models, especially with rare earths $M^{(3-4)+}$, but not usually in amide hydrolases. More often the substrate nucleophile *displaces* water (H_2O is a good leaving group from a Zn^{2+} centre), so that the electrophilicity of the tetracoordinate metal centre is maintained: then no neighbouring OH group is available, and an external nucleophile must be involved (Scheme 4.18).

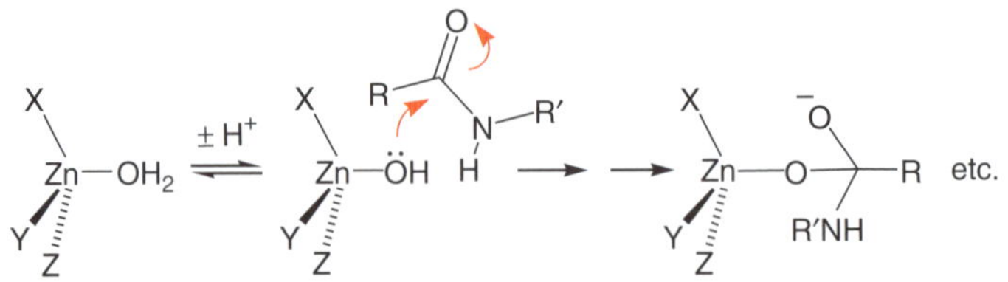

Scheme 4.16   The bound-hydroxide mechanism. The Zn centre is further available to stabilize the tetrahedral addition intermediate formed by coordination to the developing C–O⁻ anion.

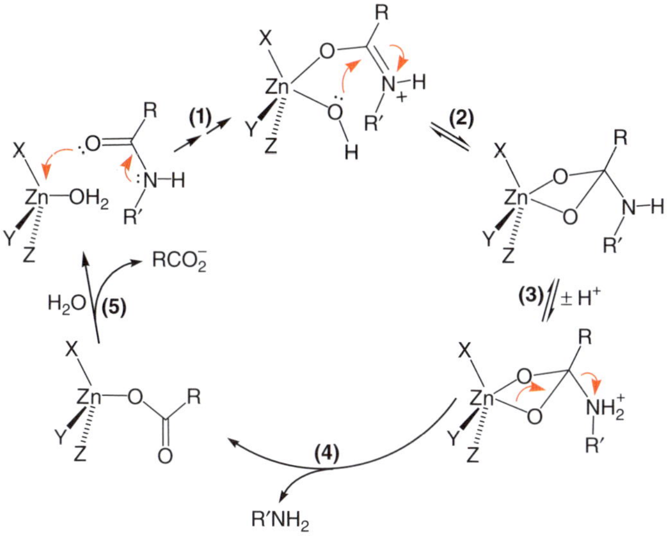

Scheme 4.17 Electrophilic activation of an amide group at a Zn^{2+} centre. (i) By ligand coordination. (Charges at the metal centre will depend on the ligands and are omitted.) For discussion of the various steps (**1–5**) see the text.

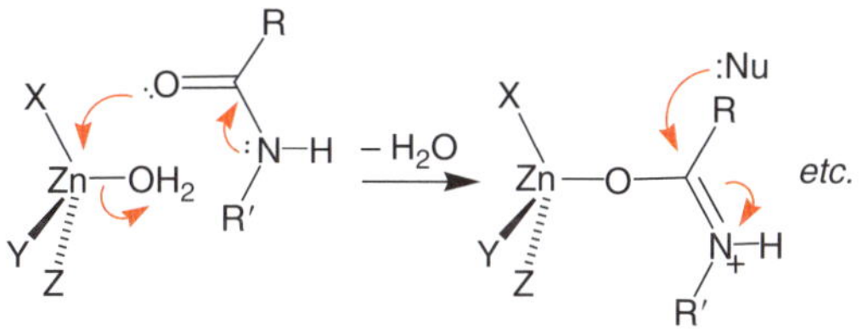

Scheme 4.18 Electrophilic activation of an amide group at a Zn^{2+} centre. (ii) By ligand exchange. An S_N2-type mechanism, with inversion at Zn is ruled out because most enzymes, and most successful models, [XYZ]-Zn-OH are sterically protected against "attack from the back" by the chelate structure connecting the three ligands (X, Y and Z).

Which mechanism is preferred for a particular reaction will depend on the ligand type and configuration, and on the substrate. The most reactive, and thus most electrophilic substrates will by definition be the least nucleophilic, so tend to favour direct attack by bound hydroxide. Typical natural substrates like amides, glycosides and phosphate ester anions (see below) are less reactive and more nucleophilic, and have more need of activation by a Lewis acid. This must be borne in mind when extrapolating from results with model systems: which in many cases only show measurable reactivity against activated substrates.[103]

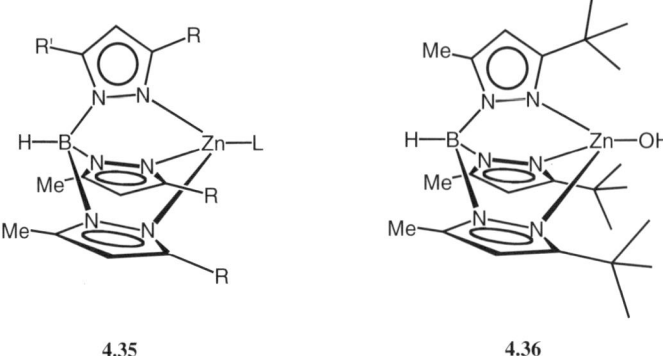

4.35   **4.36**

Scheme 4.19 Designed tridentate ligands select for the monomeric tetrahedral geometry of the hydrated zinc complex.

4.1.4.1 Enzymes

The most familiar metalloproteinases are carboxypeptidase and thermolysin, in which an active-site zinc is coordinated to two histidine imidazoles and a glutamate carboxylate. The Zn^{2+} cation in carbonic anhydrase, which catalyzes the (much easier) hydration of CO_2, is coordinated to three histidines, while cysteine (as the thiolate) is a ligand in a number of other enzymes (see Scheme 4.119). The nature and the geometry of the coordination in the [XYZ]-Zn-OH complex allow fine tuning of the properties of the Zn centre, and hence of a bound hydroxide. Thus studies using model systems show that the pK_a of zinc-bound water molecules is strongly influenced by the coordination environment: the pK_a of $[Zn(OH_2)_6]^{2+}$ is 9.0, but the pK_a of bound water depends strongly on the coordination number and charge of the complex, decreasing with decreasing coordination number and increasing positive charge; so that pK_a values as low as 7 are achievable for [XYZ]-Zn-OH complexes.

The mechanisms of action of thermolysin and carboxypeptidase are yet to be established in detail. Discussion centres around the basic mechanisms of Schemes 4.17 and 4.18, with the necessary proton transfers catalyzed by active-site carboxylate or histidine imidazole groups. Final conclusions about the mechanism will need further careful work on both the enzymes and well-designed model systems.

4.1.4.2 Model Systems with One Metal Centre

Studies with model systems turn out to be far from trivial, because the mononuclear tetrahedral zinc hydroxide structure typical of zinc enzymes is not common in small-molecule chemistry in water near pH 7.[119] In particular, the hydroxide ligand of central interest typically prefers to bridge two zinc centres. In this situation it is essential that model systems of interest are structurally fully characterized, ideally by an X-ray structure shown to persist in solution.

A successful approach to simple models with the tetrahedral [XYZ]-Zn–L structure (L = H₂O, OH, Cys-SH, *etc.*) is to incorporate the three donor groups X, Y, and Z into specially designed tridentate ligands that are designed to mimic the three protein residues that bind the active-site zinc. For example, the three nitrogen donors of tris(pyrazolyl) borate ligands select for the tetrahedral geometry (**4.35**) shown in Scheme 4.19. The substituents R and R′ on the ligands are readily varied, to modify the electronic properties of the metal centre and to provide steric protection against dimerization via either hydroxide bridges or hydrogen-bonding interactions.

Thus, the Me, t-Bu derivative **4.36**, in which pyrazole rings mimic the (more basic) imidazoles of the three histidine ligands of the Zn^{2+} centre of carbonic anhydrase (and the steric protection afforded by the conical cavity in which it sits) reacts rapidly with CO_2 to give the bicarbonate derivative.[122,123]

$$[L_3]\text{-Zn–OH} + CO_2 \rightleftharpoons [L_3]\text{-Zn–O(CO)OH}$$

The reaction is reversible, and fast on the NMR timescale at room temperature. Though observed in aprotic organic solvents rather than in water, it remains the best simple model for the same reaction catalyzed by carbonic anhydrase (which itself takes place at the bottom of a deep cavity in the protein, where access to water is strictly limited). For example, experiments using ^{17}O NMR show that **4.36**, like the enzyme, catalyzes oxygen atom exchange between CO_2 and H_2O.

Structure **4.36** models only the basic reaction of $[L_3]$-Zn–OH with CO_2. What we can learn from its properties can help in the design of true enzyme models, but no significant substrate-binding step is involved in the reaction with CO_2. At least in part for the same reason **4.36** and similar complexes do not catalyze the hydrolysis of simple amides, though they do in some cases react (stoichiometrically) with esters, and even sufficiently activated amides (*e.g.* trifluoroacetamides and β-lactams), to give the acylated derivatives (Scheme 4.20).[119,124]

For example, the triazacyclodecane complex **4.37** catalyzes the hydrolysis of methyl acetate at a significant rate above the pK_a (7.3) of the H_2O complex (Scheme 4.21). The first formed acetate complex dissociates rapidly to regenerate the catalytic OH form.[125]

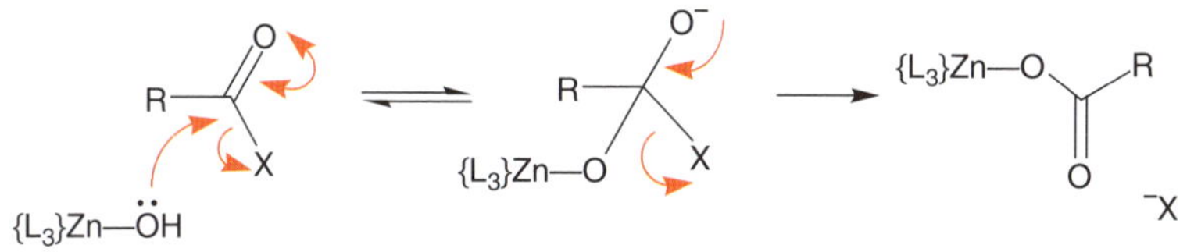

Scheme 4.20 Complexes $[L_3]$-Zn–OH are acylated by sufficiently electrophilic carboxylic acid derivatives.

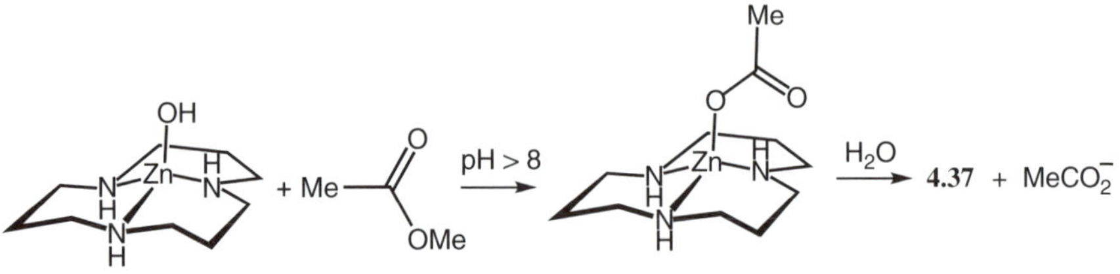

Scheme 4.21   The second-order reaction of methyl acetate with **4.37** is some 10^5 times slower than its reaction with free hydroxide anion at 25 °C, offering a modest 20–30-fold rate advantage at pH 7.3 in the absence of an enzyme.

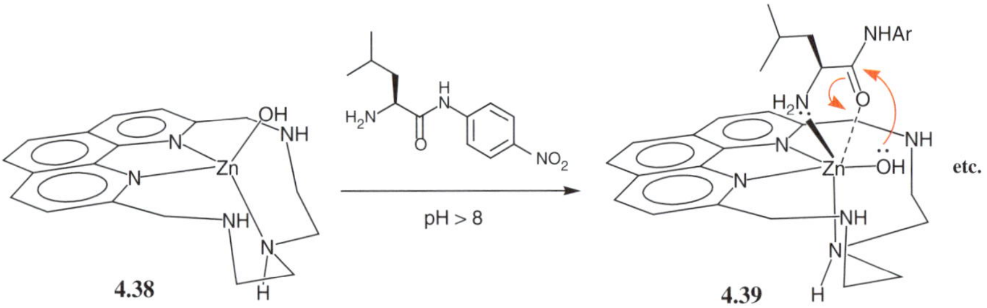

Scheme 4.22   Suggested mechanism for the hydrolysis of L-leucine *p*-nitroanilide by the zinc-bound hydroxide of complex **4.38**.

The phenanthroline complex **4.38** (Scheme 4.22) binds amino-acids and dipeptides, and catalyzes the hydrolysis of the *p*-nitroanilide of L-leucine at 35 °C.[126] Good structural information on various complexes supports a mechanism (not available to carboxy- or endopeptidases) by which the bound neutral amide, ("probably in bidentate binding mode", with a binding constant of 3.5×10^5 M) is attacked by the zinc-bound hydroxide (pK_a > 8) as shown in **4.39** (Scheme 4.22). Reaction (corresponding to a half-life of several days) is not fast, but the substrate is bound more strongly than the product leucine, and turnover is observed at pH 9.7.

4.1.4.2.1 Intramolecular Reactions.   As suggested by this result (Scheme 4.22) the simplest way to observe a measurable reaction of a Zn(OH) complex with an unactivated ester or amide group is to make it intramolecular. An instructive example occurs as part of the reaction in which the internal complex **4.40** catalyzes the hydrolysis of the activated ester *p*-nitrophenyl acetate (pNPA). The Zn-bound alkoxide anion in the complex (p$K_a = 7.4$) is a 4-times stronger nucleophile towards the activated ester *p*-nitrophenyl acetate (pNPA) than is the Zn-bound OH^- of the corresponding hydroxo-complex **4.37**,[127] just as alkoxides are generally better nucleophiles than hydroxide when free in solution. The intermediate (unactivated) acetate ester **4.41** produced – an acyl enzyme-model intermediate – is then rapidly cleaved by the neighbouring Zn-bound OH^- of **4.41**, no doubt via intra-molecular acetyl transfer to form the short-lived acetate complex **4.42** (Scheme 4.23).

In a similar reaction the unactivated amide bond in Groves' complex **4.43** (Scheme 4.24) is hydrolyzed some 1000 times faster than the reaction with external hydroxide in the absence of the metal.[128] Reactivity is rela-tively modest for intramolecular nucleophilic catalysis: compare the half-life of **4.43** of 15 hours at 70 °C with that of a few minutes at 25 °C for the imidazole-catalyzed cyclization of **4.1** (Scheme 4.2). Note that geometrical restrictions in **4.43** prevent a strong donor–acceptor interaction between the sp^2-hybridized lone-pair electrons of the planar amide $C = O$ group and the Zn centre.

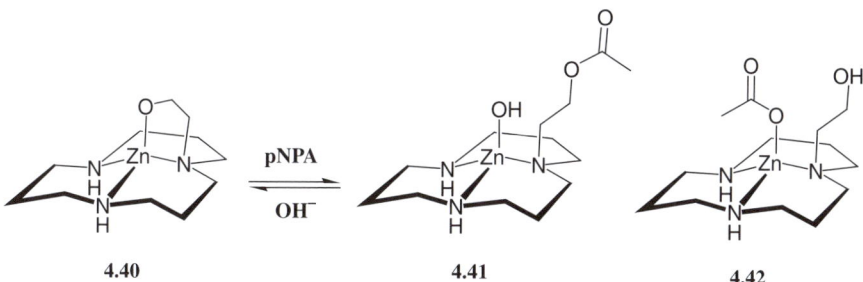

4.40   **4.41**   **4.42**

Scheme 4.23 The Zn-bound alkoxide anion in the complex **4.40** is a better nucleo-phile than bound hydroxide.

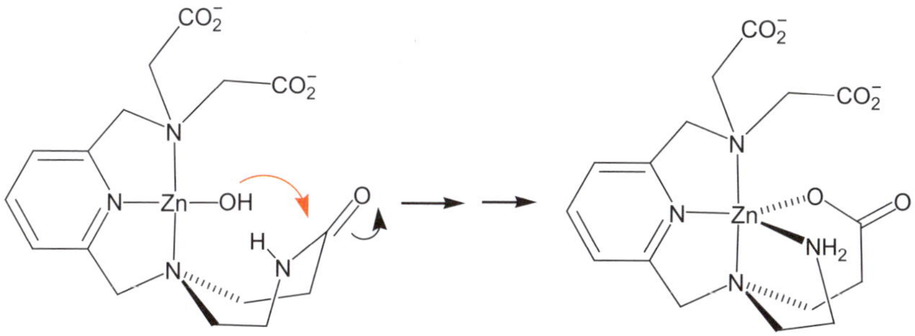

4.43

Scheme 4.24 Intramolecular nucleophilic attack on the unactivated amide group of the complex **4.43** is stereoelectronically one-dimensional. See the text.

4.1.4.2.2 Supramolecular Metalloprotease/Peptidase Models. The supramolecular hosts are generally designed to position the metal. Relatively simple examples are the twin-cyclodextrins, covalently linked by a metal-chelating ligand, designed to operate on a doubly bound substrate, as shown in Scheme 4.25. Here, the cyclodextrins bind the substrate in productive proximity to the metal cation, which acts as the catalytic "cofactor". This motif can produce impressive accelerations and reliable turnover. Thus, the hydrolysis of the extended ester substrate bound to **4.44** at pH 7 (in 80% DMSO-water, for reasons of substrate solubility) shows Michaelis–Menten kinetics, with k_{cat} = 0.45 s^{-1} and a k_{cat}/k_{uncat} of 1.45×10^7 on account of its strong binding ($K_M = 0.47\,\mu M$ even in the largely organic solvent).[129] Turnover is facilitated by the much weaker binding of the carboxylate and phenolic products, which are bound separately to different single cyclodextrins.

4.1.4.3   Model Systems with Two Metal Centres[129]

The less reactive a substrate the more it needs electrophilic activation. Adding a nucleophile as an additional ligand to a tetrahedral Zn^{2+} centre (Scheme 4.17) makes it less electrophilic and thus raises the effective pK_a of bound water: while displacing a single bound water (Scheme 4.18) removes the possibility of attack by neighbouring Zn(OH) and thus requires the involvement of an external nucleophile. Since the external nucleophile in a hydrolysis reaction must eventually be H$_2$O, and since the best way of delivering it in an enzyme-active site is often as Zn-bound hydroxide, we can understand why many enzymes hydrolyzing such unreactive substrates as amides (Section 4.1.4.3) and phosphate ester anions (Section 4.2.1.2) have evolved active sites containing two metal ions, which act together in concert. However, so far model systems for such enzymes as endopeptidases and carboxy- and aminopeptidases

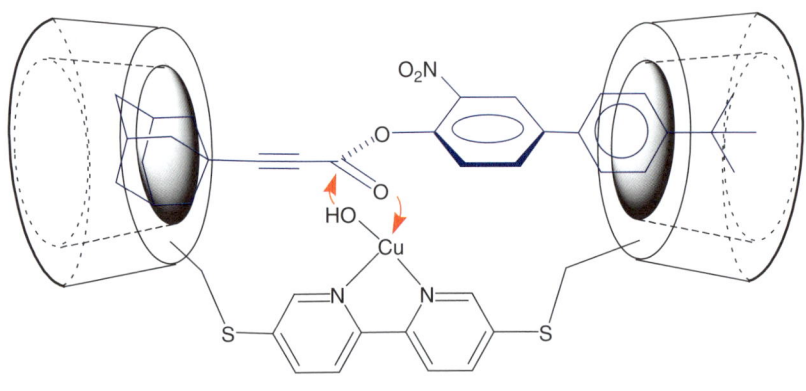

4.44

Scheme 4.25   The hydrolysis of the doubly bound ester is accelerated by a factor k_{cat}/k_{uncat} of over 10^7 in the presence of Cu^{2+}.

(exopeptidases, like carboxypeptidases but specific for the amino-terminus) are active only against activated amide substrates.

Of special interest are the β-lactamases, Nature's vigorous response to the widespread use of β-lactam antibiotics. To the chemist in this field the evolution of efficient β-lactamase enzymes – on a timescale not much longer than that needed to develop convincing enzyme models – is daunting but potentially instructive. Most known β-lactamases use the serine-protease type mechanism described in Section 4.1.1, but of interest in the present context are the class B enzymes, which may use either one or two active-site zincs. The second zinc does not appear always to make a substantial contribution to catalysis: but when it does, the Zn$_2$ enzymes are the more efficient.[130]

Model systems with two metal centres are of their nature more complicated. To operate simultaneously on the same substrate functional group the two metal centres must be close together, and are therefore typically connected by a common ligand. Thus, the arrangement found in several *β*-lactamases with two zinc centres (Scheme 4.26, **4.45**) is both structurally economical and mechanistically logical. The hydroxide bound to a tetrahedrally coordinated zinc (Zn1), of a type familiar from monozinc enzymes, is coordinated as a fifth ligand to a second Zn(2) centre, creating a structure (**4.45**) that is (roughly) trigonal bipyramidal at Zn(2). Coordination of the nitrogen centre of the lactam substrate (**4.46**, Scheme 4.26) will assist the cleavage of the C–N bond, but for a typical β-lactam antibiotic a donor interaction is also possible in the reactant state: which is not completely planar at the "amide" nitrogen.

The design of model systems containing two metal centres has been discussed by Weston.[131] The simplest is to connect two known ligands by a spacer, designed to bring the two bound metal centres into interesting proximity. A degree of flexibility is necessary, but means that even a high-resolution crystal

Scheme 4.26 Arrangements at the active centre of Zn$_2$ β-lactamases. Zn(1) is coordinated by a conserved His triad (as in the mono-Zn enzymes), and Zn(2) by conserved His and Cys (X and Y) and Asp residues and an additional water. A suggested mechanism of action (**4.46**)[119] involves coordination of the substrate at both Zn centres, with displacement of the water from Zn(2), thus setting up the C=O group for nucleophilic attack by the bound hydroxide.

structure does not guarantee the same structure in solution. The alternative "macrocyclic" strategy involves the construction (likely to require more serious synthesis) of a single molecular framework incorporating two metal binding sites, designed to hold the two metals a desired distance apart: and so with limited conformational flexibility. Convergent synthesis is preferred in both cases, with the result that the arrangements around the two metal centres, unlike those in the enzymes of interest, are generally symmetrical. Some systems based on macrocyclic spacers are discussed in the following section (4.1.4.3.1).

4.1.4.3.1 Aminopeptidase and Lactamase Models.   The high thermodynamic stability of the amide group involves n_N-$\pi^*_{C=O}$ overlap, which makes addition to the $C=O$ group more difficult: but the subsequent cleavage of the C–N bond is also unfavourable because an RNH^- anion is a very poor leaving group; which under physiological conditions demands either protonation or coordination to a Lewis acid. As a result, peptide bond cleavage catalyzed by synthetic zinc complexes, whether mono- or binuclear, so far remains a major challenge. A rare "successful" example serves to illustrate the difficulties: the hexaazadioxacyclotetracosane **4.47** (Scheme 4.27), known to bind two Zn^{2+} ions, catalyzes the hydrolysis of glycylglycine with a half-life at pH 8.37 of the order of several weeks at 70 °C.[132] Rate constants for the three reactive Zn–OH forms are some two orders of magnitude greater than expected[133] for the uncatalyzed hydrolysis reaction at this pH (by hydroxide, in the presence or absence of free Zn^{2+} ions). Not surprisingly there is no evidence of saturation kinetics, or detailed structural information that might support a mechanism reasonably represented as **4.48**.

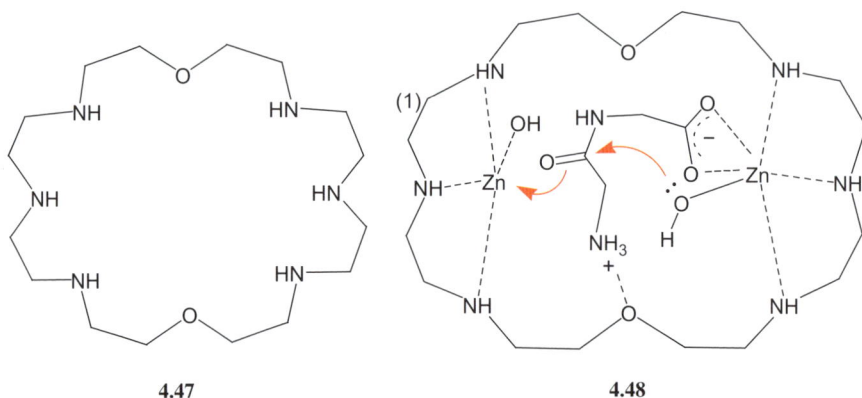

4.47   **4.48**

Scheme 4.27   Possible mechanism for the rare hydrolysis of a peptide catalyzed by a dizinc complex (of **4.47**). It is suggested that the substrate carboxylate group is bound to the nucleophilic Zn–OH centre, and the ammonium group probably also hydrogen bonded to a ring oxygen; to set up a conformation favouring the two-centre reaction shown (see the text).

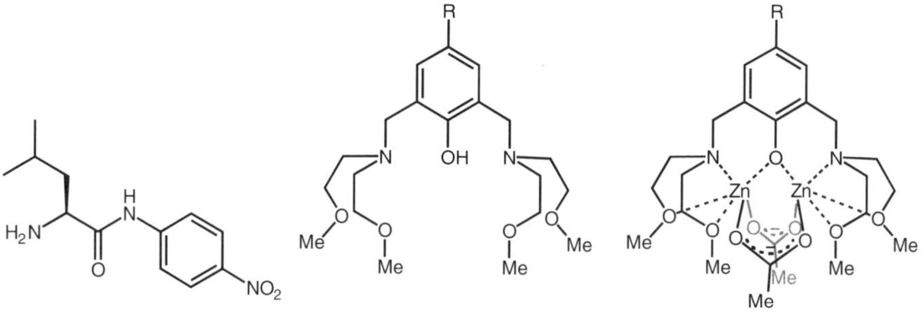

Scheme 4.28 Structures of three ligands (**4.50**, R = Me, Cl and NO₂) used in a structure–reactivity study of the hydrolysis of L-leucine *p*-nitroanilide **4.49**. Structure **4.51** indicates the geometry around the two Zn centres in the diacetate complex, as shown by X-ray crystallography (see the text).

In the absence of more informative models catalyzing the hydrolysis of unactivated amides we must learn – as usual (Section 4.1.1.3.2) from reactions of successful systems developed to catalyze the hydrolysis of activated amides.

The Sakiyama group report a short structure–reactivity study of hydrolysis of the *p*-nitroanilide **4.49** (Scheme 4.28) of L-leucine by the Zn₂ complexes of three derivatives (R = Me, Cl and NO₂) of the symmetrical system **4.50**.[134,135] A crystal structure of the methyl derivative (**4.50**, R = Me) showed distorted octahedral coordination around the two Zn centres including two bridging acetate anions (**4.51**, Scheme 4.28). The structure of the active species in the solvent used (40% aqueous DMF) is not known, but the acetate ligands are known to be readily exchanged, so that various combinations of DMF, water, substrate and product will all bind (probably increasingly strongly in that order) to the two Zn centres. The hydrolysis reactions were slow even for this activated amide substrate (though a good deal faster than the hydrolysis of the same substrate catalyzed by the monozinc complex (Scheme 4.21) discussed previously): specifically, 10 000 times slower in the case of **4.50**, R = Me than for a typical aminopeptidase reaction. But the introduction of electron withdrawing substituents R produced significant rate increases (250× for the *p*-nitro compound (**4.50**, R = NO₂), well correlated by the Hammett equation, with a high reaction constant $\rho = 2.54$. This is interpreted in terms of the increasing electrophilicity of the two zinc centres, which are reasonably presumed to bind molecules of substrate and water: which dissociates to give the reactive Zn–OH forms, with pK_as of 9–9.4 for the three compounds **4.50** (R = Me, Cl and NO₂, respectively).

Closely related to the complex **4.51** is the bis-diaminocompound **4.52** ([Zn₂L₁(μ-NO₃)(NO₃)₂]) synthesized by the Lippard group specifically to model metallo-β-lactamases (Scheme 4.29).[136] In aqueous solution one or more of the three nitrate ligands is replaced by water, and a crystal structure was obtained for the hydroxyl species [Zn₂L₁(μ-OH)-(NO₃)₂] (**4.53**). The system catalyzes the

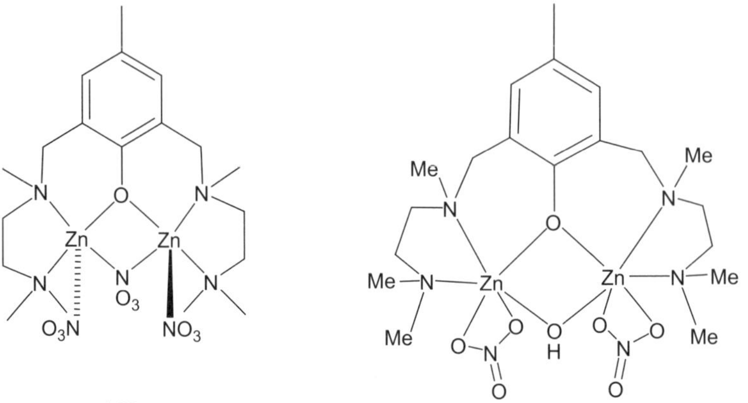

4.52   **4.53**

Scheme 4.29   Structures of Zn_2 complexes shown to catalyze the hydrolysis of nitrocefin (Scheme 4.30) and penicillin G. See the text.

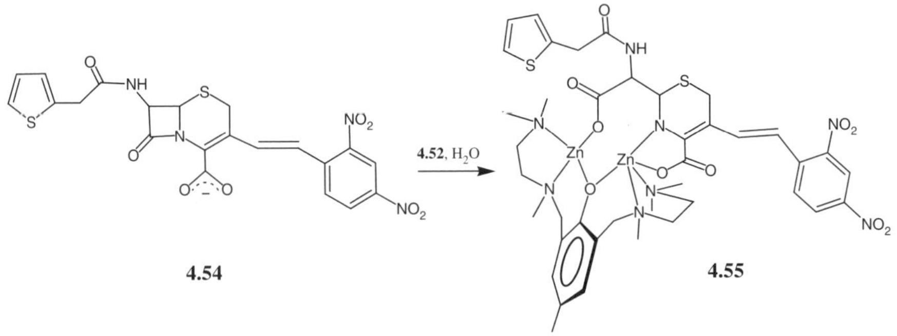

4.54   **4.55**

Scheme 4.30   The nitrocefin substrate (**4.54**) for the complexes shown in Scheme 4.29, and the structure of the reaction intermediate **4.55** (see the text).

hydrolysis of penicillin G and of nitrocefin (**4.54**, Scheme 4.30), a substrate of interest because it is used (on account of its useful chromophore) in mechanistic studies on β-lactamases. Saturation kinetics are observed, most clearly at low water concentrations in DMSO, but involving catalyst rather than substrate concentration.[136] Detailed structural and kinetic work supports a mechanism involving substrate binding via the carboxylate group of the β-lactam substrate, followed by intramolecular attack of the bridging hydroxide on the β-lactam carbonyl group.

Consistent with this mechanism, a blue intermediate, identified spectro-scopically as **4.55**, was observed when a 10-fold excess of substrate was reacted with **4.52** in wet DMSO as solvent.[137]

Although the same intermediate is observed in the hydrolysis of nitrocefin catalyzed by β-lactamase enzymes, this is an atypical substrate, with the leaving

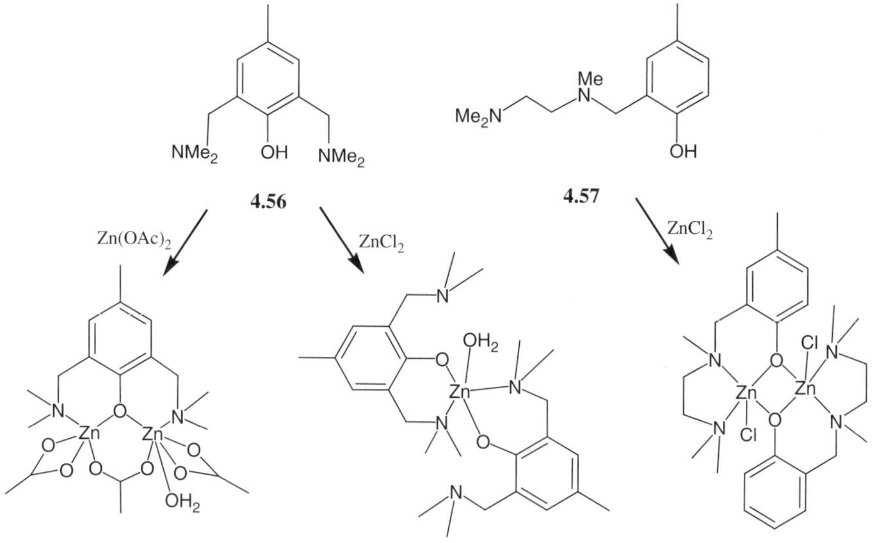

Scheme 4.31 Structures of mono- and dinuclear zinc complexes shown to catalyze the hydrolysis of β-lactam antibiotics with Michaelis-Menten kinetics.

group nitrogen strongly stabilized by delocalization into the dinitrophenyl group and so not needing protonation to assist C–N bond cleavage. However, the result does show how a Zn centre could be involved in the process. One of the conclusions from the work with this and a second similar complex was that the second zinc centre does not contribute significantly to catalysis, since similar monozinc complexes are just as reactive in the hydrolysis of β-lactams.[136] This appears to be true also in some enzyme reactions, though improved models are likely to tell us otherwise.

A recent report describes a set of four relatively simple mono- and dinuclear zinc complexes that are rather efficient catalysts for the hydrolysis of β-lactam antibiotics – and show Michaelis–Menten kinetics.[138] We have seen (see Section 2.3) that saturation kinetics of this form is a defining feature of catalysis by enzymes, so the most efficient of these complexes qualifies as the most complete simple model for a β-lactamase. The models are based on ligands **4.56**, (Scheme 4.31) which reacts with Zn(OAc)$_2$.6H$_2$O and ZnCl$_2$, respectively, to give the di- and monozinc complexes **4.58** and **4.59**; and **4.57**, which with ZnCl$_2$ gives the dimeric complex **4.60**.

All three complexes (at 2.12×10^{-5} M) catalyze the hydrolysis of oxacillin (Scheme 4.32), in pH 7.5 HEPES buffer containing 10% of DMSO, with half-lives of 1.5–4 h, compared with over 100 hours for a control reaction. The reactions are first order in the catalyst and follow Michaelis–Menten kinetics.[138] The derived parameters are compared in Table 4.2 with those obtained under the same conditions for the same reaction catalyzed by a natural β-lactamase from *B. Cereus*. Noteworthy, if unsurprising, is that the Zn$_2$ complexes

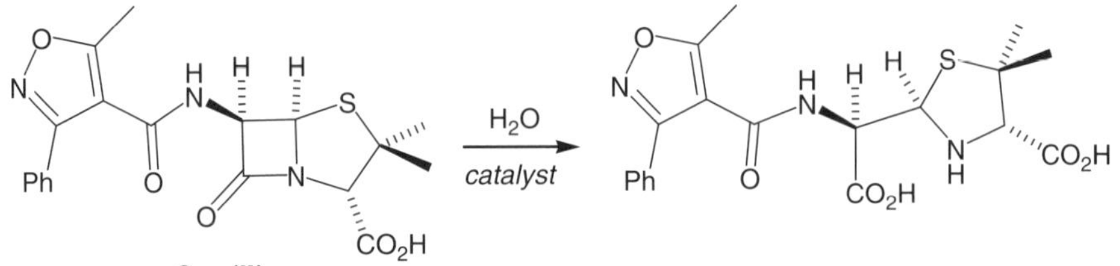

Oxacillin

Scheme 4.32   The hydrolysis of oxacillin is catalyzed by β-lactamase models **4.58–4.60** (see Table 4.2).

Table 4.2   Michaelis–Menten parameters compared, for the hydrolysis of oxacillin (Scheme 4.32) catalyzed by models **4.58–4.60** and a β-lactamase from *B. Cereus*.

Catalyst[a]	V_{max} $(\mu M\,min^{-1})$	K_M (μM)	k_{cat} (min^{-1})	k_{cat}/K_M $(M^{-1}min^{-1})$
4.58 (Zn$_2$)	28.4	8.94	2.83	3.17×10^5
4.59 (mono-Zn)	83.3	34.3	8.33	2.43×10^5
4.60 (Zn$_2$)	9.08	3.43	0.91	2.64×10^5
Enzyme (Zn$_2$)	$(31.6)^a$	6.97	717	10.3×10^7

Notes. *a* The different catalyst concentrations of 21.2 μM for **4.58–4.60** and 44.2 nM for the enzyme mean that the V_{max} values are not directly comparable.

and the (Zn$_2$) enzyme show significantly lower K_M values, *i.e.* bind the substrate more tightly: more significant is that the values are closely similar, adding validity to the models. k_{cat} and the catalytic efficiency parameter k_{cat}/K_M are (again not surprisingly) substantially higher for the enzyme: but by a factor of only 86 in k_{cat} for the mono-Zn complex **4.59**. However, its high K_M means that **4.59** is (marginally) the least efficient catalyst in the table.

The acetate ligands of the most efficient model compound **4.58**, (which is less *efficient* than the enzyme by a factor of 325) are expected to dissociate in solution, perhaps exchanging with the carboxylate group of the substrate, so the exact solution structures, and thus also the mechanism of action of the most efficient of these models, are not known. But the pH dependence shows that reaction of **4.58** depends on a group with pK_a of about 8.3, as expected for a Zn-bound water reacting with substrate bound to a second Zn centre. As discussed in this, and indeed most of the papers cited in this section, the Zn–Zn distance is an important parameter in catalytic Zn$_2$ systems, including enzymes.[119,131]

4.2   Phosphoryl Transfer

Phosphoryl group (or phosphate) transfer is another of the commonest biological reactions, and is catalyzed by a very large number of enzymes. An excellent overview of the enzymes and the chemistry involved is available.[139] The chemistry is more diverse than carboxylic acyl transfer because two mechanistically very different classes of phosphate ester are of the first importance in biology. Phosphomonoesterases and kinases catalyze the hydrolysis and formation of monoesters, while phosphodiesterases and various types of nucleases handle the hydrolysis of diesters. Both classes of substrates are typically extremely unreactive, so present a fascinating mechanistic challenge to the enzymes involved. They rise to this challenge in ways that will be familiar from our discussions of carboxylic acyl transfer, generally involving transfer of the phosphoryl group, and thus P–O rather than C–O cleavage. A third class of substrate, the man-made phosphate triesters, has appeared in the biosphere only recently, but new bacterial enzymes have already evolved to hydrolyze them; much as happened with the β-lactamases described in the previous section.

Phosphate mono- and diesters, and inorganic phosphate, are negatively charged under physiological conditions. They are present in solution near pH 7 as the diester anions and as rapidly equilibrating mixtures of the mono- and dianions of monoesters: inorganic phosphate (trianion) and phosphoric acid itself are present in significant concentrations only at very high and very low pH, respectively. (Scheme 4.33).

The negative charges have a major influence on reactivity. The ionization of the acid form of a phosphodiester, for example, clearly reduces the electrophilicity of the P centre. Activation of the anion by protonation would be a thermodynamically unfavourable way to start a reaction that needs to be rapid near

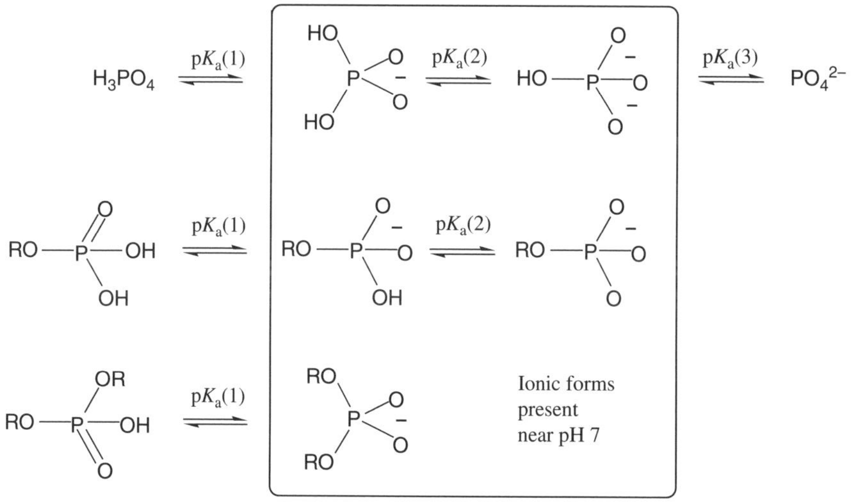

Scheme 4.33 Ionization of phosphoric acid and its esters. pK_as vary to some extent with the R group(s), but are typically close to 2 for $pK_a(1)$, 7 for $pK_a(2)$ and 12 for $pK_a(3)$. Note that all the P–O bonds of a given anion are equivalent: so all are drawn here as single bonds, rather than identifying one of them arbitrarily as $P=O$. A negative charge drawn between two oxygens indicates that it is shared, *i.e.* that both the P–O bonds concerned have significant double-bond character.

pH 7, but the negatively charged oxygens of phosphate anions are obvious targets for favourable binding interactions with more abundant cations. Thus, the active sites of enzymes catalyzing phosphate-transfer reactions are always positively charged, and commonly use metal cations to complement the functional groups available on amino-acid side-chains.

In contrast to the situation with phosphate diesters the second negative charge on the dianion of a phosphate monoester can be activating. This fundamental difference between mono- and diesters – replacing the P–OH group of $ROP(OH)O_2^-$ by a second P–OR – means that the reactions of the two are best discussed separately.

4.2.1 Phosphoryl Group Transfer from Monoesters

Protonation, by definition, makes a dianion overall more electrophilic. However, the reactivity of phosphate monoester dianions $ROPO_3^{2-}$ depends particularly strongly on the OR group. As a result, if RO^- is a good enough leaving group (*i.e.* the weak conjugate base of a relatively strong acid ROH) the dianion becomes more reactive in phosphate-transfer reactions. This is because the P–OR bond is significantly weakened by n_O–σ^*_{P-OR} interactions (Scheme 4.34, **A**). The process is (by definition) further advanced in the transition state leading to P–OR cleavage.

The PO_3^- (metaphosphate) group at the centre of Scheme 4.34 is a high-energy species and not generally a discrete intermediate in a true $S_N1(P)$ reaction: any actual bond breaking in the sense $P^{(\delta+)} \ldots {}^{(-)}OR$, corresponding to the canonical forms written in Scheme 4.34(**A**), normally leads to recombination, simply regenerating the reactant ester. However, at or near this point on the reaction coordinate the phosphate group can be transferred to a nucleophile

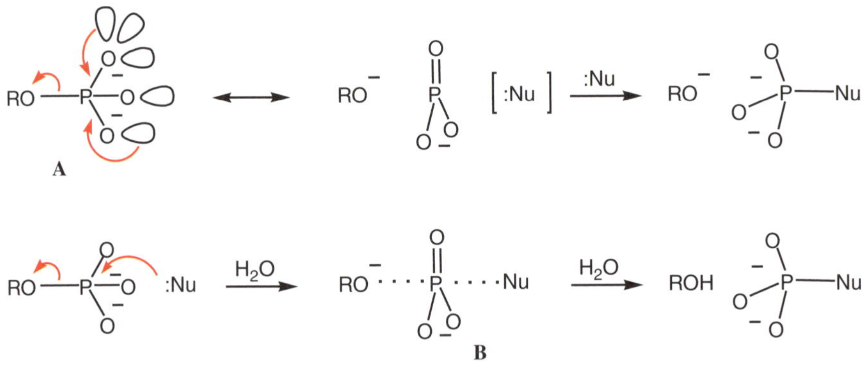

Scheme 4.34 Phosphate transfer from a monoester dianion. The P–OR bond is significantly weakened by the n_O–σ^*_{P-OR} interactions shown (**A**): but reaction almost invariably involves inversion at phosphorus, thus indicating "in-line" displacement and the participation of a nucleophile (the $S_N2(P)$ mechanism). Only weak bonding to the nucleophile is needed because the bond to the leaving group is almost broken in the "loose" transition state **B**. (See the text).

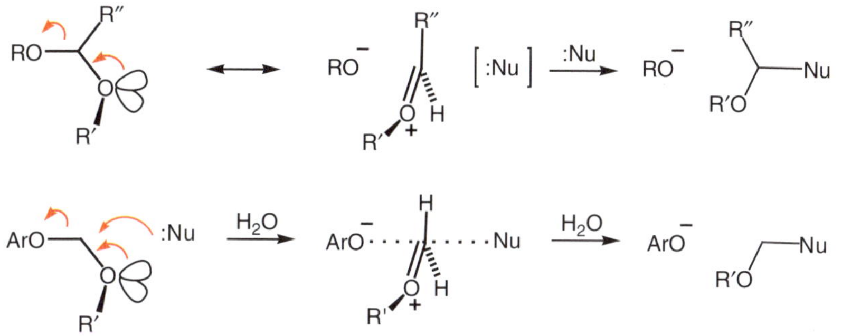

Scheme 4.35 The C–OR bond of an acetal is much more easily cleaved in the sense C$^+$–$^-$OR than that of an alcohol or ether, because the oxocarbocation so formed is stabilized by electron donation from the remaining oxygen. Thus, the S$_N$1 mechanism is normal for the cleavage of acetals ROCHR″–OR′. However, this stabilization is insufficient in the case of acetals ROCH$_2$–OR′ derived from formaldehyde (R″ = H), where in the presence of a nucleophile (present in the encounter complex) the observed mechanism is best described as (enforced) S$_N$2, with much S$_N$1 character.[142]

[:Nu] already present in the encounter complex: in a mechanism best described as S$_N$2(P) with much S$_N$1(P) (or dissociative*) character, and thus minimal associative character in the transition state (Scheme 4.34(B)). In fact, for a good enough leaving group the rate may become entirely independent of the nucleophilicity of the incoming reactant.[141] This mechanism has much in common with the "borderline" S$_N$2 reactions observed with acetals derived from formaldehyde (Scheme 4.35).

The high sensitivity to the pK_a of the leaving group characteristic of the uncatalyzed hydrolysis of the dianions of phosphate monoesters (cleavage is some 17 times faster per unit decrease in the pK_a of the (conjugate acid of) the leaving group, compared with a factor of less than 2 for the reaction of the monoanions)[143] suggests that efficient phosphate transfer from monoesters is likely to involve the PO$_3$$^{2-}$ group. This applies even to the hydrolysis of the monoanions. The monoanion is the most reactive ionic form of a monoalkyl phosphate, which is hydrolyzed faster than the corresponding dianion. (It is also much faster than the anion of a diester (RO)$_2$PO$_2$$^-$, *i.e.* this is not an effect of the increased electrophilicity of the phosphorus centre). The explanation is that reaction involves the pre-equilibrium formation of the tautomer **T** (Scheme 4.36), making the leaving group the neutral alcohol rather than the alkoxide anion. Though this pre-equilibrium is highly unfavourable (by a factor of some 10^{10}), the high sensitivity to the pK_a of the leaving group makes **T** over 10^{20}

---

*In a dissociative transition state bond order decreases in going from the ground state to the transition state (the sum of the bonding between the incoming nucleophile and outgoing leaving group is less than one). An associative transition state is defined as one in which bond order *increases* in going from the ground state to the transition state (i.e. the sum of the bond orders from P to the incoming nucleophile and outgoing leaving group is greater than one).[140]

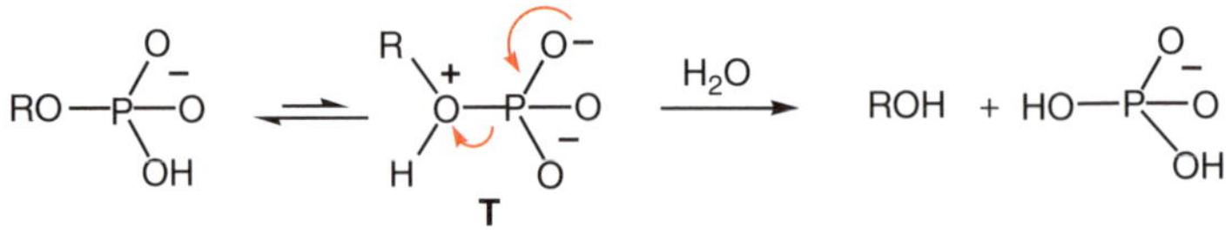

Scheme 4.36   The (uncatalyzed) hydrolysis of the monoanion of an alkyl phosphate monoester involves the unfavourable pre-equilibrium formation of a PO_3^{2-} derivative **T**, with a very good (ROH) leaving group.

times more reactive than $RO-PO_3^{2-}$, thus more than compensating for its low concentration.[143]

4.2.1.1   Enzymes I. Phosphoryl Transfer Without Metals: PTPases

Phosphatases catalyze the transfer of the phosphoryl group from phosphate monoesters to water, to give inorganic phosphate. They use two classes of mechanism similar in principle to those described in Section 4.1 for carboxylic acyl transfer: involving either transfer to an active-site nucleophile, followed by hydrolysis of the phosphoryl enzyme intermediate; or direct transfer to water. A reliable diagnostic test of mechanism is the stereochemistry of the overall reaction (monoesters chiral at the phosphorus centre are available, using the three different isotopes of oxygen, ^{16}O, ^{17}O and ^{18}O). Inversion at phosphorus indicates direct transfer to water, while retention of stereochemistry is taken to result from two consecutive inversions, as expected for the two-step transfer via a phosphoryl enzyme intermediate. Enzymes using the direct transfer route typically use a metal-bound hydroxide as the nucleophile, with assistance from a second metal centre. Some phosphatases using the two-step mechanism also use two active-site metals, but important classes of enzyme involved in the dephosphorylation of phosphoproteins use none.

Protein phosphorylation is a key regulatory process, which controls the individual activities of many enzymes and receptors. The target side-chain nuclophiles are generally the hydroxyl groups of specific serine, threonine and tyrosine residues. These are phosphorylated by kinase enzymes, using ATP as the phosphate donor, and dephosphorylated by the protein phosphatases. These enzymes are typically specific for tyrosine, or for serine and/or threonine; though some dual-specific enzymes hydrolyze phosphates of all three.

Protein tyrosine phosphatases (PTPases) are of special interest because they have evolved a simple but effective mechanism (Scheme 4.37) which does not use metal cations. The same mechanism is found in the dual-specific protein phos-phorylases, with which they share a common active-site sequence $Cys.(X)_5.Arg$ (with the catalytic Cys and Arg residues separated by exactly five amino-acid residues), but otherwise no sequence similarity. The mechanism has been established as a double displacement involving the anion of the cysteine residue as the nucleophile; so that the phosphoryl enzyme is a phosphorothiolate ester.

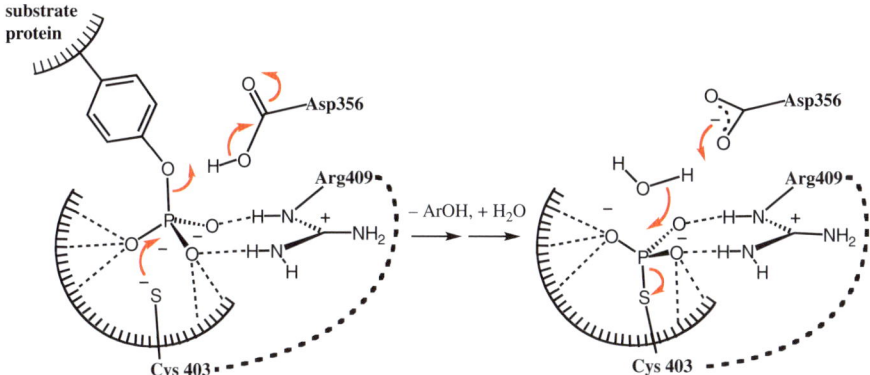

Scheme 4.37   Reaction mechanism typical of protein tyrosine phosphatases. General acid catalysis by the conserved Asp356 is concerted with nucleophilic attack by the thiolate anion of Cys403. (Representation based on Hoff *et al.*[145])

The hydrolysis of this phosphoryl enzyme intermediate in the second step of the reaction is mechanistically the reverse of the first, with the Asp residue involved as a general acid in step 1 now available to act as a general base in step 2,[144] when the tyrosine OH is replaced by water (Scheme 4.37).

Detailed mechanistic examination of this reaction has shown that the reactive ionic forms are the thiolate anion of the substrate and the dianion of the phosphate monoester. Thus, these enzymes take advantage of the intrinsic reactivity of the monoester dianion.[145,146] Evidently, the enzyme is able to stabilize the loose transition-state characteristic of this sort of reaction (**B** in Scheme 4.34), using a pattern of hydrogen-bonding interactions specific for its charge distribution and trigonal-bipyramidal geometry.[147,148]

4.2.1.1.1 Models.   No simple model is available for the attack of a thiolate on the dianion of a phosphate monoester, though the special, cationic environment of the bound substrate may be expected to reduce any electrostatic repulsion inherent in the attack of an anion on a dianion. Significantly, the first simple system shown to support the reactions of (oxy)anion nucleophiles with the dianions of an aryl phosphate monoester involves assistance by a cationic general acid (Scheme 4.38).[149]

4.2.1.1.2 Intramolecular Models.   Catalysis of the hydrolysis of phosphate monoesters by neighbouring groups observed in simple systems generally involves reactions of the type shown in Scheme 4.38, in which the departure of a leaving group from the PO_3^{2-} is assisted by a neighbouring general acid. Nucleophilic attack on the PO_3^{2-} group by external neutral nucleophiles is observed for good, *i.e.* weakly basic (pK_a 7 or less) leaving groups: intramolecular nucleophilic catalysis is not possible because of the requirement for

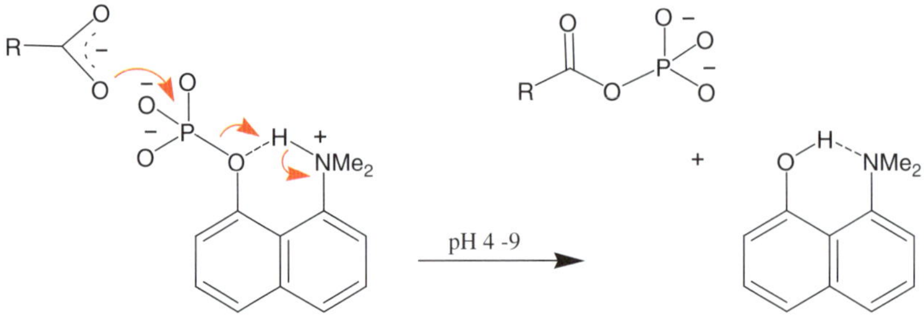

pH 4 -9

4.61

Scheme 4.38 Oxyanions attack the phosphorus centre of the PO_3^{2-} group of the naphthyl phosphate **4.61** with efficient general acid catalysis from the neighbouring dimethylammonium group.[149]

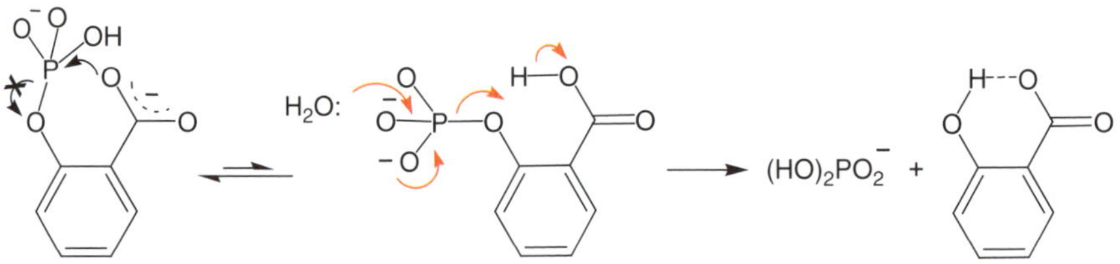

4.62

Scheme 4.39 The mechanism of hydrolysis of the dianion of salicyl phosphate **4.62** (the monoanion is less reactive and the fully ionized trianion totally unreactive).

in-line displacement at the phosphorus centre (see below, Scheme 4.44). For example, the hydrolysis of the dianion of salicyl phosphate **4.62** (Scheme 4.39) involves the less favoured ionic form with adjacent PO_3^{2-} and COOH groups. This enables general acid catalysis by the neighbouring COOH group efficient enough to overcome the initial unfavourable equilibrium, with a strong intramolecular hydrogen bond developing in the salicylic anion leaving group,[150] as involved also in the reaction of **4.61** in Scheme 4.38.

4.2.1.1.3 Supramolecular Models. Designing a supramolecular catalyst for phosphate transfer from a phosphate monoester is a serious challenge. The in-line displacement involved in the $S_N2(P)$ process (Scheme 4.34, above) places the nucleophile and the leaving group far enough apart that a system capable of handling both at the same time has to be a big one. The simplest, and still the most successful system of this sort is Lehn's cryptand (Scheme 4.40),[151] which catalyzes key biological phosphate-transfer reactions of ATP *in vitro*, by way of an intermediate phosphoramidate. Such polyammonium macrocycles bind polyphosphate anions by a combination of electrostatic

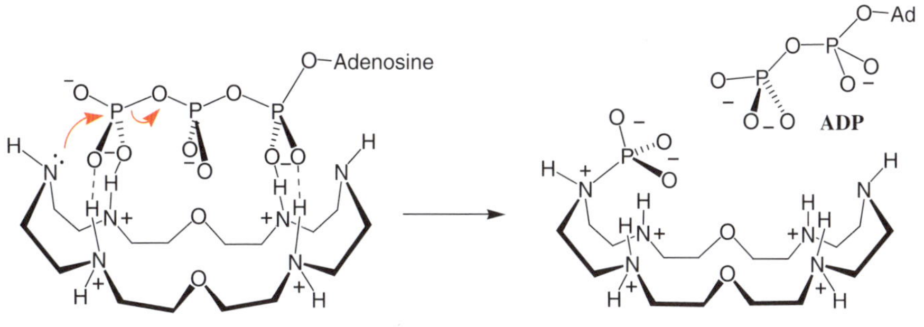

4.63 **4.64**

Scheme 4.40 The hydrolysis of ATP is catalyzed by the hexaazadioxamacrocycle tetracation shown (**4.63**) by nucleophilic catalysis involving one of the unprotonated nitrogens. The short-lived phosphoramidate inter-mediate **4.64** is hydrolyzed to inorganic phosphate, or can transfer phosphate to another nucleophile.

interactions and multiple hydrogen bonding in water. At pH 7 the cryptand tetracation (**4.63**) binds the triphosphate group of ATP in close proximity to a (partially) unprotonated nitrogen nucleophile. This displaces the ADP trianion – a good leaving group that needs little or no external assistance – in a reac-tion that follows Michaelis–Menten kinetics ($k_{cat} = 1.07 \times 10^{-3}\,\mathrm{s}^{-1}$, $K_M = 10^{-4}$ M at 70 °C). The hydrolysis of ATP – a natural metabolite that is itself an activated substrate – is accelerated under these conditions by a factor of a few hundred.[152]

4.2.1.2 Enzymes II. Metalloenzymes

Most phosphatases catalyzing the hydrolysis of phosphate monoesters use two or even three active-site metal cations. Enzymes catalyzing direct transfer to water typically use a metal-bound hydroxide as the nucleophile, with assistance from the second metal centre, and some phosphatases using active-site metals also catalyze the two-step mechanism. The best known and most intensively studied of these are the ubiquitous alkaline phosphatases, and especially the enzyme from *E.Coli*, which catalyze the hydrolysis of the dianions of a broad range of phosphate monoesters. In addition to two Zn cations the active site of an alkaline phosphatase also contains a conserved Mg^{2+} ion, thought to be involved, through a Mg-bound hydroxide, in the generation of the active nucleophile, which is the alkoxide anion derived from Ser-102 (Scheme 4.41). This oxygen is also coordinated to a Zn cation, and the combined effect is to lower the pK_a of the serine OH to about 5. The substrate is bound to the same Zn centre, and also by hydrogen bonding to the guanidinium side-chain of the conserved Arg166. This arrangement allows in-line attack of the serine–O^- on phosphorus, with the (presumably concerted) displacement of the RO^- leaving

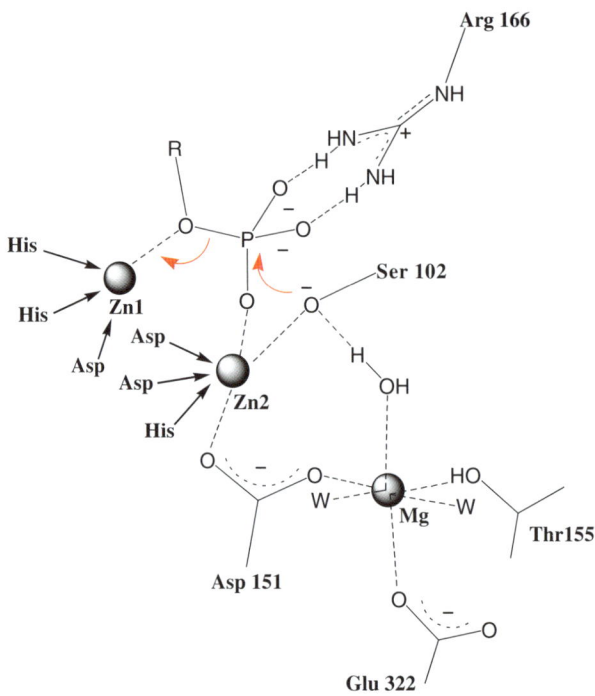

Scheme 4.41   Suggested mechanism for the first step of the hydrolysis of an alkyl
phosphate dianion ROPO$_3^{2-}$ by an alkaline phosphatase. The roles of
the two Zn centres are reversed in the hydrolysis of the serine phosphate
formed, with Zn1–OH now the nucleophile and coordination to Zn2
stabilizing the departing serine–O$^-$. The double-displacement process
accounts for the observed retention of configuration at phosphorus.

group, assisted by coordination to the second Zn centre. The charge distribution on the dianion of a phosphate ester can put a significant amount of
negative charge on the R–O–P bridging (leaving group) oxygen, as shown by
the significant hydrogen bonding of the corresponding oxygen in **4.61** (Scheme
4.38), which raises the pK_a of the dimethylammonium group from less than 5 to
over 9.

   As is often the case, much of the early mechanistic work on alkaline phosphatases used the *p*-nitrophenyl derivative as substrate for these obligingly
accommodating enzymes, and recent work has shown that results with
p-nitrophenyl phosphate were complicated by product inhibition (by inorganic
phosphate, with an inhibition constant K_i close to 1 μM[140]). The convincing
conclusion is that the alkaline phosphatases too take advantage of the
intrinsic reactivity of monoester dianions, with reaction proceeding via loose,
primarily dissociative transition states of the sort described above (**B**,
Scheme 4.34): characterized by extensive breaking of the bond to the leaving
group.[140]

4.2.1.2.1 Models. Only a few zinc-based complexes have been examined as potential models for alkaline phosphatase, typically as minor parts of investigations where carefully constructed metal-containing systems turned out to perform more effectively against phosphate diesters. For example, a systematic study of spacer length and geometry on a series of symmetrical systems bearing a pair of triazacycloalkane-ligands known to bind Zn^{2+} found only modest catalysis of the hydrolysis of the *p*-nitrophenyl phosphate dianion, with a rate just 6 times faster than background in the presence of 0.227 mM of the Zn_2-complex of the most efficient catalyst **4.65** (Scheme 4.42).[153] The system evidently produces a modestly activated hydroxide-equivalent, but is not designed to bind a phosphate dianion.

A more successful system, **4.66**,[154] with no less than 8 amino-groups, binds two Zn cations to give a Zn_2 complex **4.67** that is actually selective for monoester dianions (and extended polyphosphate derivatives, like ATP). Crystal structures define the (solid-state) structures of the Zn_2 complex **4.67** and the product complex **4.69**. The zinc centres are pentacoordinate, with distorted trigonal bipyramidal geometry, which presumably binds the substrate dianion via a bridging interaction more or less as shown as **4.68** (Scheme 4.43). The hydroxyl group on the propanol bridge was designed to act as the nucleophile

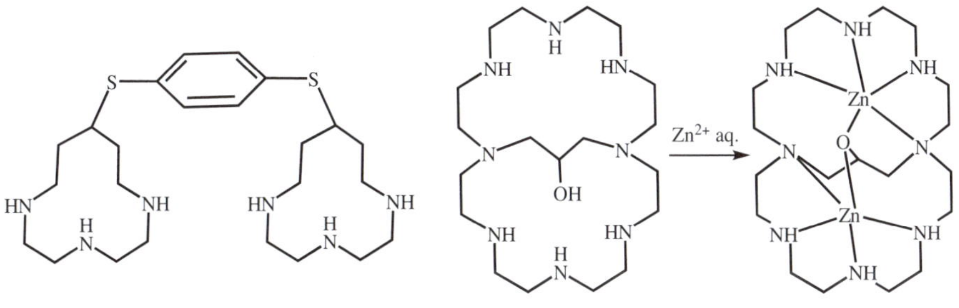

Scheme 4.42 Symmetrical systems designed to model the first step of the alkaline phosphatase reaction. See the text.

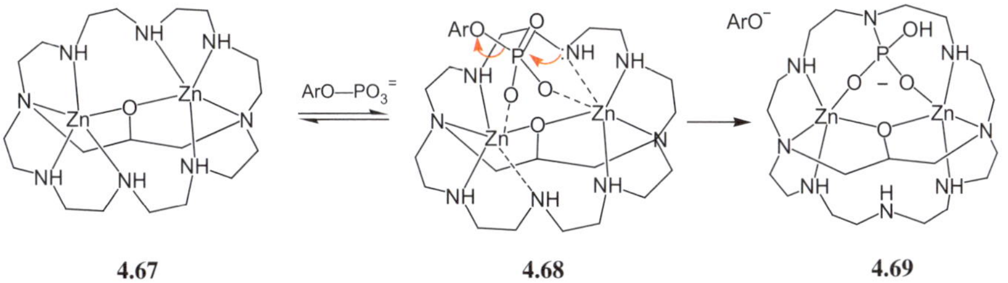

Scheme 4.43 Suggested mechanism of action of the Zn_2 model system **4.67** (Scheme 4.42). The structures of **4.67** and the product **4.69** are based on crystal-structure determinations, and the substrate $ArOPO_3^{2-}$ is 4-nitrophenyl phosphate (see the text).[154]

to effect the cleavage of the bound substrate, but the crystal structure of the stable product shows that this oxygen remains bound to the two zinc centres, reducing its reactivity, so that the preferred nucleophile is an amine nitrogen.

The observed rate of release of 4-nitrophenol from 4-nitrophenyl phosphate (10 mM) at 35 °C in the presence of **4.67** (5 mM) shows a bell-shaped pH dependence, with an optimum near pH 6 and a rate maximum around $7.8 \times 10^{-6} \, s^{-1}$, corresponding to an 8-fold acceleration above background and a second-order rate constant (corresponding to a k_{cat}/K_M under the initial rate conditions) of $1.52 \times 10^{-3} \, mol^{-1} \, s^{-1}$. This is not bad for a model, the combination of intramolecularity (**4.68** in Scheme 4.43) and any activation of the substrate delivering a 4000–5000-fold acceleration compared with the second-order rate constant for the intermolecular reaction of an amine with 4-nitrophenyl phosphate. Though it is less impressive when compared with alkaline phosphatase, which is a particularly efficient enzyme with k_{cat}/K_M of the order of $10^7 \, mol^{-1} \, s^{-1}$ for the same unnatural substrate.[155]

This chemistry is reminiscent of the reaction of the Lehn cryptand **4.63** (Scheme 4.40),[151] which catalyzes the cleavage of ATP^{4-} to yield ADP^{3-} and the phosphoramidate **4.64**, observed only as short-lived intermediate. The Zn_2 complex **4.67** also cleaves ATP^{4-}, to yield ADP^{3-} and a phosphoramidate (**4.69**), while the metal-free cryptand **4.66** promotes its hydrolysis, at about the same rate, to ADP^{3-} and inorganic phosphate. It seems clear that phosphoramidates are intermediates in all these reactions: which are rapidly hydrolyzed in the Zn-free systems, but stabilized by the interactions with the Zn centres in **4.69**.

This discussion has been limited to Zn_2 enzymes and models. Models using more highly charged cations can be expected to bind and activate phosphate monoesters more strongly, and to favour more associative mechanisms. There is evidence that this may be the case for the purple acid phosphatase enzymes, which use at least one Fe^{3+} in combination with a second metal dication.[119,139,156]

4.2.2   Phosphoryl Group Transfer from Phosphodiesters

Phosphodiesters exist under physiological conditions exclusively as the anions. The single negative charge (shared by two oxygens: see Scheme 4.33) makes the phosphorus centre less electrophilic than that of a comparable triester, and also ineligible for the special, predominantly dissociative mechanism available for the cleavage of monoester dianions (Scheme 4.44: compare Scheme 4.34, above). Consequently, diesters are generally the least reactive of the extraordinarily unreactive esters of phosphoric acid. (Important exceptions are cyclic diesters in which the [O–P–O] unit is part of a 5-membered ring: see Scheme 4.47.)

Reactivity for diesters depends significantly on the nucleophile, as well as (still much more strongly) on the leaving group, indicating that bonding at P is tighter in the transition state than for the corresponding reactions of monoester dianions.[139] And reactions with oxyanion nucleophiles, especially alkaline hydrolysis, are reasonably readily observed; though reactivity is somewhat reduced by

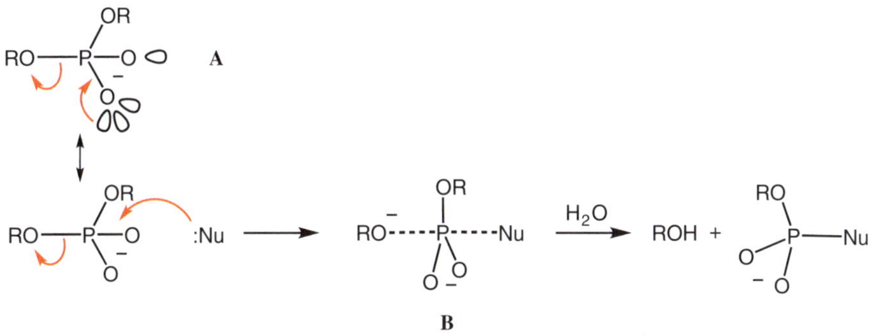

Scheme 4.44 Outline of the phosphoryl transfer process involving a phosphodiester. The single negative charge and reduced charge density on the non-bridging oxygens compared with the (substantially more strongly basic) dianion of a phosphate monoester mean that n_O–σ^*_{P-OR} interactions (**A**) are much weaker. Conversely, pentacovalent addition intermediates are more accessible, and the trigonal-pyramidal transition structure **B** may lead to a pentacovalent phosphorane with a significant lifetime in specially favourable cases. Unless this is the case the $S_N2(P)$ (in-line) process reliably involves inversion of configuration.

electrostatic repulsion, as expected for reactions between anions.[157] It is likely that this effect operates primarily at the level of the formation of the encounter complex, because intramolecular attack by oxyanions on phosphodiester phosphorus is exceptionally efficient compared with comparable reactions of triesters (Scheme 4.45).[8]

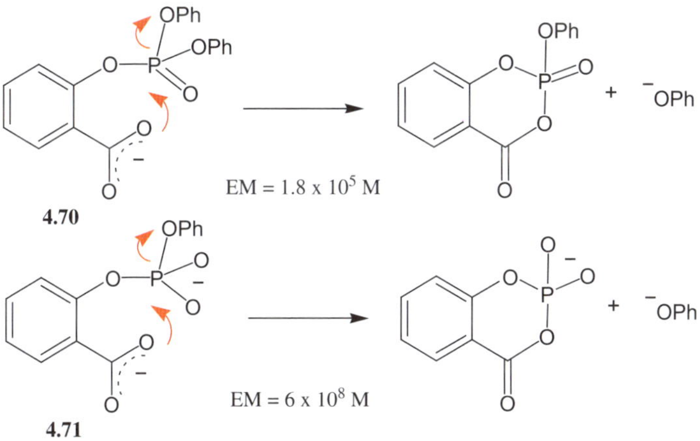

Scheme 4.45 Effective molarities (see Section 1.4.1) compared for the otherwise identical reactions of a phosphate di- and triester. The comparison reaction in each case is the equivalent intermolecular process involving the diffusion together of the ester and the carboxylate anion to form the encounter complex.[8]

4.2.2.1   *Intramolecular Reactions*

The preference for inversion of configuration observed for $S_N2(P)$ reactions of phosphodiester anions, and its explanation, are nicely illustrated by the systems shown in Scheme 4.45. The diester **4.71** reacts exclusively by in-line displacement of the exocyclic phenoxide leaving group. No endocyclic displacement of the salicylate anion is observed, even though it is the better leaving group. Two closely related explanations (Scheme 4.46) are (a) that the reaction is concerted, preferring the in-line pathway for the usual reasons: or (b) that it goes by way of a trigonal-bipyramidal, phosphorane intermediate **4.72**, which maintains the apical-apical nucleophile-in and leaving group-out arrangement for good stereoelectronic reasons. (Specifically, the preference for the most electronegative substituents on a phosphorane to be in apical positions, which means that the two oxygen anions of **4.72** must remain equatorial.)

A phosphorane intermediate (*e.g.* **4.73**) formed from the addition of a nucleophile to a triester has only a single oxygen anion substituent, which can remain equatorial as the remaining two pairs of substituents exchange apical and equatorial positions: by the process known as pseudorotation.[158] The salicylate oxygen of **4.74** then becomes apical, and thus able to depart, generating the mixed anhydride **4.75**: which is hydrolyzed under the conditions to salicylate and diphenyl phosphate. By varying the exocyclic leaving group it could be shown that the pseudorotation barrier, estimated as $> 5\,\mathrm{kcal}\,(21\,\mathrm{kJ})$ mol^{-1} for the reaction of the diester **4.71**, is not significant in the reaction of the triester; because the product ratio is independent of the geometry of the displacement, and depends only on the leaving group.[159]

The possibility of pseudorotation adds an important dimension to the phosphate transfer chemistry of diesters. Note that the basic requirement, that there should be only a single (or of course no) negatively charged oxygen in a product-determining phosphorane, can also be met by the neutral, protonated forms of diesters (or, in principle, even monoesters).

4.2.2.1.1 Intramolecular Attack by OH. Biologically the most important intramolecular reaction of phosphodiesters is the substrate-assisted cleavage of RNA. Naturally occurring 3′-phosphate esters of ribonucleosides have a potent nucleophile – the ribose 2′-hydroxy group – built into the substrate, perfectly placed to act as an intramolecular nucleophile by way of a favourable 5-membered cyclic transition state (Scheme 4.47). The geometry of attack by this oxygen on phosphodiester phosphorus is well defined, and the process conveniently studied as an intramolecular reaction in simple systems. The effective molarity of the hydroxyl group in the ribose system **4.76** is of the order of 10^7 M, compared with about 10^4 M for the corresponding reaction of an ethylene glycol derivative.[8]

At high pH the intramolecular reaction of Scheme 4.47 is specific base catalyzed, involving the pre-equilibrium formation of the alkoxide anion (the pK_a of the OH group of **4.76** is between 13–14), and this reaction accounts for the ready alkaline hydrolysis of RNA. The first step is cyclization to form the phosphorane

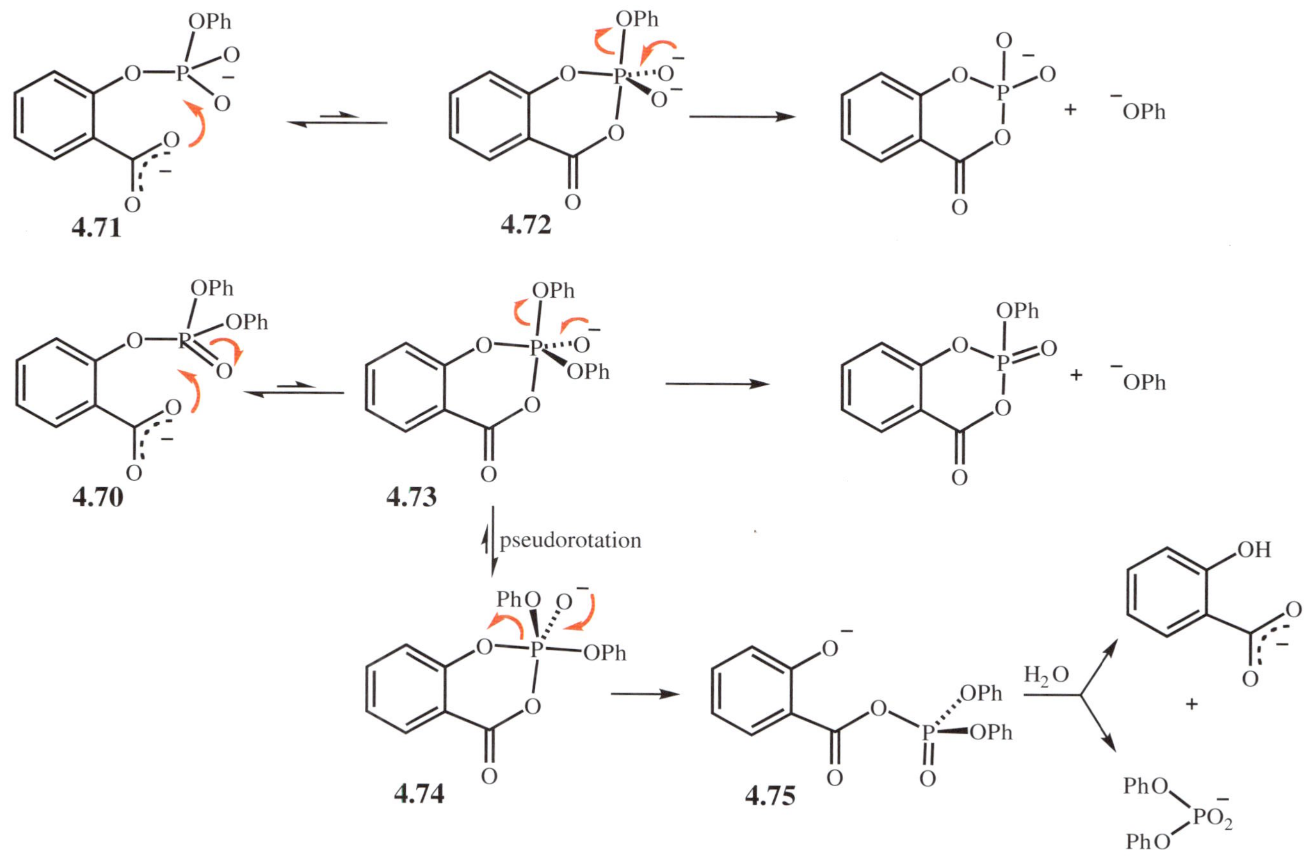

Scheme 4.46 The contrast between the exclusively exocyclic displacement of phenoxide from the phenyl salicyl phosphate dianion **4.71** and the similar reaction of the corresponding triester **4.70**; which gives a mixture of exocyclic and endocyclic displacements that depends only on the leaving group capabilities of the aryloxide anions.

dianion **4.79** (Scheme 4.48). There is good kinetic evidence that this step is rate determining for leaving groups RO$^-$ less basic (*i.e.* better) than that the 2'-O$^-$ (the competing leaving group from **4.79**). For the most basic (alkoxide) leaving groups (pK_a of ROH > 13) the rate becomes much more sensitive to basicity, as the cleavage of the P'–OR bond becomes rate determining.[161]

Under physiological conditions near pH 7 the reaction is general base catalyzed: a conclusion that is relatively straightforward for model compounds with good leaving groups R'O$^-$, but less so for reactions of systems, like RNA or simpler oligoribonucleotides with poor, alkoxide leaving groups; because the general base-catalyzed formation of a phosphorane intermediate and its general acid-catalyzed breakdown are kinetically equivalent (Scheme 4.49: see Section 1.3.3).

The two mechanisms can be distinguished in model systems. The breakdown of the phosphorane **4.81** (Scheme 4.50) must be rate determining in the intramolecular cyclization of the dianion of the diester **4.80**, because the nucleophile, a phenolate anion, is a much better leaving group than methoxide. Rapid in-line displacement leads exclusively to the cyclic phosphodiester **4.82**, and the

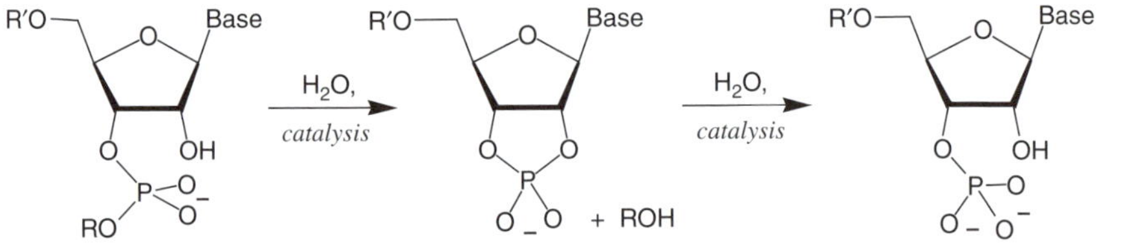

4.76   **4.77**   **4.78**
                                                                              (+ Nucleoside-2'-phosphate)

Scheme 4.47   The neighbouring hydroxyl group is an effective nucleophile for the cleavage of 3'-phosphate esters of ribonucleosides. The (relatively rapid) nonenzymic hydrolysis of the cyclic phosphate ester **4.77** produces a mixture of 2'- and 3'- derivatives.[160]

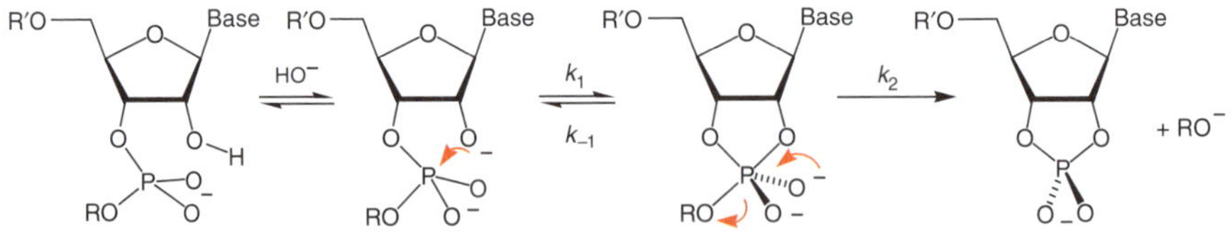

4.79

Scheme 4.48   The alkaline hydrolysis of model ribonucleotide diesters, and presumably of oligo- and polyribonucleotides generally, goes by way of a transient phosphorane dianion intermediate **4.79**. This intermediate does not accumulate but breaks down rapidly to products or to regenerate the reactant anion. Breakdown to products becomes rate determining when $k_{-1} > k_2$.[161]

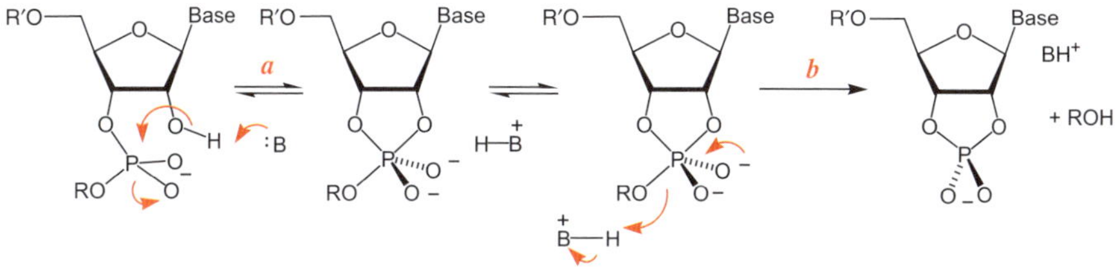

Scheme 4.49 The general acid–base catalyzed cleavage of a ribonucleotide. The general base-catalyzed formation (step *a*) and general acid-catalyzed breakdown (step *b*) of the phosphorane intermediate **4.79** (Scheme 4.48) both involve the elements of the substrate anion plus the general base. The important mechanistic difference is the (kinetically inaccessible) location of the proton.

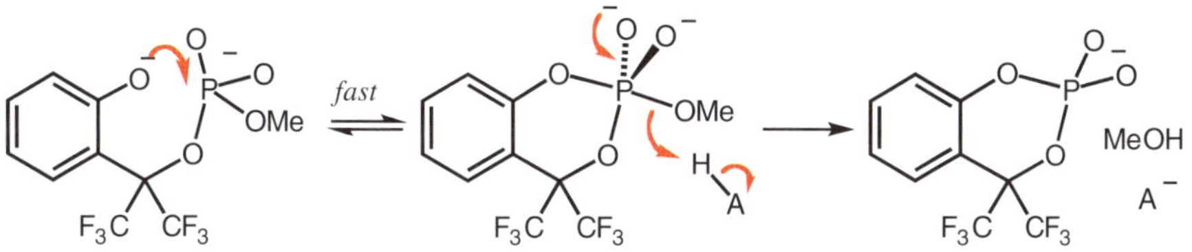

4.80   **4.81**   **4.82**

Scheme 4.50 General acid catalysis of the cyclization of the diester dianion **4.80** can be assigned unambiguously to proton transfer to the departing methoxide leaving group.

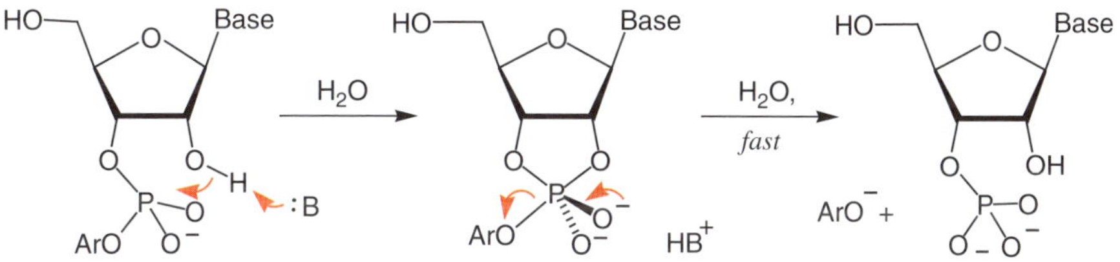

4.83

Scheme 4.51 General base catalysis of the formation of the phosphorane dianion is rate determining for the phosphodiester **4.83**, with the good leaving group $ArO^- = p$-nitrophenoxide.

observed general acid catalysis can be assigned unambiguously to the cleavage of the bond to the leaving group.

The Brønsted α of -0.33 observed for this general acid catalysis complements perfectly (if no doubt fortuitously) the Brønsted β of 0.67 measured for the general base-catalyzed cyclization of the more obviously relevant model system **4.83** (Scheme 4.51),[162] which reacts via a phosphorane in which the exocyclic *p*-nitrophenoxide is a much better leaving group than the ribose 2′-oxygen, making its formation cleanly rate determining.

Many related mechanistic investigations have looked at general acid–base catalysis of the cyclization-cleavage reactions of RNA itself, or of ribonucleotide diesters designed to answer specific questions. For various reasons unambiguous answers have rarely been forthcoming: despite their relatively high reactivity (for phosphodiesters) these systems are, nevertheless, still very unreactive in real time, so that data of the highest quality are not easily obtained. Furthermore, oligomers (with more than a single phosphodiester link) are better models for RNA, but different internucleotide bonds must be expected to react at least quantitatively differently. A (relatively) simple compromise solution is to study a "chimeric" oligomer, with a single ribonucleotide link in an otherwise oligodeoxynucleotide chain. For example, the hydrolysis of TTUTT, which has a half-life of weeks at 80 °C near pH 7,[163] shows modest catalysis by both imidazole and imidazolium constituents of imidazole buffers. The products also vary, depending on the pH, with isomerization to the 2′-phosphate competing with cyclization to the cyclic phosphate at lower pH (Scheme 4.52): strong evidence for the intermediacy of the phosphorane monoanions, which can interconvert via pseudorotation (Ψ). Excellent reviews of the extensive work in this area are available.[160,164]

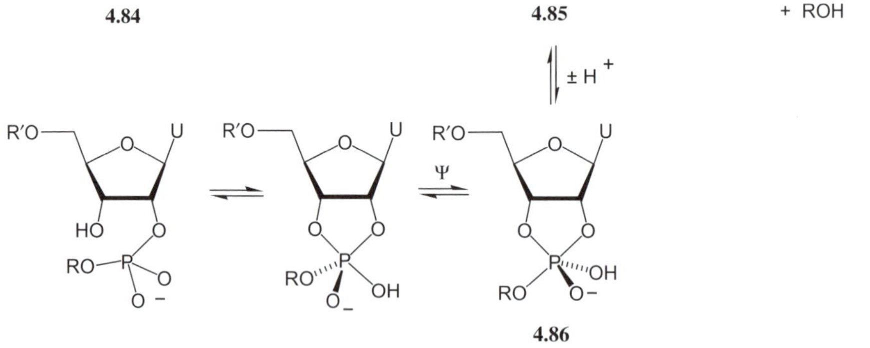

Scheme 4.52   Isomerization by way of the phosphorane monoanion **4.86** competes with cleavage at acidic pH. There is convincing kinetic evidence for the transient intermediate dianion **4.85** in the alkaline hydrolysis of uridine 3′-phosphate esters (Scheme 4.48).[161]

4.2.2.2 Enzymes I. Phosphoryl Group Transfer without Metals

The large numbers of enzymes involved in the cleavage of phosphodiesters can be broadly classified mechanistically in terms of the involvement of metal cations in catalysis, and the intrinsic reactivity of their substrates.[165] The two criteria are largely interdependent: phosphoryl transfer reactions of typical, unactivated phosphodiesters like DNA generally involve active-site metal ions; while the reactions of intrinsically reactive systems like RNA, with a neighbouring hydroxyl group acting as the immediate nucleophile, and 5-membered cyclic diesters, often do not. The four classes thus defined are summarized briefly below in terms of their metal requirements.

Relatively few enzymes are capable of cleaving the P–OR bond of a typically unreactive phosphodiester without the aid of a metal. So those known to do so – and their number is growing – are of special interest. The key to a successful substitution reaction at a given unreactive phosphorus centre is always the delivery of the primary nucleophile. In the absence of a metal cation this is typically not water but a nucleophilic group from an active site amino-acid side-chain: a histidine imidazole in enzymes belonging to the extensive phospholipase D superfamily, which includes a number of bacterial DNAses,[166,167] and the hydroxyl group of a tyrosine in the DNA topoisomerases.[168] Little is known about the detailed mechanisms of the topoisomerases, which are involved in a good deal more than "simple" bond making and breaking, though the key chemical step is the formation of a protein-tyrosine phosphate diester at the 3′-terminus of the DNA chain. The hydrolysis of this intermediate, which is itself a seriously unreactive diester, can itself be catalyzed by an enzyme belonging to the phospholipase D superfamily.[169]

This phospholipase D superfamily is characterized by an extensive common-sequence motif, HisXLys(X)$_4$Asp(X)$_6$GlySerXAsn, which occurs twice in the (typically dual-domain or dimeric) holoenzyme, thus providing two histidine residues in optimal positions for the in-line double-displacement processes of the overall reaction. The basic mechanism suggested for this group of enzymes, and for the "ubiquitous acid endonuclease" DNAse II [170] is outlined in Scheme 4.53.

No simple model is available for this mechanism. Closest is the highly efficient intramolecular general acid catalysis observed for the attack of various nucleophiles on the diester **4.88** (Scheme 4.54) by the neighbouring dialkylammonium group, which in this system has a pK_a close to that (around 7) of the imidazolium cation.[171] Reactivity depends moderately strongly on the basicity of the nucleophile, and strongly on the developing intramolecular hydrogen bond, which is known to be strong in the final product naphthol (the pK_a of its OH group is raised to 14.9 in water).

In fact a simple bifunctional model for the mechanism of Scheme 4.53 is not a practical proposition: building an intramolecular nucleophile into a system like **4.88** offering efficient intramolecular general acid catalysis presents a major challenge because the S$_N$2(P) process (Scheme 4.53) demands in-line displacement. The only simple solution – not exactly an enzyme model (!) – would be to put the nucleophile in the second ester group. Thus, the "substrate-assisted"

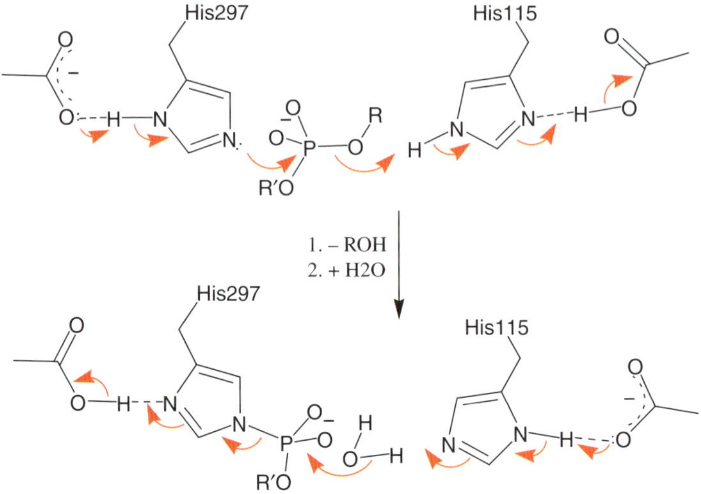

4.87

Scheme 4.53   Enzymes of the phospholipase D superfamily use two active-site histidines to catalyze the cleavage of unreactive phosphodiesters by a nucleophilic mechanism involving a phosphoryl-enzyme intermediate **4.87**. The intermediate is rapidly hydrolyzed by the reverse of the cleavage mechanism, with water replacing the departing nucleotide ROH.

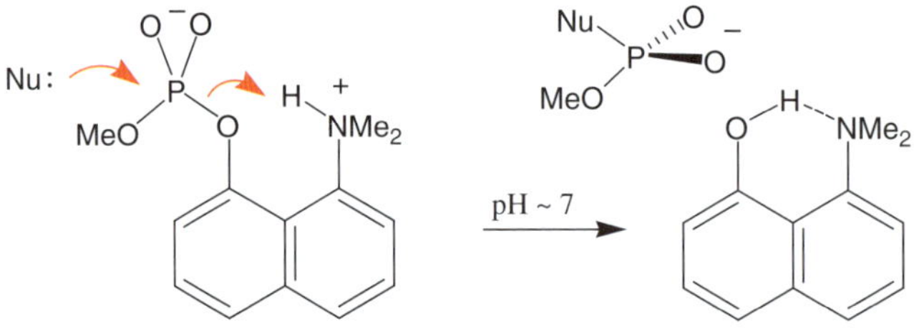

4.88

Scheme 4.54   The dialkyl ammonium group of **4.88** acts as a general acid to catalyze the attack of nucleophiles on the phosphorus centre of the neighbouring, intrinsically unreactive phosphodiester, by factors of up to 10^6 in rate.[171]

hydrolysis of the bis-salicyl diester **4.89** (Scheme 4.55, half-life 10.2 min at 39 °C) is some 10^{10} times faster at the pH 4.5 maximum of the bell-shaped pH–rate profile than that of diphenyl phosphate under the same conditions.[172]

4.2.2.2.1 Transfer to Neighbouring OH.   The many different ribonucleases all[ii] take advantage of the neighbouring 2′-hydroxyl group to hydrolyze RNA by way of the cyclic 2′,3′-phosphate. The familiar "ribonuclease

---

[ii] With the interesting exception of RNAse H (see Section 4.2.2.3).

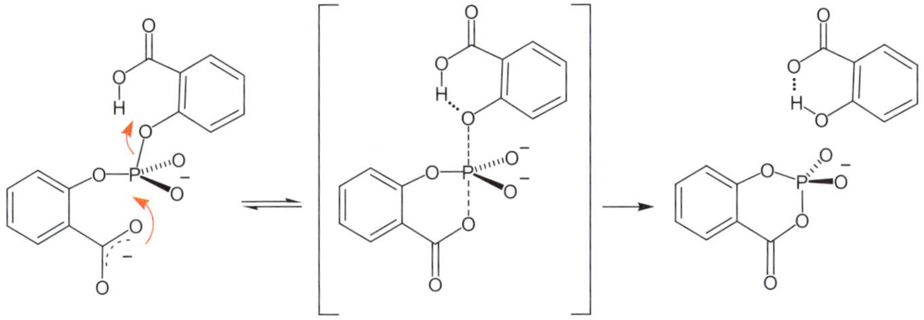

Scheme 4.55 The hydrolysis of the monoanion **4.89** of bis-salicyl phosphate involves catalysis by both carboxy groups. Nucleophilic catalysis is dominant, compared with the contribution from general acid catalysis: perhaps because reaction involves the pre-equilibrium formation of the pentavalent species **4.90** as an intermediate.[172]

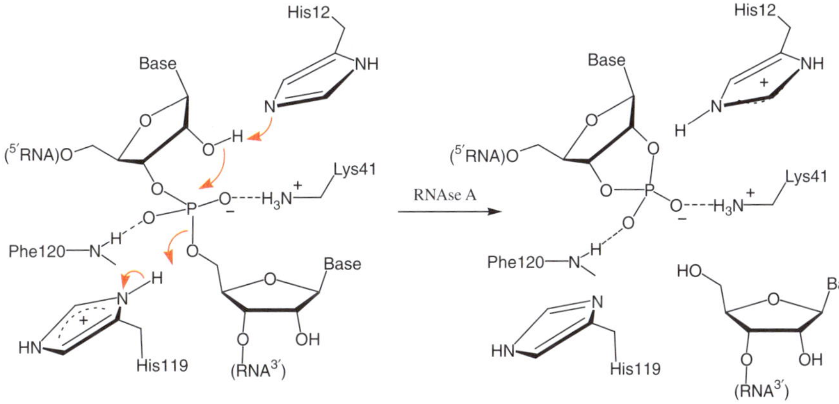

Scheme 4.56 Outline mechanism of action of ribonuclease A. The developing negative charge on the pentacovalent transition state is stabilized by electrostatic and hydrogen-bonding interactions with the NH_3^+ group of Lys41 and the backbone amide group of Phe120. See the text.

mechanism" shown in Scheme 4.56, based on extensive work on ribonuclease A ("the most studied enzyme of the twentieth century"[173]), remains the preferred choice; having survived some rigorous questioning based on a great deal of work on model systems.[160,164] The active site of the endoribonuclease RNAse A brings the bound substrate into contact with two histidine imidazoles, which act highly efficiently as general base and general acid to catalyze the effectively concerted attack of the substrate 2′-OH group and the in-line departure of the 5′-terminal OH of the leaving ribonucleotide, respectively.

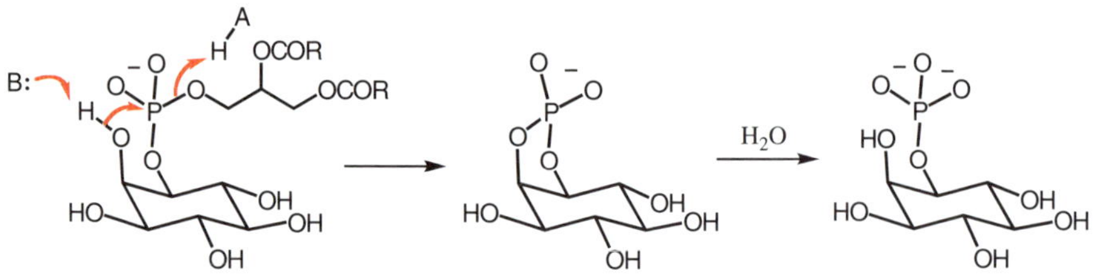

4.91

Scheme 4.57   The reaction catalyzed by phosphatidylinositol-specific phospholipase
C. See the text.

The efficiency of general acid–base catalysis in the ribonuclease reaction can
be estimated from comparisons with mutant enzymes in which the catalytic
histidines 12 and 119 are replaced by alanines. Rates are reduced 10^{4-5}-fold
in each case compared with wild type.[173]

The initial cyclic 2′, 3′-phosphate product (Scheme 4.56) is mostly released by
the enzyme (though RNAse A can catalyze its hydrolysis). Product inhibition is
always a potential problem in the cleavage of polymeric substrates, for which
two large structures differ only by the presence or absence of a single covalent
bond: and thus a common problem also for enzyme mimics. Endoribonucleases
take advantage of their substrate-assisted single-step cleavage mechanism, and
release both oligomeric product chains to maximize turnover.

Generally similar considerations apply to the reaction catalyzed by the
bacterial enzyme phosphatidylinositol-specific phospholipase C, which also
involves substrate-assisted cleavage by a neighbouring OH group, of the bond
to a diacyl glycerol. This reaction too goes by way of a 5-membered cyclic
transition state (Scheme 4.57), with the hydrolysis of the cyclic phosphate **4.91**
produced much slower than its formation. The proposed mechanism of the
first step is similar to that suggested for ribonuclease (Scheme 4.56), with two
histidine imidazoles acting together as a general acid-base pair, probably
positioned by aspartate and aspartic acid residues, with an arginine guanidinium group stabilizing the pentacovalent transition state by electrostatic and
hydrogen-bonding interactions.[174,175]

4.2.2.2.2 Supramolecular Models. Phosphate diesters are extraordinarily
unreactive in water at pH 7 and so perhaps the most challenging targets of
the major metabolites for the model designer. Artificial sequence-specific
restriction enzymes would be of practical utility, and base pairing clearly
offers the possibility of sequence-specific binding: though no obvious prospect
of how to achieve the enormous rate accelerations of diester cleavage that
practical utility would demand. RNA, on the other hand, is already an activated substrate, by virtue of the nucleophilic 2′-OH group adjacent to each
phosphate linkage. As a result, RNAse mimics have been popular targets.

Breslow constructed three simple models for the RNAse A active site by equipping β-cyclodextrin, like the enzyme active site, with two imidazole groups. These were attached to two sugar $6'$-CH$_2$ carbons at the narrow end of the cavity in the three possible configurations (Scheme 4.58), to set up three geometrically different bifunctional catalytic systems.

All three compounds catalyzed the hydrolysis of the (activated, non-natural) substrate **4.92** (Scheme 4.59), showing saturation kinetics and bell-shaped pH–rate profiles with pH optima at pH 6–7, just like the enzyme RNAse A. Thus, the reaction, like the enzyme, is considered to involve one imidazole as the free base and the second as the conjugate acid, presumably as general base and general acid: while deuterium kinetic isotope effect measurements indicate that their dual involvement is concerted.[176] The superiority (by a factor of 5 in k_{cat}) of the **AB** isomer as a catalyst introduces a simple geometric factor to be taken account of in the discussion of mechanism.

Viable models that perform sequence-specific cleavage of a polynucleotide without involving metals have been reported by van Boom [177] and by Komiyama.[178] The common strategy involves recognition derived from stable

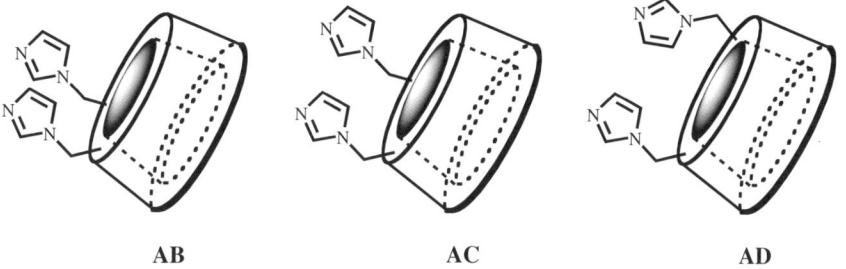

AB   **AC**   **AD**

Scheme 4.58   The three positional isomers of β-cyclodextrin-6,6′-bis-(1-imidazole). **AB** (imidazoles on adjacent sugar residues), **AC** and **AD**. See the text.

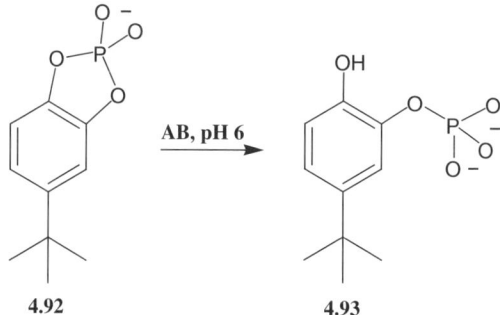

AB, pH 6

4.92   **4.93**

Scheme 4.59   The hydrolysis of the cyclic phosphate **4.92** is catalyzed by all three cyclodextrin bis-imidazoles **AB**, **AC** and **AD** (Scheme 4.58). The **AB** isomer is the most effective, with $k_{cat}/K_M = 7.8 \, \text{mol}^{-1}\text{s}^{-1}$ at the pH optimum and 25 °C.[176]

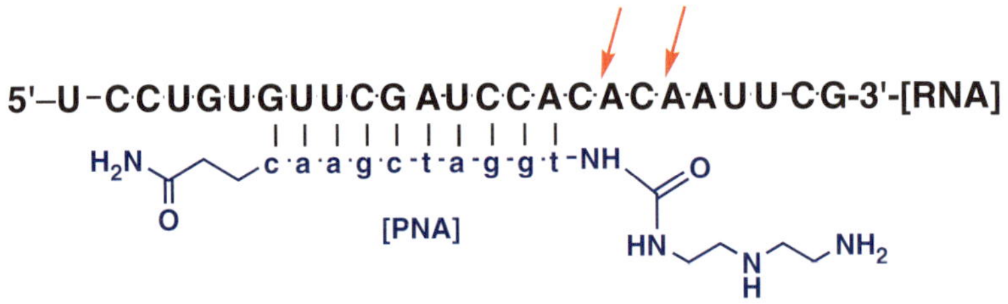

5'–U‑C‑C‑U‑G‑U‑G‑U‑U‑C‑G A‑U‑C‑C‑A‑C‑A‑C‑A‑A‑U‑U‑C‑G‑3'‑[RNA]

H₂N.   c‑a‑a‑g‑c‑t‑a‑g‑g‑t‑NH

[PNA]

Scheme 4.60   van Boom's catalyst (in blue) cleaves its target RNA substrate at two centres (red arrows) close to the 5'‑terminus of the bound PNA sequence, producing 19- and 17-mers (ratio $\sim$2:1) by cleaving the P–O(3') bonds.

catalyst oligonucleotides complementary to sequences in (25 and 30-mer) RNA substrates. The base-pairing recognition elements in the Komiyama and van Boom systems (Scheme 4.60) were a 19-nucleotide sequence of DNA and a PNA[iii] 10-mer, respectively. Each system had the same diethylenetriamine attached to its 5'-terminus, capable of acting as a general acid and/or base to catalyze cleavage by any ribonucleotide 2'-OH group within reach. The van Boom system (Scheme 4.60) is less selective but cleaves its substrate faster (with a half-time of 8 h at pH 7 and 40 °C).

A number of simpler artificial RNAses are discussed in a recent critical review.[179]

Enzymes do not have to rely on the relatively inflexible base-pairing approach to recognition even of specific nucleotide sequences, and protein RNA- and DNA-recognition domains are well known. Thus, the restriction endonuclease BfiI, which cleaves DNA without benefit of catalytic metal ions, appears to have evolved by domain fusion of a DNA-recognition element and an N-terminal nonspecific nuclease. This sort of fusion, of a common catalytic core (see Scheme 4.53) to separate domains for substrate recognition may account for the evolution of many different enzymes in the phospholipase D family.[166] It offers a potential route also to artificial enzymes, depending on the successful integration of effective synthetic catalytic units.[180]

4.2.2.3   *Enzymes II. Metalloenzymes*

Like most phosphatases catalyzing the hydrolysis of phosphate monoesters (Section 4.2.1.1), most phosphodiesterases also use two or even three active-site metal cations (typically Zn, Ca or Mg); making possible direct phosphoryl group transfer to water in the shape of metal-bound hydroxide, with assistance from a second metal centre.[181] Some enzymes even use the same mechanism to cleave RNA. These make up the ribonuclease H family, endonucleases that bind RNA/DNA hybrids and hydrolyze overlapping single strand RNA, using two Mg^{2+} cations. In such cases the important role of 2'-hydroxyl-group lies in substrate binding and recognition rather than catalysis.[182]

[iii] PNA (peptide nucleic acid) binds RNA more strongly than does DNA.

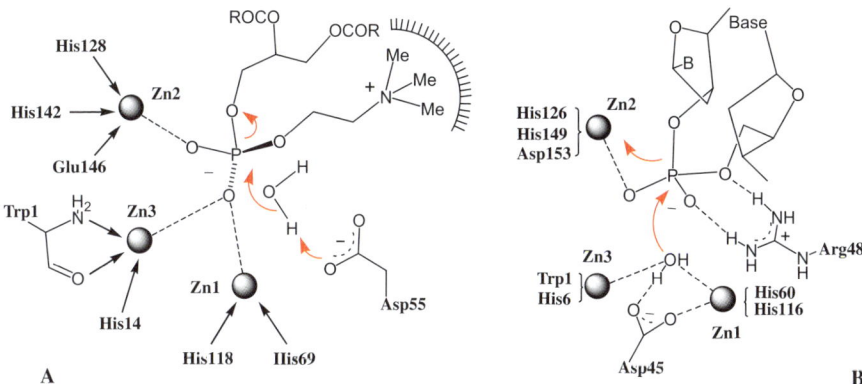

Scheme 4.61 A. Suggested part-mechanism for the hydrolysis of phosphatidyl choline catalyzed by phospholipase C from *B. Cereus*.[131] **B** A closely similar 3-zinc motif is found in P1 nuclease, which catalyzes the hydrolysis of single-stranded DNA and RNA.[184]

Metallophosphodiesterases that cleave unactivated esters without substrate assistance fall mechanistically, it is suggested, into two classes:[181] those that use only the metal ions and attached ligands for catalysis, and those that also involve active-site groups from amino-acid side-chains. The first class includes some nucleases (including members of the ribonuclease H family) and ribozymes (not surprisingly since they have no active-site groups of obvious consequence; though see below, Sections 4.6.4 and 5.3.1) as well as numerous DNA and RNA polymerases and recombinases.[182,183] When the catalytic action takes place in the active site of a protein enzyme it seems unlikely that active-site groups will not be involved at some level, and there is no clear-cut dividing line between the two classes at the present time. As more detailed structural and mechanistic evidence appears for the participation of more groups, especially for the most difficult phosphodiester cleavage reactions, the perceived borderline is likely to shift.

For example: apart from ribonuclease H the best-studied examples of the first class of enzymes, using only the metal, are the structurally similar P1 nucleases, which catalyze the hydrolysis of single-stranded DNA (or RNA) by cleaving the P–O(3′) bond with inversion of configuration; and the phospholipase C that hydrolyzes phosphatidylcholine to phosphocholine and the 1,2-diacyl glycerol (Scheme 4.61).[iv] The P1 nuclease reaction involves nucleophilic attack by a Zn-bound water, assumed to react as the bound hydroxide, with Asp45 helping to "properly orient" it: Zn2 plays a key role in binding and activating the phosphate and stabilizing the leaving O3′-oxanion leaving group.[184] An arginine residue is also present, in position to stabilize the developing negative charge in the transition structure. The phospholipase C has

---

[iv] Not to be confused with the phosphatidylinositol-specific phospholipase C discussed above (Scheme 4.57), which uses no active-site metal.

in common with the P1 nuclease a 3-metal motif (similar to that described in Scheme 4.41 for *E. Coli* alkaline phosphatase), and was at one time thought to use the same mechanism. However, recent X-ray structural evidence shows the latter enzyme has no Zn-bound water, and it has become clear that Asp55 is involved as a general base.[185] The departure of the leaving group is also expected to be assisted by a general acid (which could be a Zn-bond water), though none has so far been identified.

There can be no doubt that active-site groups from the protein are involved in catalysis, as nucleophiles, in cases where the stereochemistry at phosphorus is retained: necessarily the result of a double inversion of configuration. Examples are the snake venom and bovine intestinal diesterases, which both use a threonine OH as the primary nucleophile, and a single Zn cation.[139]

4.2.2.3.1 Supramolecular Models: RNA Cleavage. The two or three-metal motifs found in the active sites of many phosphodiesterases have been fertile sources of inspiration for model builders. At the simplest level it is useful to characterize the properties of metal-bound water/hydroxide in simple systems, thus establishing the rules of engagement for two and three-metals held in close proximity to each other, and identifying any special factors apparent in the enzyme systems. The examples in this section have been chosen to illustrate the main conclusions of the extensive body of work with model systems, which has been extensively reviewed:[119,131,186,187] and to indicate the directions of recent research.

The earliest models were inspired by the ribonuclease reaction. The most popular substrate (for the usual reasons: see Section 4.1.1.3) has been 2-hydroxypropyl *p*-nitrophenyl phosphate HPNP, just about the simplest possible model for a ribonucleotide **RN** (Scheme 4.62). The parent system **RN** is both a closer model and intrinsically more reactive (For the same leaving group, the *p*-nitrophenyl ester **RN** (R = *p*-nitrophenyl, R' = H, Base = uracil) is hydrolyzed up to 10^4 times faster than HPNP at high pH [188]).

As discussed in Section 4.1.1.3, *p*-nitrophenyl esters are imperfect models for their alkyl counterparts, in particular because the good leaving group needs no general acid catalyst to help it on its way; so that pentacovalent species are

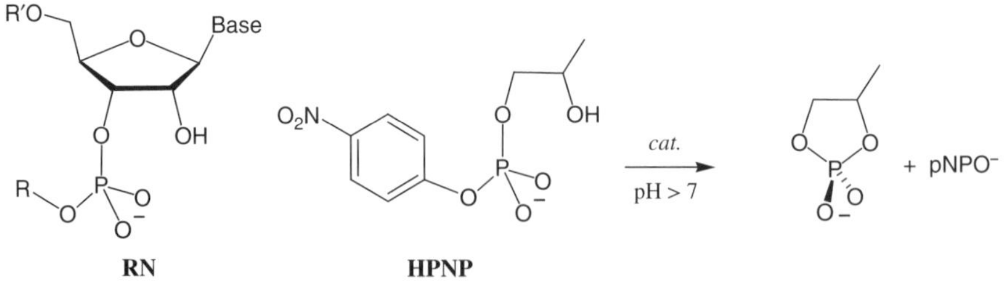

Scheme 4.62   *p*-Nitrophenyl 2-hydroxypropyl phosphate (**HPNP**) is a simple and convenient model for a ribonucleotide (**RN**).

unlikely to be stable enough (as the anions) to be involved as intermediates in phosphate-transfer reactions involving *p*-nitrophenyl phosphodiesters. Thus, though reactions of HPNP are convenient, and useful in comparing reactivities and geometries, the conclusions from these studies must always be seen in a limiting context. For example: the two-zinc system **4.94** (Scheme 4.63), designed specifically for the distance between the two metal centres to match that of relevant bimetallic enzymes, showed a satisfactory 80-fold rate enhancement over the monozinc analogue **4.95** for the reaction with HPNP; but had no effect on the cleavage of RNA (as represented by the ribodinucleotide UpU).[189] The authors concluded that "the appropriate metal-metal distance may be necessary, but is not sufficient" for efficient catalysis. (Yet another example of the broad, general conclusion that many factors are always likely to be involved in an efficient enzyme reaction.)

By contrast, the (significantly higher) reactivity against HPNP of the di- and trizinc complexes **4.96-Zn₂** and **4.97-Zn₃** (Scheme 4.64) is carried over to its reactions with dinucleotides. For these systems Michaelis–Menten kinetics are observed, and k_{cat} values for the reactions with HPNP at 25°C and GpG at 50 °C are comparable.[187]

Much recent work has been concerned with this sort of comparison between the reactions of HPNP and ribonucleotide models with alkoxyl leaving groups.

The mechanisms of the di- and trimetal-system catalyzed cleavages of 2-hydroxyalkyl-phosphodiesters depend on the metal and the leaving group, but generally involve binding of the phosphate anion to one metal centre, thus activating the substrate towards attack on P, and of the substrate OH group to a second, making it a stronger acid. Consequently, the $S_N2(P)$ process in models typically involves the substrate oxyanion as a nucleophile, as shown for **4.96** in Scheme 4.65. Although the calixarenes, like the cyclodextrins, have a central hydrophobic cavity, it seems unlikely that there is specific binding of substrate HPNP in these and similar systems, apart from the $PO_2^- - Zn^{2+}$ ligand–metal interaction. However, there is some evidence that a uracil or guanine base may bind to a second or third metal centre.[190] Thus, the simplest

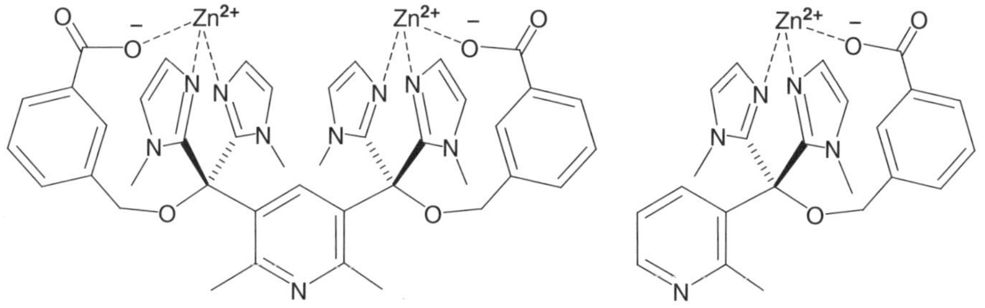

4.94   **4.95**

Scheme 4.63   Anslyn and coworkers' binuclear zinc complex **4.94**, containing "all natural" ligands (see the text), is a more effective catalyst for the hydrolysis of HPNP than **4.95** (see the text).[131]

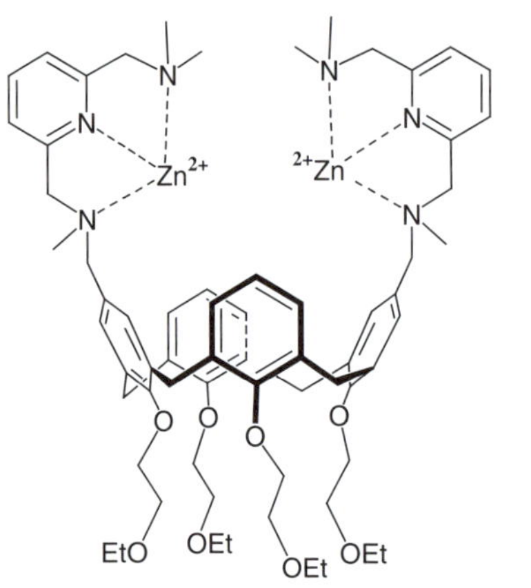

4.96-Zn₂   **4.97-Zn₃**

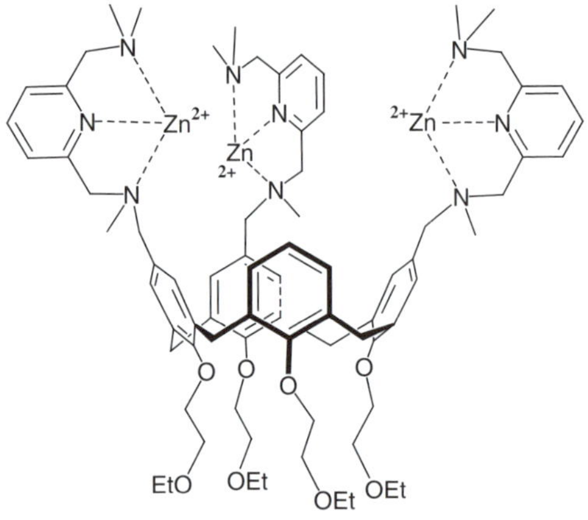

Scheme 4.64   Calix[4]arenes **4.96** and **4.97** were equipped with two and three effective metal-binding ligands.

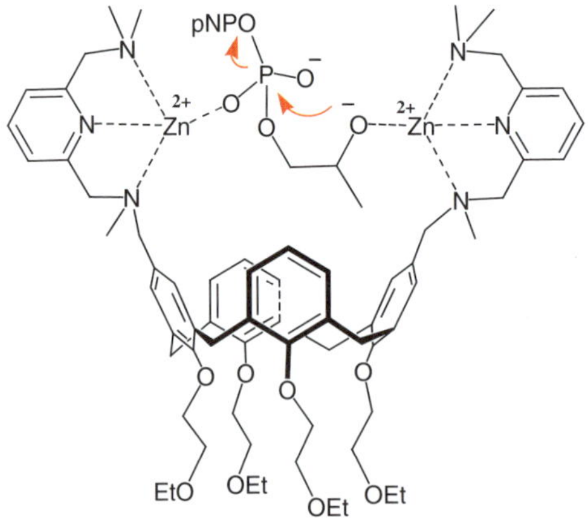

4.96 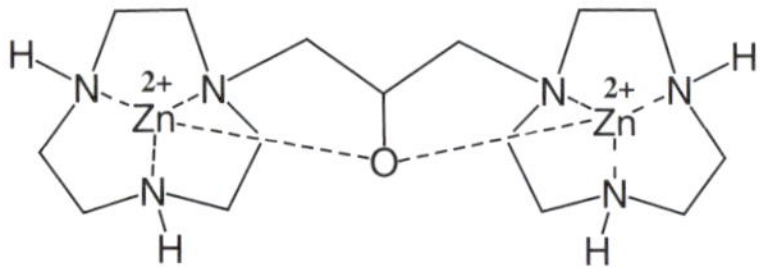   **4.99**

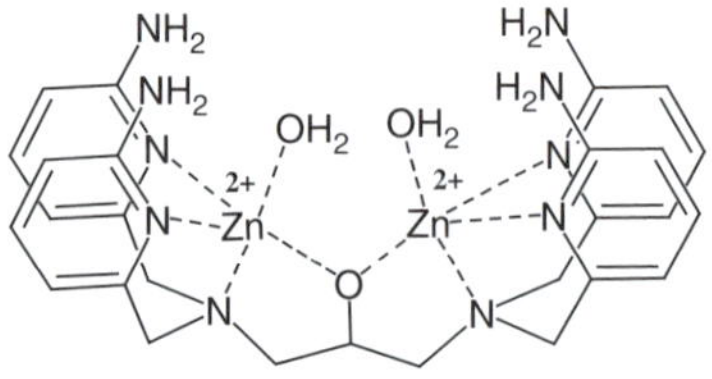

Scheme 4.65   Suggested mechanism of cleavage of HPNP (pNP = *p*-nitrophenyl) by the dizinc complex **4.96.Zn₂** (based on Molenveld *et al.*[187]) The Zn₂ complexes **4.98** and especially **4.99**, without hydrophobic cavities, are more effective catalysts for the hydrolysis of HPNP, and also cleave oligoribonucleotide bonds (see the text).

dizinc system **4.98** (Scheme 4.65), which has both zinc centres chelated to its secondary alkoxide group, is a useful catalyst for the hydrolysis of both HPNP and the ribodinucleotide UpU (Table 4.3).[191]

The introduction of a second or third metal centre may increase the efficiency of catalysis in a given system, but there is evidence that the addition of hydrogen-bond donors – potential general acids and bases – can be more

Table 4.3 Catalytic proficiencies for metal-complex catalysis of the hydrolysis of HPNP and UpU by Zn_x complexes, in water at 25 °C.[187,191,192]

Catalyst	Substrate	k_{cat} (s^{-1})	k_{cat}/k_{uncat}	K_M (mM)	Catalytic [a] proficiency	ΔG_{TS} (cal/mol)	Ref.
4.96-Zn₂	**HPNP**	7.7×10^{-4}	1.4×10^7	0.018	1.5×10^9	-12.6	187
4.97-Zn₃	**HPNP**	2.4×10^{-3}	8.8×10^4	0.83	1.0×10^8	-11.0	187
4.98-Zn₂	**HPNP**	4.1×10^{-3}	1.2×10^5	16	**1.1×10^7**	**-9.6**	191
4.99-Zn₂	**HPNP**	1.7×10^{-2}	$\sim 10^6$	0.32	3.3×10^9	-13.0	192
4.97-Zn₃	**UpU**[b]	1.1×10^{-4}	1.1×10^4	0.34[b]	3.2×10^7	-7.5	187
4.98-Zn₂	**UpU**				**7×10^6**	**-9.3**	191

Notes. [a]K_{TS}, calculated as $[(k_{cat}/K_M)/k_{uncat}]$ or (for figures in **bold**, as $[(k_{cat}/K_M)/k_2]$. [b]Measured at 50 °C in 35% EtOH-water at pH 8.0.

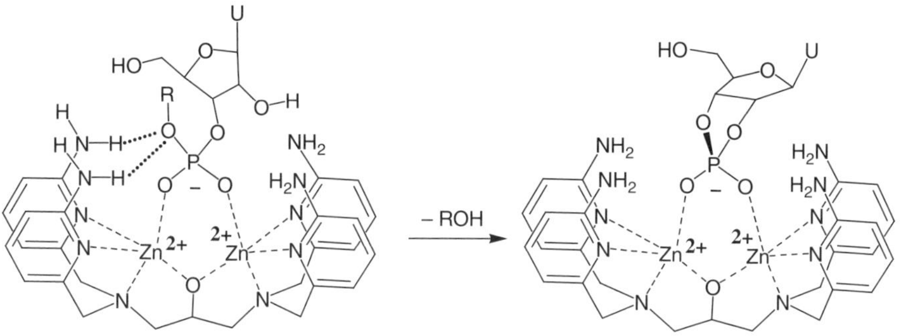

4.100   **4.101**

Scheme 4.66 The aminopyridyl hydrogen-bond donor groups of **4.99** (Scheme 4.65) are well placed to interact with both nucleophilic and leaving OH groups (**4.100**) of a bound 2-hydroxyalkyl phosphodiester substrate. Reactions catalyzed by **4.99** are up to 1000 times faster than the system lacking the four *ortho*-amino groups.[192]

effective still. Thus the most effective catalyst reported to date for the cleavage of HPNP is **4.99**:[192] a 0.2 mM solution reduces the half-life of (0.05 mm) HPNP to about a minute at pH 7.4 and 25 °C – corresponding to a 10^6-fold rate acceleration, and a catalytic proficiency of almost 10^9 (Table 4.3). The system shows Michaelis–Menten kinetics and efficient turnover, rapidly converting a 10-fold excess of HPNP into the propylene cyclic phosphate (Scheme 4.62). Furthermore, **4.99** catalyzes the hydrolysis of UpU with similar efficiency (**4.100**, Scheme 4.66), a 1 mM solution giving a similar 10^6- fold rate acceleration near pH 7.[192]

4.2.2.3.2 Supramolecular Models: DNA-Cleavage.

The problems presented by the familiar combination of unreactive substrate and inefficient catalysis come to a head in the quest for artificial DNAses. Even the standard, imperfect, activated model phosphodiester substrate, bis-p-nitrophenyl phosphate (BNPP), is stubbornly unresponsive, with a half-life estimated at 1300 years at 25 °C and pH 7.[186] Oligodeoxyribonucleotides, as dialkyl phosphates, are of course much less reactive. The saving grace for the modeler has been the discovery that candidate artificial DNAses can be tested conveniently against plasmid DNA: which offers target natural internucleotide P–O bonds that are not only on average more reactive than those of short oligonucleotides, but come in polynucleotide bundles containing many thousands of them. The cleavage of a single bond can be detected by simple gel electrophoresis, increasing the sensitivity of the assay by a factor of the order of 10^4.

No simple supramolecular model is known that is capable of binding and hydrolyzing an unactivated phosphodiester without involving a metal, and systems incorporating a single metal cation are generally inefficient. (An exception is the highly electrophilic Ce^{4+}, which Komiyama has shown to catalyze the P–O (rather than the oxidative) cleavage of deoxyribonucleotides only 20–40 times more slowly than that of corresponding ribonucleotides.[193] (The half-life at pH 7 of TpT (**4.102**, Scheme **4.67**) in the presence of 10 mM $Ce(NH_4)(NO_3)_6$ is 3.6 hours at 50 °C.) The phosphate anion is bound to the cerium tetracation in place of one or two of its (8 or 9) solvating waters; one of which, as bound hydroxide, is thought to act as an intramolecular nucleophile (Scheme 4.67): though less effectively (4-membered ring) than the 3'-OH of a ribonucleotide.

Cerium is not involved in natural enzyme chemistry, and aqueous solutions of lanthanides are typically toxic to biological systems; but their high intrinsic reactivity make them interesting components for synthetic phosphodiester cleaving systems (including potential artificial restriction enzymes[194]). Especially since binding and catalysis have been shown to be greatly further enhanced in methanol, where a combination of complex formation and

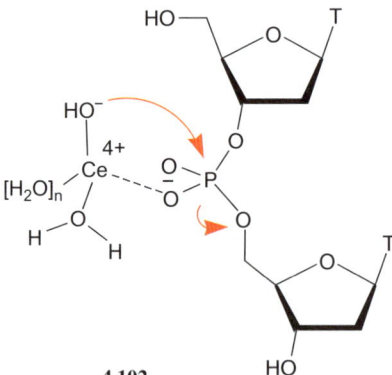

4.102

Scheme 4.67   Suggested mechanism for the Ce^{4+}-catalyzed cleavage of a deoxyribodinucleotide. The metal provides powerful electrophilic activation of both the phosphodiester and bound waters.[193]

medium effects can support very large accelerations. For example, the cleavage of HPNP is catalyzed by La^{3+} in methanol, in the presence of one equivalent of methoxide, by a factor of 10^{10}, over a million times greater than observed for catalysis by La^{3+} in water.[195] This system, not surprisingly, is also highly active against simple activated phosphodiesters such as methyl *p*-nitrophenyl phosphate. It involves the formation of various catalytically active species, and especially lanthanide dimers, which can involve two metal centres in the catalytic mechanism. The lower dielectric constant of an alcohol medium strongly favors ion association, though of course biological systems must have more sophisticated and specific control mechanisms. So the truest models combine specific binding sites for appropriate commonly available metals with those for the target substrate – or better – the transition state for its reaction of interest.

We discuss two representative, relatively efficient systems based on single Zn^{2+} centres. A 3.6 mM solution in water of the histidine-based system **4.103** (Scheme 4.68) cleaves BNPP with a 36 000-fold acceleration over background at the pH optimum near pH 7. Structure **4.103** also shows significant reactivity against plasmid DNA, a 3 µM solution catalyzing complete cleavage to linear DNA in 24 hours at 50 °C.[196] The active nucleophile is probably the Zn-hydroxide, though a potential kinetic ambiguity is not resolved. Reaction with the plasmid is over 100 times faster than the cleavage under the same conditions of the dinucleotide TpT, which offers fewer binding possibilities: but neither reaction shows saturation kinetics.[197]

Clearly a successful small-molecule catalyst for DNA cleavage must take advantage of the binding possibilities offered by the polynucleotide structure for reasons of catalytic efficiency as well as sequence selectivity. Perhaps the simplest binding mode is intercalation: anthraquinones are known intercalators and derived nuclease models **4.104** bind as expected with K_M values of the order of 10^{-4} M. The compound with the longest spacer (**4.104**, $n = 3$) is the most reactive, a 5 mM solution cleaving a bond of plasmid DNA with a half-life of 42 h at 37 °C.[186]

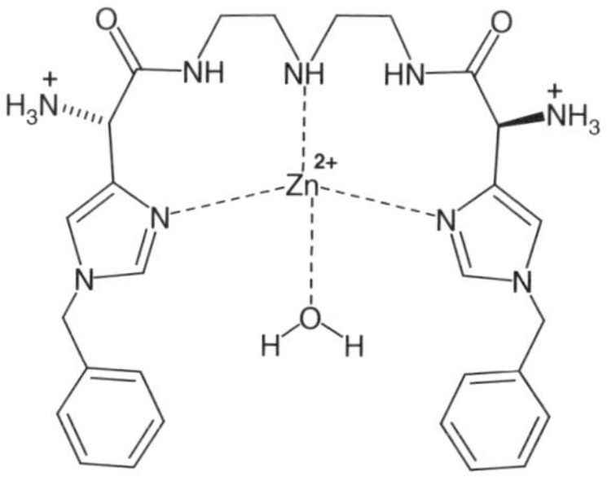

4.103 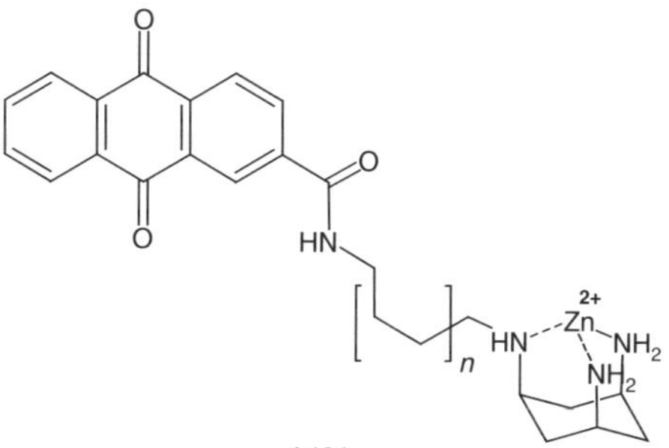   4.104

Scheme 4.68   The Zn(II) complex **4.103** catalyzes the cleavage of P–O bonds in both BNPP and nucleotide substrates. The anthraquinone derivative **4.104** ($n = 1$–3) is specific for the DNA superstructure because it intercalates to bind.[186]

At the other extreme are the zinc-finger nucleases, which combine a natural or engineered specific DNA-binding domain with the nonspecific DNA cleavage domain of a suitable restriction enzyme.[198] Comparisons with available small-molecule model systems based on two or three metal centres suggest that biologically based systems currently look more promising, certainly for use *in vivo*.

4.3 Glycosyl Transfer

Glycosyl transfer is the signature reaction of the carbohydrates, the third great class of biological structures. The basic reaction (Scheme 4.69) is less complicated than acyl or phosphoryl transfer, because it typically involves the one-step nucleophilic substitution of a neutral substrate group without ionizations in the pH region. As for acyl and phosphate transfer, many thousands of enzymes catalyze this reaction: making and breaking bonds to glycosidic centres not just in saccharides but in nucleotides, nucleotide coenzymes, glycolipids and numerous glycoproteins (which include antibodies and some hormones). Like phosphorylation, glycosylation is also a common and flexible recognition motif, and glycoproteins, glycosylated on serine, threonine or tyrosine OH or asparagine NH_2 side-chains, are important in intracellular signaling.

Stripped to its essentials, glycosyl transfer is an S_N2 reaction at an acetal centre, a reaction unknown in classical acetal chemistry. Acetals are well known to be stable under neutral or basic conditions, but hydrolyzed by acid, reacting by well-understood A1, or S_N1-type mechanisms (Scheme 4.70).

We know that enzymes, working at or near neutral pH, typically use general rather than specific acid catalysis, but it was not until this generalization was reinforced by mechanistic conclusions based on the first crystal structure of an enzyme that chemists identified the mechanism in a simple (glycoside) system

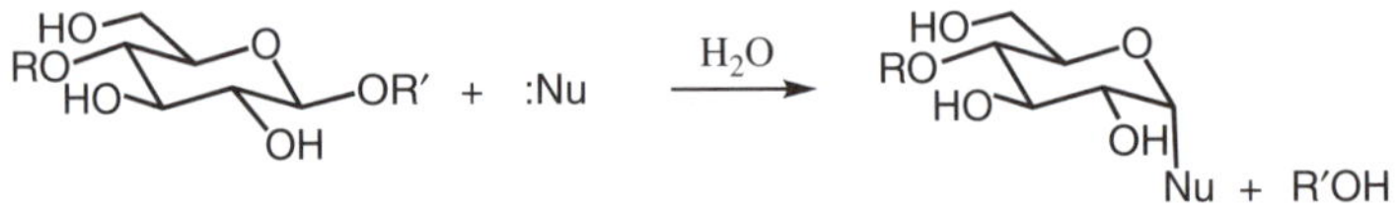

Scheme 4.69 The basic glycosyl transfer reaction. An acceptor nucleophile displaces the leaving group $R'O^-$, usually with inversion of configuration at the anomeric centre (see the text).

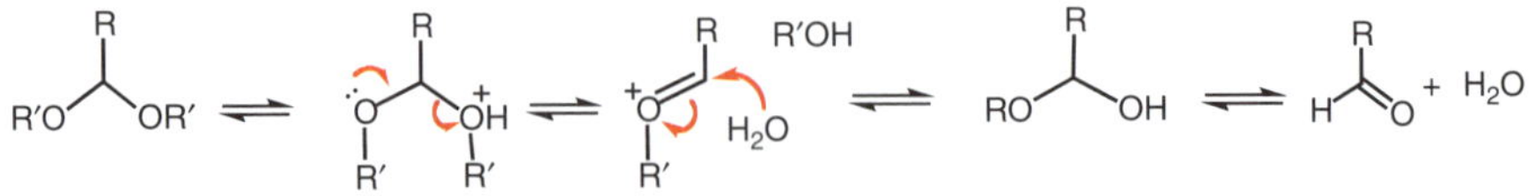

Scheme 4.70 The acid-catalyzed hydrolysis of an acetal is specific acid catalyzed. Protonation converts one of the two poor potential leaving groups RO^- into ROH, a very good one.

(see Section 1.9). General acid catalysis of acetal hydrolysis is not easily observed in simple systems, and this first observation involved an intramolecular reaction (Scheme 4.71).

S_N2 reactions are also not easily observed at acetal centres, but the mechanism has been characterized in acetals derived from formaldehyde, where the S_N1-type mechanism of Scheme 4.70 is least favoured. The reaction has much in common with the "borderline" S_N2 (P) reaction of phosphate monoester dianions discussed above (Section 4.2.1). As in the S_N2 (P) reactions reactivity shows notably low sensitivity to the basicity of the nucleophile, while depending strongly on the leaving group; even the absolute reactivities are closely similar for a given good leaving group.

The hydrolysis of a formaldehyde acetal **4.106** with a good leaving group is catalyzed by weak nucleophiles like carboxylate anions by way of a "loose" or "S_N1-like" transition state (**4.107** in Scheme 4.72), characterized by weak bonding to both nucleophile and leaving group, with positive charge accumulating on the central carbon and especially its adjacent oxygen. The secondary deuterium isotope effect ($10 \pm 5\%$ per deuteron of the CD_2 group, depending on the nucleophile) falls in the range expected for an S_N1 process.[199] Water behaves no differently from other nucleophiles: the conclusion is that the methoxymethyl cation (**4.107**, $R' = Me$ in Scheme 4.72, with the dashed bonds deleted) has too short a lifetime in water to exist as a full intermediate: so that the S_N2 mechanism is "enforced".[142]

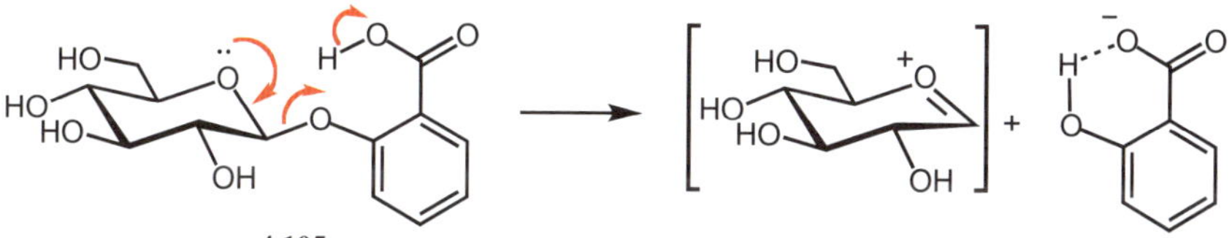

4.105

Scheme 4.71 Intramolecular general acid catalysis accounts for the rapid hydrolysis of 2-carboxyphenyl β-D-glucoside.[25] (From Section 1.9.)

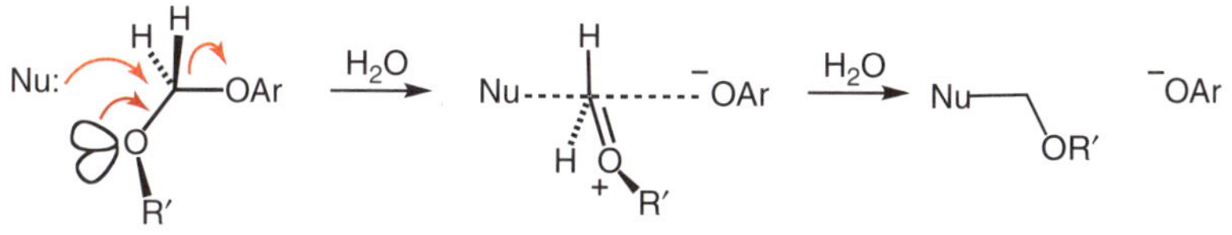

4.106   **4.107**

Scheme 4.72 Nucleophilic substitution at an acetal centre. For a good leaving group ArO^- the $n_O-\sigma^*_{C-OR}$ interaction shown in **4.106** weakens the C–OR bond (comparable alkyl aryl ethers are unreactive), and bond formation to the nucleophile in the transition state **4.107** leads to inversion of configuration. Only weak bonding to the nucleophile is needed because the bond to the leaving group is largely broken in the "loose" transition state (see the text).

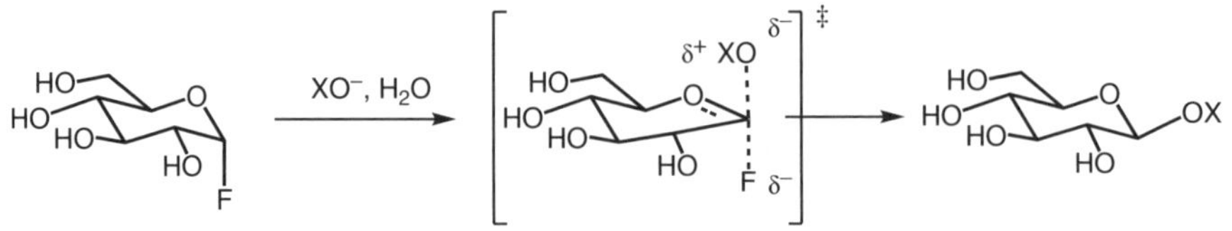

4.108   **4.109‡**

Scheme 4.73   Nucleophilic substitution at a glycosyl centre. There is a large amount of
bond breaking and little bond making in the dissociative transition state
4.109‡ (see the text).

Though these are very simple models, similar considerations apply generally
to substitutions at glycosidic centres. A careful examination of the displacement
of fluoride (a leaving group good enough for reactions with nucleophiles to be
followed at 30 °C in water) from alpha-D-glucopyranosyl fluoride (**4.108**)
revealed well-defined second-order reactions with anionic nucleophiles (Scheme
4.73). The reactions, which proceed with 100% inversion of configuration at
the anomeric centre and also show notably low sensitivity to the basicity of the
nucleophile, were interpreted as enforced concerted substitutions, with the
glycosyl cation, like the methoxymethyl cation, too short-lived to exist in
the presence of a strong nucleophile.[200] Once again there is a large amount of
positive charge development at the reaction centre in the transition state
(**4.109‡**, Scheme 4.73). This charge interacts favourably with anions, but not
with uncharged amine nucleophiles.[200] This is consistent with the observation
that the nucleophiles involved in the active sites of enzymes catalyzing glycosyl
transfer reactions are generally full or developing oxyanions.

The n_O–σ^*_{C-OR} interaction shown in **4.106** introduces a stereoelectronic
dimension to reactivity and ground-state stability. It is strongest when a donor
electron pair on oxygen is antiperiplanar to the C–OR bond, making gauche
(**4.110**) the preferred conformation about the central C–OR bonds of an acyclic
acetal.[201] This is the case for the conformationally constrained tetrahydropranyl
acetals **4.111** when the exocyclic C–OR' bond is axial (Scheme 4.74): if the acetal
is symmetrical, as in **4.110**, or the conformation gauche about both of them, as
in both **4.110** and **4.111a**, both C–OR bonds are strengthened. This selective
stabilization of axial isomers opposes the normal steric preference of substituents on saturated six-membered rings for the equatorial configuration, and
accounts for the preferred axial conformations of tetrahydropranyl acetals
4.111. It is the basis of the anomeric effect, the high proportion of axial anomers
found for glycopyranosides (*e.g.* **4.112** in Scheme 4.74).[202]

In the equatorial derivatives **4.111e** and **4.112e** the C–OR bond is necessarily
antiperiplanar to the in-ring C–O bond. For the tetrahydropyranyl system
equilibration between axial and equatorial isomers can involve *either* ring
inversion *or* reversible bond breaking and making at the acetal centre; but ring
inversion is unfavourable for hexopyranoses (the equatorial preference of the 6-
CH_2OH group is generally decisive) and equilibration generally requires cleavage of one (usually the exocyclic one) of the acetal C–O bonds.

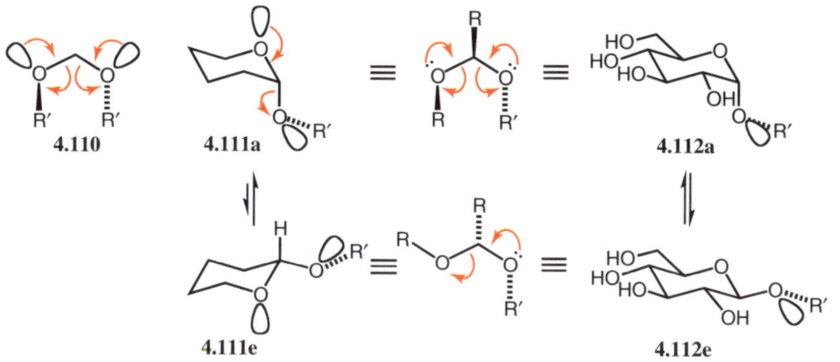

Scheme 4.74 Other things being equal the preferred conformation about the central C–O bonds of an acetal is gauche. This is achieved in conformationally unconstrained acetals (**4.110**) and axial tetrahydropyranyl acetals (**4.111a**), including glycosides (**4.112a**).

Cleavage of the C–OR′ bond of an acetal has an absolute requirement for $n_O-\sigma^*_{C-OR'}$ electron donation from the ring oxygen: the activation energy for the reaction can be almost 20 kcal mol^{-1} higher if the leaving group is rigidly fixed equatorial.[203] However, activation energies for conformational changes are far lower, and the $n_O-\sigma^*_{C-OR'}$ orbital overlap necessary for cleavage can develop as the conformation changes. Thus, the equatorial acetal **4.113e** (Scheme 4.75, Ar = 4-nitrophenyl), nominally conformationally fixed by the *trans*-bridgehead, is readily hydrolyzed to the oxocarbocation **4.114**, no doubt via a twist-boat conformation like **4.113e′** with a lone pair manoeuvering into a position antiperiplanar to the breaking C–OAr bond. The equatorial acetal is hydrolyzed a few times faster than its axial anomer **4.113a**, as expected if there is no stereoelectronic impediment to the reaction (the axial anomer is less reactive because it is more stable, a result of the anomeric effect[204]).

In effect the conformationally flexible system finds its way round the potential stereoelectronic barrier. That the barrier is real and substantial is shown by the stability of the rigid bicyclic system **4.116** (Scheme 4.75), which is hydrolyzed some 10^{13} times more slowly than (estimated for) the corresponding axial system **4.117**.[203] β-glycosidases, enzymes that catalyze the hydrolysis of saccharides with equatorial leaving groups, must also find their way around this potential stereoelectronic barrier: either by binding their substrate in a non-ground-state conformation or by incorporating the necessary conformational changes in to the catalytic process.

4.3.1 Simple Models

The design of models to characterize, or to answer particular questions about the mechanisms of glycosyl transfer reactions must make the usual compromises between simplicity, convenience and potential for giving useful information. Most informative should be reactions of glycosides themselves, but alkyl

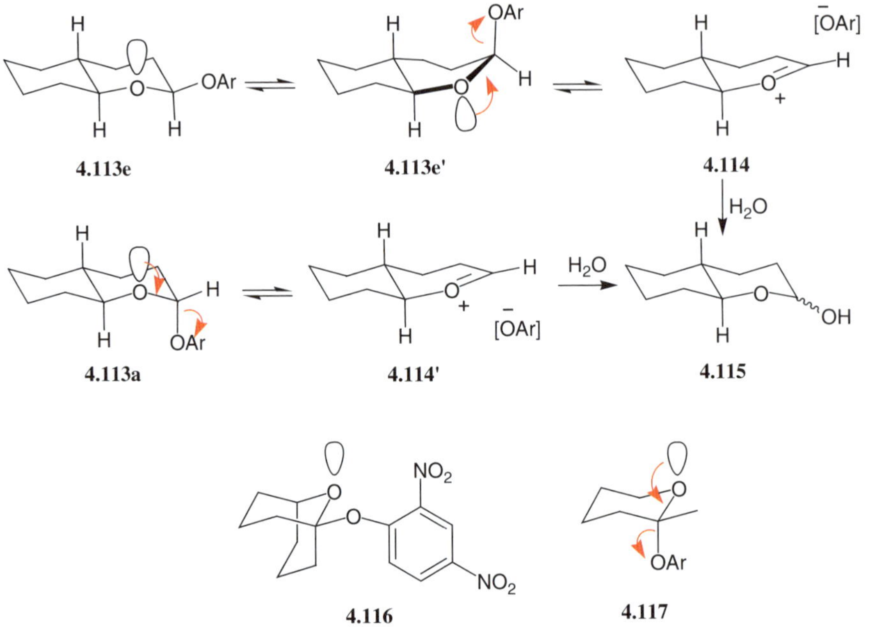

4.113e   **4.113e'**   **4.114**

4.113a   **4.114'**   **4.115**

4.116   **4.117**

Scheme 4.75   Stereoelectronic control of acetal cleavage depends on conformational flexibility (see the text).

glycosides are exceedingly unreactive in water near pH 7: methyl β-D-glucoside, for example, has an estimated half-life of millions of years near pH 7, making simple alkyl glucosides even less reactive than phosphodiesters.[205] So, the convenience criterion once again dictates the use of activated systems. The simple formaldehyde and tetrahydropyranyl acetals **4.110** and **4.111** discussed above (Schemes 4.74 and 4.75) are more reactive than simple glucosides by factors of 10^4 and 10^7, respectively, and the factor is even larger (up to 10^{10}) for acetals $PhCH(OR)_2$ derived from benzaldehyde. These last systems are obviously not close models for glycosides, but they are of interest because the hydrolysis of derivatives of tertiary alcohols, almost uniquely for alkyl acetals, is subject to general acid catalysis by carboxylic acids (Scheme 4.76).[206]

The key to understanding the balance between general acid catalysis and the acid-catalysis mechanisms usually observed in these systems is the stability of the two intermediates involved in the specific acid-catalyzed mechanism (SAC): the conjugate acid **4.110.H$^+$** of the reactant and the oxocarbocation **4.118** (Scheme 4.76). If the leaving group oxygen is very weakly basic, and the carbocation very stable, the C–OR$'$ bond may break before proton transfer is complete, so that the two processes become concerted and conditions for general acid catalysis (the mechanism as well as the second-order rate law) are fulfilled. The balance can thus be shifted towards general acid catalysis in a given system by making the leaving group oxygen sufficiently weakly basic (as for example in the *p*-nitrophenyl tetrahydropyranyl acetal **4.119**,[207] Scheme 4.77) or the cation

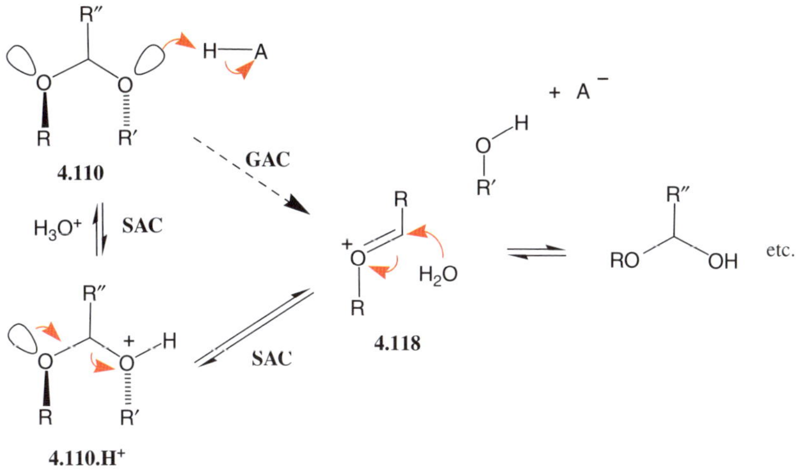

Scheme 4.76 General acid catalysis of acetal hydrolysis must always compete with the specific acid-catalysis mechanism: which usually wins at acidic pH (see the text).

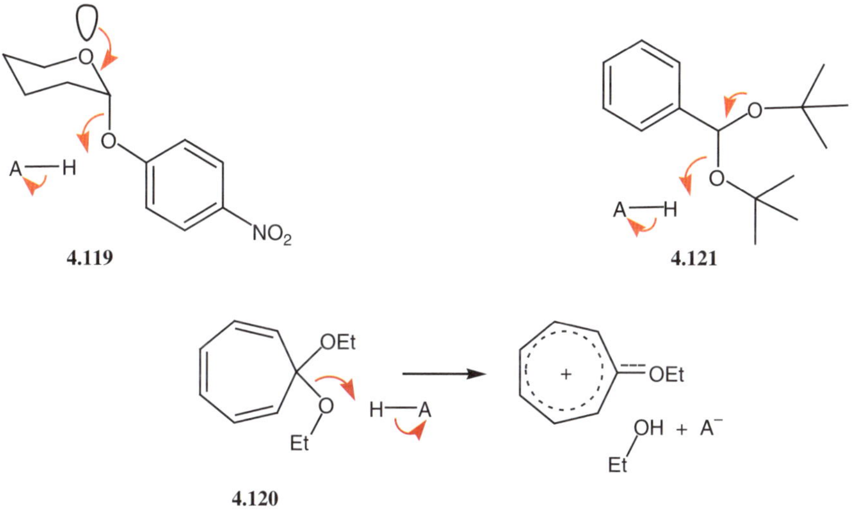

Scheme 4.77 Simple model systems showing general acid catalysis of acetal hydrolysis (see the text).

exceptionally stable, as in the case of the tropone acetal **4.120**.[208] In the case of the acetals $PhCH(OR)_2$ derived from tertiary alcohols mentioned above, the release of steric strain involved in breaking one of the C–OBu^t bonds (**4.121**) evidently gives GAC the advantage over the two-step process.

The hydrolysis reactions of alkyl glycosides are very slow near pH 7 because the leaving groups are poor, and because the carbocation intermediates that

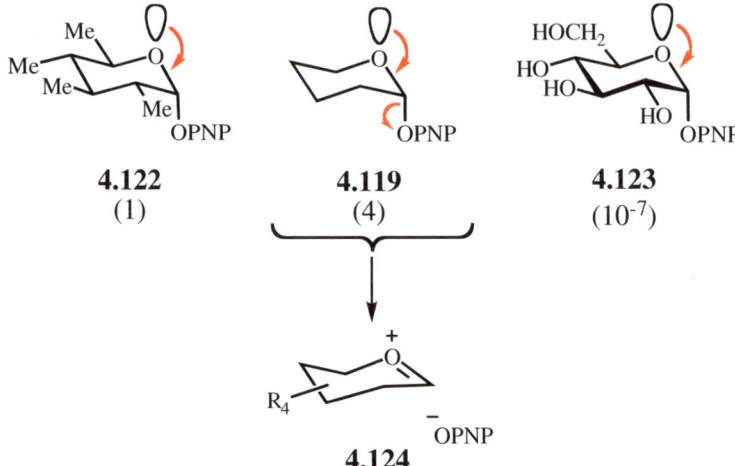

Scheme 4.78   The spontaneous hydrolysis reactions of *p*-nitrophenyl tetrahydropyranyl acetals are slowed by both hydroxyl and methyl substitution. Relative rates (at 39 °C in 20% dioxan-water) are shown in brackets.

would be involved in spontaneous (S_N1-type) or specific acid-catalyzed reactions are high-energy species. This is primarily the cumulative electronic effect of the four electron-withdrawing OH groups on a typical hexopyranoside. However, replacing four equatorial hydrogens of the tetrahydropyranyl ring by methyl groups (**4.122**, Scheme 4.78) also slows the reaction: by a factor of only four, but this must conceal a larger effect because methyl substitution normally stabilizes carbocations.[209] This larger effect is torsional: equatorial substituents stabilize the chair conformation, and twisting the chair into the half-chair conformation of the oxocarbocation (**4.124**), which must be planar at C(1), introduces torsional strain. Methyl groups are sterically larger than OH, and the torsional effect of the four equatorial OH groups of a glucoside (**4.123**) is estimated to contribute a factor of up to 50 to the low reactivity of the system.

4.3.2   Enzymes

The thousands[v] of enzymes catalyzing glycosyl transfer (Scheme 4.69, above) fall into two main classes: the hydrolases, which catalyze the transfer to water and deal with a specific leaving group, however poor; and transferases, involved in metabolic and synthetic processes, and often using activated glycosyl phosphates to glycosylate their specific nucleophile. The thousands of glycoside hydrolases (glycosidases) have been classified into over a hundred different *families* of enzymes: and the hundredth family of transferases will be no doubt

---

[v] A dedicated, regularly updated web site (http://www.cazy.org/) "describes the families of structurally related catalytic and carbohydrate-binding modules (or functional domains) of enzymes that degrade, modify, or create glycosidic bonds" and offers an instructive overview of the impressive breadth and current activity in this area.

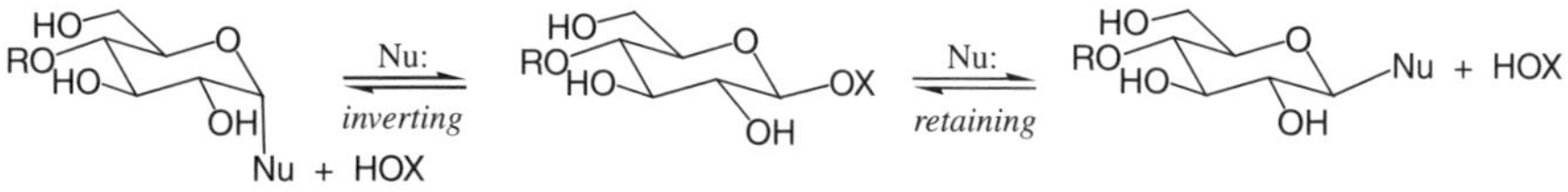

HO ― O Nu: HO ― O Nu: HO ― O
RO ― HO ― ⇌ RO ― HO ― ―OX ⇌ RO ― HO ― ―Nu + HOX
OH *inverting* OH *retaining* OH
Nu + HOX

Scheme 4.79 Glycosyl transfer can involve inversion or retention of configuration at the anomeric (glycosidic) centre.

identified before very long. The classification of so many enzymes is a complicated business (especially as their reactions may be reversible), and may be based either on structure (families defined on the basis of sequence similarities turn out to have similar three-dimensional structures) or function, or both. However, there is for both classes a clear-cut mechanistic distinction, between retaining and inverting enzymes, depending on the relative configurations of reactant and product at the anomeric centre (Scheme 4.79).

An authoritative in-depth review of the chemistry of enzyme-catalyzed glycosyl transfer[210] is a good starting point for further reading.

4.3.2.1 Glycoside Hydrolases

The typical glycoside hydrolase mechanism involves two catalytic groups from the enzyme-active site, a general acid to assist departure of the leaving group and a second, basic, group that may act directly as the nucleophile, or as a general base to assist its attack. (Glycoside centres being uncharged, metals are involved in their reactions only very rarely.) If the immediate nucleophile is water this completes the reaction, resulting in inversion of configuration: if it is an active-site group a second hydrolytic step is required and the overall result is retention of configuration. The two catalytic groups are most often aspartate or glutamate carboxyls, and the distance between them provides a useful criterion of mechanism.[211] When they are close (typically around 5 Å apart) the glycoside group of the substrate fits neatly between them on binding. A greater distance, of the order of 9 Å, leaves room for an intervening water molecule. The two mechanisms are sketched out in Scheme 4.80.

Note that in the case of a glycoside hydrolase the choice of inverting or retaining mechanism is mechanism- rather than product-driven (the products are in equilibrium with each other under physiological conditions).

The textbook example of a glycosidase is lysozyme. The first crystal structure to be determined for any enzyme was that of the hen egg white enzyme,[212] and the structural investigations gave valuable information about how the polysaccharide substrate must bind, with the target glycosidic centre within reach of two, and only two, potential catalytic groups. These, glutamic acid-35 and aspartate-52, are known to be involved as the COOH and COO⁻ forms, respectively: much as represented in the retaining enzyme mechanism shown in Scheme 4.80. Convincing experimental evidence for the glycosyl enzyme intermediate (**4.125** in Scheme 4.80) also came from (much more recent) work with lysozyme.[213] Not surprisingly the intermediate does not accumulate on the

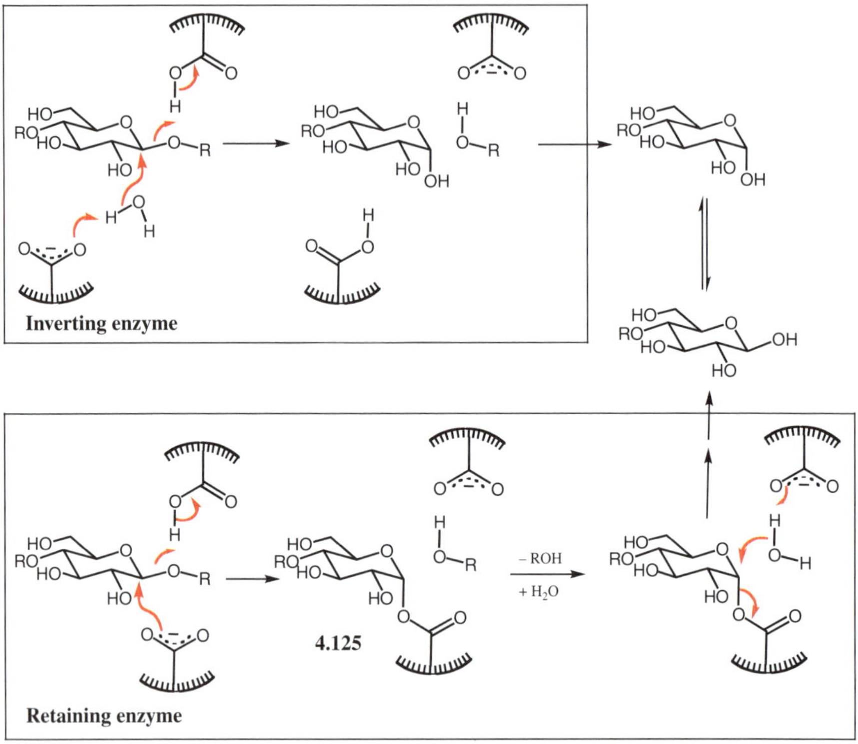

Scheme 4.80 Outline mechanisms for retaining and inverting glycosidases. The hydrolysis of the intermediate glycosyl enzyme **4.125** in the retaining enzyme mechanism is mechanistically simply the reverse of the initial cleavage step, and nucleophiles other than water may react at this stage.

enzyme, but could be persuaded to do so by mutating the active-site glutamic acid-35 to glutamine. This removes the general base catalyzing the attack of water on the intermediate (Scheme 4.80), which thus survives longer in the modified active site. However, in the wild-type enzyme this group is involved, as the general acid, to catalyze the initial formation of the intermediate, so its formation needed to be artificially accelerated. This was done by using fluorine, providing a leaving group good enough to depart without assistance from a general acid. Thus when the E35Q mutant lysozyme was incubated with NAG$_2$-fluoride the steady-state concentration of a new intermediate, with mass number consistent with glycosyl-enzyme A (Scheme 4.81) was high enough to be observed by ESI-MS (electrospray ionization mass spectrometry).

Further stabilization of the intermediate, and further corroboration, were obtained by replacing the group in the 2-position acetamido (AcNH) in the natural and most model substrates) of the substrate disaccharide by fluorine. Since fluorine is much more strongly electron withdrawing than AcNH, it strongly disfavors reactions at the glycosidic centre, so that glycosyl-enzyme B

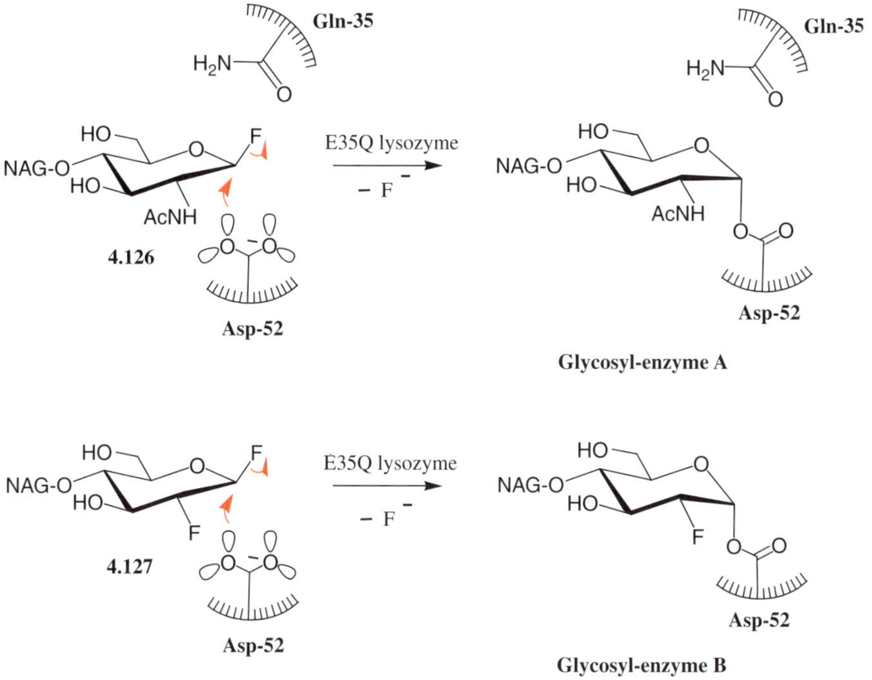

Scheme 4.81 Experiments with mutant E35Q lysozyme confirm the intermediacy of a glycosyl enzyme (see the text).[213]

can be observed by ESI-MS in the reaction of the wild-type enzyme. Glycosyl-enzyme B lives long enough in the mutant E35Q enzyme for its structure to be confirmed by a crystal-structure determination.[213]

The 2-acetamidogroup plays an active role in the reactions of a small group of glycosidases that use substrate assistance in the hydrolysis mechanism. Enzymes in these few of the many families of retaining glycosidases lack the enzyme nucleophile needed for the usual retaining mechanism illustrated in Schemes 4.80 and 4.81. It turns out that its place is taken by the 2-acetamido-group,[214] which is well placed to displace a neighbouring equatorial leaving group, via a 5-membered cyclic transition state, to form an oxazolinium intermediate **4.128** (Scheme 4.82). This is rapidly hydrolyzed, in the same way as the glycosyl enzyme intermediates **4.125** shown in Scheme 4.80.

4.3.2.2 Glycoside Transferases

Since glycoside hydrolases are simply transferases using water as the acceptor nucleophile at the same glycosidic centres, close mechanistic similarities between the two classes of enzyme are to be expected. (Though a new factor when the leaving group is a nucleoside diphosphate can be the involvement of a metal cation, as a Lewis acid, in place of the general acid.) Some retaining

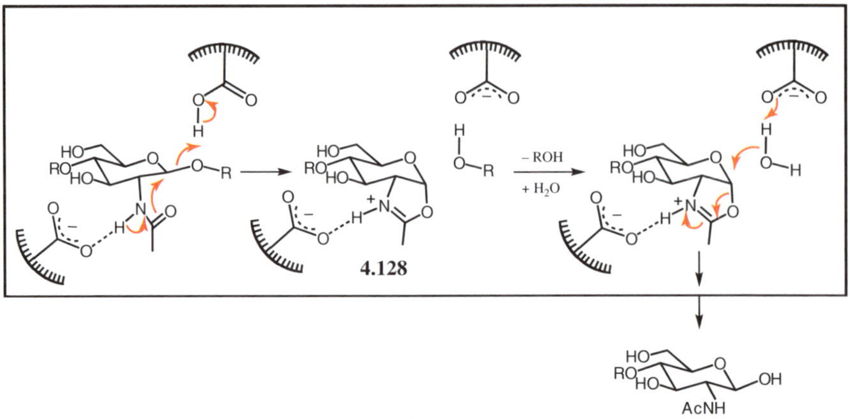

Scheme 4.82 Substrate-assisted hydrolysis of some 2-acetylamino-glycosides involves intramolecular nucleophilic catalysis by the amide group. The oxazolinium intermediate (**4.128**) formed is probably stabilized by hydrogen bonding to a conserved aspartate.[215]

glycoside hydrolases can themselves act as transferases, with the nucleophilic water replaced in the final step of the retaining mechanism (Scheme 4.80) by other nucleophiles. So, constitutional transferases are certainly expected to use the same mechanism, with the nucleophile a specific OH group of an acceptor sugar. This appears to be generally true for enzymes catalyzing exchanges between oligo- or polysaccharides,[216] though despite considerable effort,[217] no convincing identification of a catalytic nucleophile or evidence for a covalent intermediate has been obtained, This lack is not necessarily negative evidence, but has led to the discussion of alternative mechanisms, notably an "S_Ni-like" process, in which retention of configuration results from bond formation to the attacking nucleophile taking place from the same face as the leaving group departure (Scheme 4.83). The process demands a high degree of cationic character at the anomeric centre, and approach of the incoming nucleophile along much the same trajectory as the departure of the leaving group, excluding a concerted displacement process. There is evidence from the acid-catalyzed alcoholysis of glycosyl fluorides that glycosyl transfer can proceed with some retention of configuration without the intervention of ion-pair intermediates even in simple systems,[218] but the active site of an enzyme is liberally provided with nucleophilic centres, and it would seem likely that a default nucleophile, in the shape of a side-chain (or even main-chain) amide group, would have become involved if such a high-energy species as a glycosyl cation were developing. However, an intermediate (*e.g.* **4.129** in Scheme 4.83) formed in this way might well be very short lived – and its identification difficult.

The "S_Ni-like" mechanism for glycosyl transfer was originally proposed for glycogen phosphorylase, a glycosyltransferase that catalyzes the reversible transfer to inorganic phosphate of a glucosyl group from glycogen, the

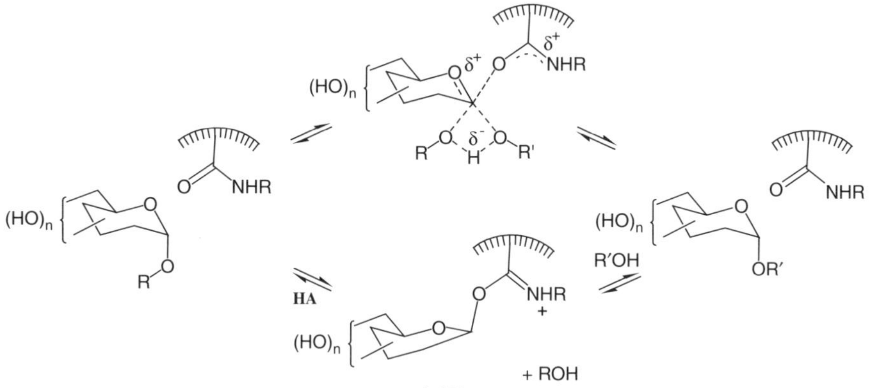

4.129

Scheme 4.83 Suggested outline mechanism (upper route) for an "S_Ni-like" mechanism for glycosyl transfer with retention of configuration.[219] (The exchange of nucleophile and leaving group cannot be a synchronous concerted process: see the text). The alternative double-displacement mechanism using an amide group as the primary nucleophile, would involve a reactive intermediate **4.129**.

polymeric storage form of glucose in mammals; giving glucose-1-phosphate with net retention of configuration. (This enzyme is remarkable less for its lack of an obvious active-site nucleophile than for its totally unexpected – and still unexplained – requirement for pyridoxal phosphate as a cofactor.[220]) Recent work on another system has provided the most convincing explanations so far for these observations.

The active-site residues of the trehalose-6-phosphate synthase, OtsA, are essentially superimposable on those of glycogen synthase. The two enzymes evidently use the same mechanism, and the OtsA structure[221] makes a useful addition to the still meagre collection of structures of retaining glycosyl-transferases. OtsA catalyzes the synthesis of trehalose-6-phosphate (**4.130**) using UDP-Glc (**4.131**) as the donor and the α-anomer of glucose-6-phosphate (**4.132**) as the acceptor. The reaction occurs with net retention at the reacting anomeric centre and results in the formation of the double glycosidic linkage of trehalose (Scheme 4.84).

In this enzyme also there is no obvious active-site nucleophile to indicate a double-displacement pathway, and the "S_Ni-like" mechanism is considered by the authors to be the strongest candidate. Once again the best-placed potential nucleophile is a main-chain amide group, closely equivalent to one in glycogen phosphorylase.[220] The available evidence does not support conclusively either the double displacement or the S_Ni-like mechanism. This fascinating mechanistic problem will most likely be solved by work with enzymes: perhaps complemented by the experiments with models that have characterized the nucleophilic reactions of amide groups at glycosidic centres (above, and Section 4.3.3).

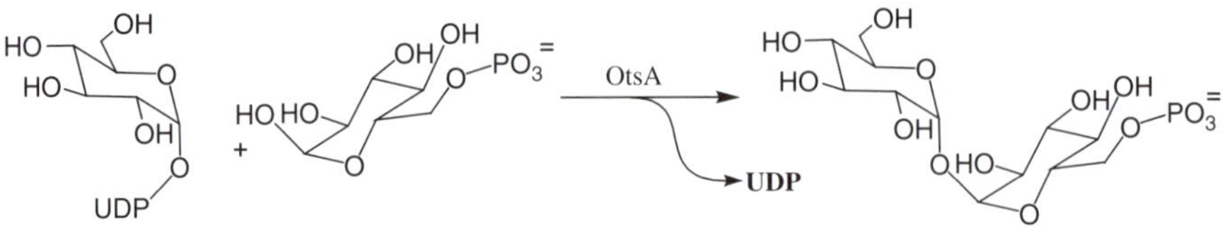

4.131   **4.132**   **4.130**

Scheme 4.84   The synthesis of trehalose-6-phosphate (**4.130**) is catalyzed by the synthase OtsA.

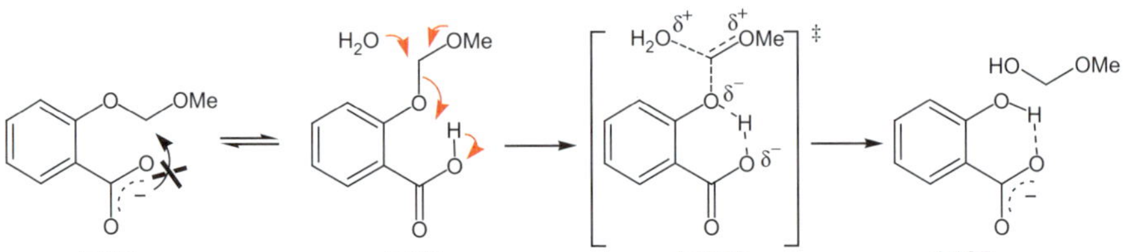

4.133   **4.134**   **4.134‡**   **4.135**

Scheme 4.85   The carboxyl group of 2-methoxymethoxybenzoic acid acts as a general acid to catalyze the hydrolysis of the neighbouring acetal. No significant direct cyclization reaction is observed for either the carboxyl or the carboxylate form.

4.3.3   Intramolecular Models

One of the important benefits of intramolecular models is the possibility of observing reactions of substrate groups that are unactivated, and thus closer to those in natural substrates of the enzymes of interest. Unfortunately, the big accelerations are primarily associated with nucleophilic, cyclization reactions, and displacements at acetal centres are notably insensitive to the nucleophile, typically depending primarily on assistance from general acids. Thus, the carboxyl group *ortho* to the methoxymethyl acetal (**4.134**, Scheme 4.85) of salicylic acid acts as a general acid [222] rather than as a nucleophile, as it would towards almost all the other functional groups we have discussed so far.

A model designed specifically to maximize the nucleophilic potential of the carboxylate group illustrates some of the problems. Carboxylate group oxygens have two geometrically different pairs of lone pairs of electrons, and it has been suggested that these will have different basicities and thus nucleophilicities (though any difference, other than steric, is now thought to be small[223]). Intramolecular reactions will normally use the E-lone pair for simple steric reasons (*e.g.* in **4.133**, above) while enzymes can bring functional groups together at 180° if necessary, and so may be expected to use a z-lone pair, as in **4.136** (Scheme 4.86), also for simple steric reasons. The design of the *p*-nitrophenyl riboside **4.137** ensured that intramolecular attack by the carboxylate group will involve a z-lone pair, and the release of *p*-nitrophenoxide is indeed 860 times faster (at 100 °C near pH 7) than from the reference compound **G. 4.139**, without the carboxyl group.[224]

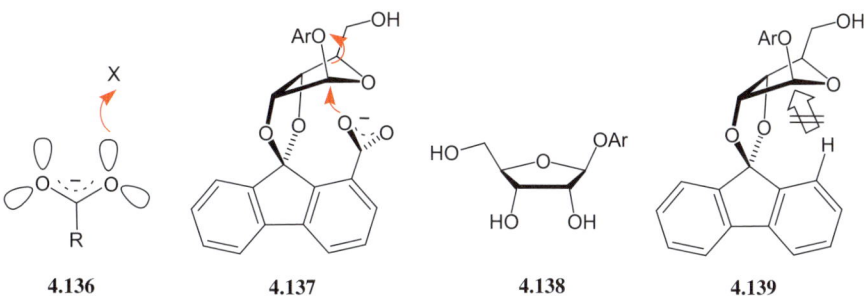

Scheme 4.86 Carboxylate groups using a Z-lone pair (**4.136** and **4.137**) to attack electrophiles: and the three *p*-nitrophenyl ribosides (**4.137, 138,** and **139**, Ar = *p*-nitrophenyl) described in the text.

However, this factor is less an indication of the high reactivity of **4.137** than of the low reactivity of the "reference compound" **4.139**: it turns out that **4.137** is hydrolyzed only 2–3 times faster than the parent riboside **4.138**, while both are several hundred times more reactive than **4.139**, the model system lacking the carboxy group. The explanation seems clear:[224] the rigid structure of **4.139** prevents easy access of nucleophilic water from the back (barred arrow), which is the required approach for an S_N2-type mechanism. This observation is also clearly inconsistent with significant reactivity involving S_N1 or S_Ni mechanisms (see below).

A second problem for the design of intramolecular models is that intramolecular general acid catalysis is itself typically seriously inefficient, with effective molarities less than 10 M in the great majority of systems tested (Section 1.4.1).[8] Salicylic acid derivatives like **4.134**, with effective molarities (EMs) of the order of 10^4 M, are exceptions. So a major objective of work in the area has been to identify the basis of this efficient catalysis: since general acid/base catalysis is common to almost all enzyme reactions, where it can reasonably be assumed to be highly efficient.

The common factor in efficient general acid catalysis is the formation of a strong intramolecular hydrogen bond as part of the proton-transfer process. A strong intramolecular hydrogen bond is a well-known property of the salicylate monoanion **4.135**, which it stabilizes by some 5 kcal (over 20 kJ) mol^{-1} in water. The extensive amount of bond breaking in the transition state for displacements at acetal centres (Schemes 4.72 and 4.73) means that it is very much product-like: so that features that stabilize the product stabilize the transition state significantly also. Transition-state stabilization is of course the basis of catalysis, and must be optimized in the most efficient systems. There is no reason in principle why an intramolecular hydrogen bond should not be stronger in the transition state than the product. Hydrogen bonds are strongest when the pK_as of donor and acceptor are matched. They are not equal in either reactant or product of Scheme 4.85, but can become equal during the course of the cleavage of the C–O bond: at the point when the right amount of negative charge has developed on the leaving group oxygen. This will happen at or close to the transition state **4.134‡**.

The strength of an intramolecular hydrogen bond also depends on its geometry, in particular the spacing between the donor and acceptor centres and the bond angle at the proton. Other things being equal linear hydrogen bonds are strongest, and lone-pair orbital directionalities mean that the six-membered ring geometry of the salicylate anion **4.135** or the transition state **4.134‡** leading to it, though favourable, is not optimal. So, a basic design feature for improved efficiency is a linear intramolecular hydrogen bond.

This reasoning led to the benzofuran **4.140** (Scheme 4.87).[225] This system, designed specifically to support a strong, linear intramolecular hydrogen bond in the product **4.141**, shows the most efficient intramolecular general acid catalysis yet achieved for the departure of a phenolic leaving group. The combination of functional groups and geometry is now efficient enough to translate successfully to an aliphatic system: the rate of loss of the tertiary alcohol leaving group from **4.142** corresponds to an acceleration of 10^{10} compared with the system lacking the COOH group. In terms of transition-state stabilization this acceleration is worth over 14 kcal (58 kJ)/mol.[226] The crystal structure of the product anion **4.143** revealed the expected intramolecular hydrogen bond, with an O–H . . . O bond angle of 175.1° and an O . . . O distance of 2.71 Å.

Thus, intramolecular general acid catalysis can match intramolecular nucleophilic catalysis in efficiency in a properly designed system, and putting the two together successfully in a concerted process might be expected to lead to catalysis that approaches enzyme efficiencies. However, it has not so far been possible to achieve this goal in any acetal-cleavage reaction. For example, both nucleophile and general acid would appear to be well-placed in the glycoside **4.144** (Scheme 4.88),[227] but the reaction is only 7 times faster that the hydrolysis of the salicyl glucoside (**4.105** in Scheme 4.71). Evidently, nucleophile and general acid are not efficiently matched electronically in this system.

Nucleophile and general acid might be expected to be better matched in a series of disalicyl acetals,[228] including the benzaldehyde derivative **4.146**: which is reactive enough to be studied at 25 °C, and shows a bell-shaped pH–rate profile defining the reaction of the monoanion as over 10^9 times more reactive than its dimethyl ester **4.147**. However, almost all this acceleration can be

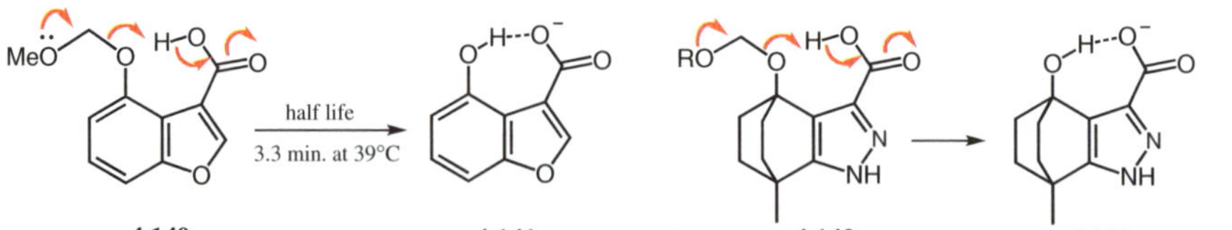

Scheme 4.87 Systems showing highly efficient intramolecular general acid catalysis. The saturated tricyclic system **4.142** is designed to have a hydrogen-bond geometry as close as possible to that of the aromatic system **4.140**.

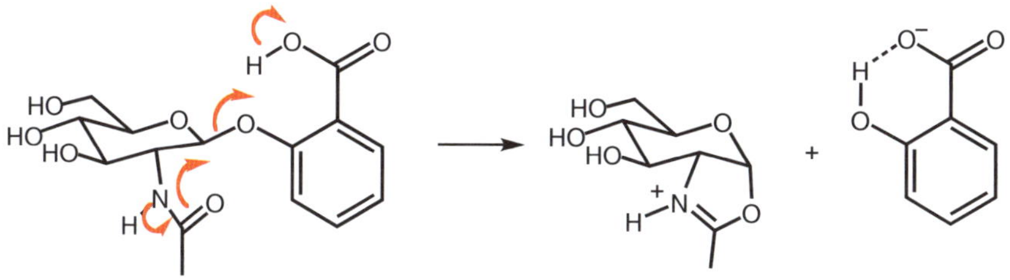

4.144   **4.145**

Scheme 4.88   Both intramolecular nucleophilic and general acid catalysis appear to be involved in the hydrolysis of glycoside **4.144** (see the text). The presumed intermediate **4.145** is too short lived to be detected.

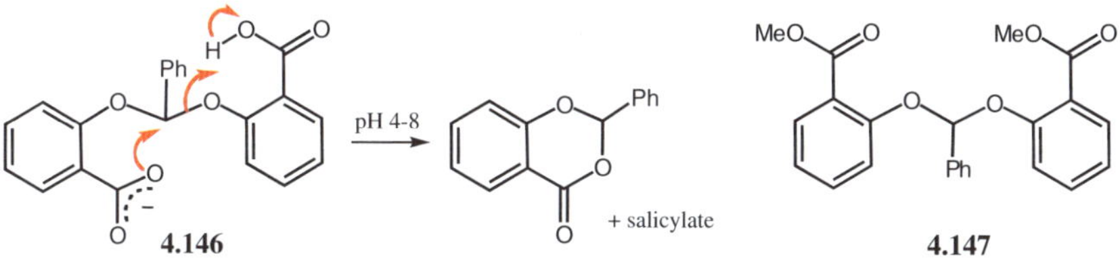

4.146   **4.147**

Scheme 4.89   The carboxylate anion acts as the nucleophile in the hydrolysis of the disalicyl acetal **4.146**, but makes a relatively minor contribution to the rate of the reaction.

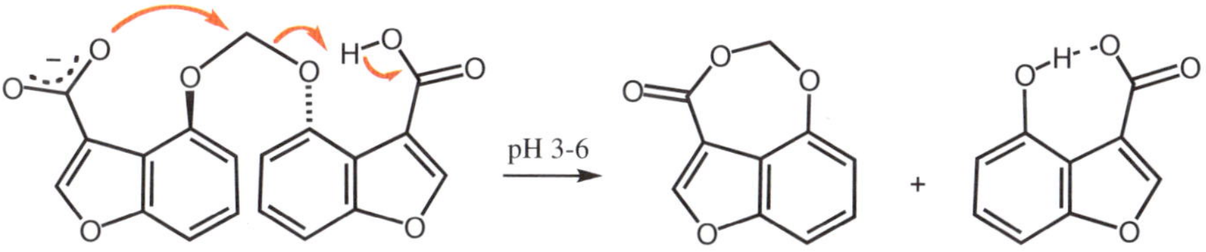

4.148   **4.149**

Scheme 4.90   Both nucleophile and general acid contribute significantly to the cyclization of the formaldehyde acetal **4.148**, although nucleophilic catalysis involves the formation of a (less favourable) seven-membered ring.

attributed to the COOH group acting as a general acid, with the contribution of the carboxylate group apparently worth only about one order of magnitude.

The best balance between nucleophile and general acid in the series of acetals with phenol leaving groups is found in the system **4.148** (Scheme 4.90), based on the compound **4.140** (Scheme 4.87) showing the most efficient general acid catalysis. The carboxylate group in **4.148** contributes about two orders of magnitude to the total acceleration of 10^5, as compared with the rate expected for the specific acid-catalyzed reaction.[229]

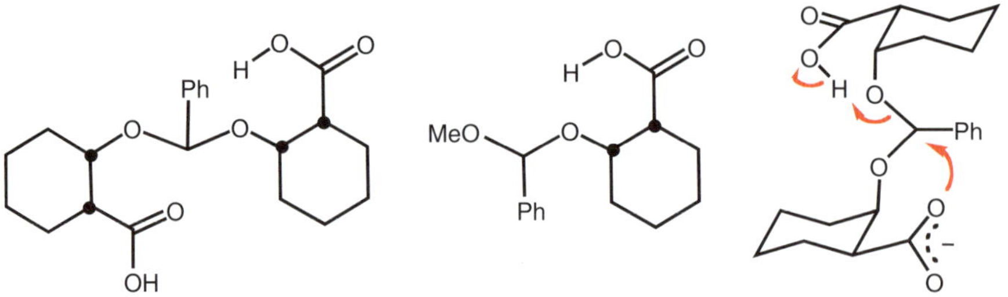

4.150   **4.151**   **4.152**

Scheme 4.91   Intramolecular general acid catalysis of the hydrolysis of the acetal
                   group is not observed in the reaction of **4.151**, but the introduction of
                   the second, CO_2^- group in **4.150** leads to effective catalysis involving
                   contributions from both groups (**4.152**).

Perhaps the most informative model system of this sort is the acetal **4.150**
(Scheme 4.91), based on system **4.146** (Scheme 4.89) but with two aliphatic
alcohols as potential leaving groups. As usual for aliphatic leaving groups, no
intramolecular general acid catalysis is observable in the hydrolysis of the
methyl acetal **4.151** (the anion of course also shows no reaction), but the
hydrolysis of the monoanion of **4.150** (in 50% dioxan-water at 30 °C) shows a
bell-shaped pH–rate profile between pH 4–7 and a rate 4×10^4 times faster than
that of the diethyl ester under the same conditions.[230] Evidently, general acid
catalysis is only made possible by the presence of the carboxylate anion, most
likely acting as a nucleophile as shown in **4.152**. Here, the two catalytic groups
appear to show true synergy: neither is significantly catalytic by itself, but
increasing bond formation to the nucleophile results in sufficient negative
charge development on the leaving group oxygen for general acid catalysis to
become viable. Or – alternatively – incipient hydrogen bonding from the carboxyl group to oxygen makes it a good enough leaving group for nucleophilic
attack to become viable. Most likely both factors are involved.

4.3.4   Enzyme Mimics

Despite the biological importance of glycosyl transfer the development of
artificial enzymes capable of catalyzing the process is still at a very early
stage. Most published work involves cyclodextrins as receptors, but the
recognition element is typically a non-natural leaving group, rather than the
carbohydrate component of the substrate: and the reaction its departure. The
hydrophobic cavity of β-cyclodextrin has the right dimensions to bind aromatic
rings (see Section 4.1.1.3.1). For example, it binds the *p*-nitrophenyl group of
the tetrahydropyranyl acetal (**4.153**), but as a result stabilizes the compound
towards hydrolysis (a small acceleration is observed in the presence of
α-cyclodextrin at neutral pH).[231] The rate of hydrolysis of the corresponding
β-D-glucoside is not significantly affected by binding (**4.154**) to β-cyclodextrin
itself (Scheme 4.92): which makes the cyclodextrin a perfect template for the

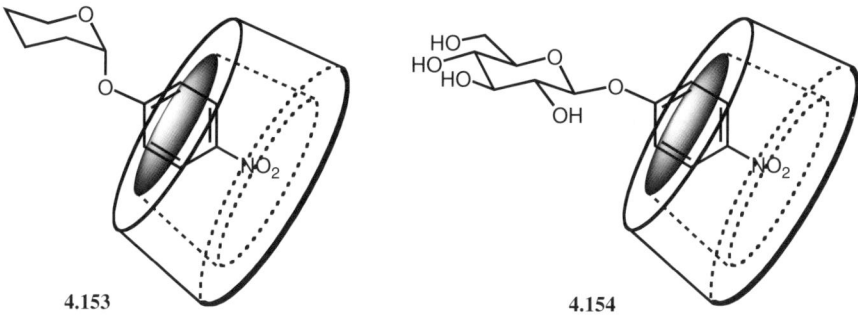

4.153   **4.154**

Scheme 4.92   Cyclodextrins are known to bind aromatic rings (see Section 4.1.1.3.1). Unless otherwise stated the standard representation used here indicates specifically β-cyclodextrin.

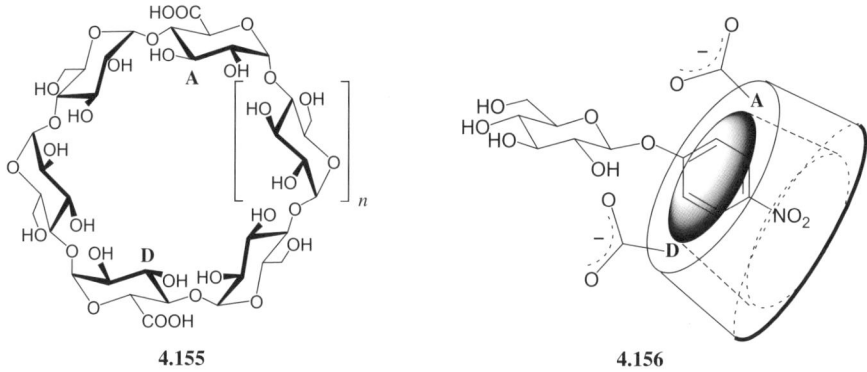

4.155   **4.156**

Scheme 4.93   Dicarboxylic acids derived from α- and β-cyclodextrin (**4.155**, $n = 0$ and 1, respectively) bind ($K_M = 2.4$ and 13.4 mM) and catalyze the hydrolysis of *p*-nitrophenyl β-D-glucoside with $k_{cat} = 1.46 \times 10^{-7}$ and $3.34 \times 10^{-7}\,\mathrm{s}^{-1}$, respectively. The substrate is thought to bind much as shown in **4.156**.

placement of potential catalytic groups designed to cleave the bond to the leaving group; as in compounds **4.156–4.159** (Schemes 4.93 and 4.94).

The first combination of potential catalytic groups to come to mind is of course a pair of carboxyls, ideally placed 5–6 Å apart, as in a typical retaining glycosidase. The distance between two groups at the 6-positions of rings A and D of a cyclodextrin is conveniently 6.5 and 5.0 Å for the β- and α-cyclodextrin rings **4.155** ($n = 1$ and 0, respectively) in Scheme 4.93.[232]

Both model systems **4.155** catalyze the hydrolysis of the activated substrate *p*-nitrophenyl β-D-glucoside. The reactions show Michaelis–Menten kinetics, but are very slow ($k_{cat}/k_{uncat} = 35$ and 18 at 59 °C for catalysis by the β- and α-cyclodextrin (**4.155**, $n = 1$ and 0, respectively) and the mechanisms are not well understood. In particular, catalysis in observed only at pH > 6.4, where both carboxyl groups are ionized, so that neither appears to act as a general

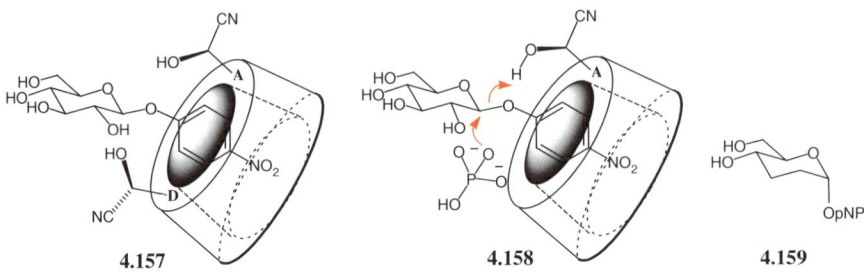

Scheme 4.94   The hydrolysis of *p*-nitrophenyl β-D-glucoside is catalyzed by cyanohydrins derived from α- and β-cyclodextrin, with k_{cat}/k_{uncat} values up to
almost 8000. The rate enhancement is only some six times less for the
system with only one cyanohydrin group.[233] No catalysis is observed
for the dideoxy derivative **4.159** (see the text).

acid. Weak nucleophilic catalysis is a possibility, or some sort of electrostatic
stabilization of the positive charge developing in the transition state.[232]

More effective catalysis appears, remarkably, in geometrically similar systems where the carboxyl groups of the cyclodextrins **4.155** are replaced by
cyanohydrins derived from the corresponding aldehydes (Scheme 4.94).[233,234]
Catalysis follows Michaelis–Menten kinetics and k_{cat}/k_{uncat} values up to many
thousands can be observed. Substrate binding involves the aromatic leaving
group, as shown (Scheme 4.94): the hydrolyses of (both α- and β-anomers of) *p*-
nitrophenyl hexopyranosides of different sugars are catalyzed with rates comparable to that for the glucoside shown in the Scheme. One interesting exception that shows no significant catalysis is the 2,3-dideoxyglucoside **4.159**
(Scheme 4.94). The dideoxy system is intrinsically about a thousand times more
reactive than the parent glucoside, but presumably binds differently enough not
to be within productive reach of the catalytic apparatus. Catalysis is also
inhibited by inert molecules known to bind strongly to cyclodextrins.

Thus, although catalysis is not highly efficient the cyanohydrin is qualitatively a convincing enzyme mimic, which shows some turnover. However, here
too the mechanism of catalysis is not well understood. Catalysis is enhanced by
phosphate buffer, depending, but not linearly, on the concentration of the
phosphate dianion. This suggests a possible nucleophilic role for phosphate
(**4.158**), possibly itself loosely bound by multiple hydrogen bonding to OH
groups of the substrate and the 6-CH$_2$OH groups of the host (but clearly not in
the cavity because α- and β-cyclodextrins **4.157** show similar activities). The
role of the cyanohydrin is somewhat better defined by results with the series of
β-cyclodextrins equipped with similar hydroxyalkyl groups shown in Scheme
4.95. Catalytic efficiency clearly depends on the pK_a of the hydroxyl group (the
cyanohydrin is the strongest acid in the series, with a pK_a close to 11,[235] sug-
gestively close to the pK_as of the cyclodextrin OH groups), consistent with it
acting as a general acid. A rough estimate using four measured k_{cat} values gives
a Brønsted $\alpha = 0.75 \pm 0.15$, interestingly similar to the value of 0.69 measured

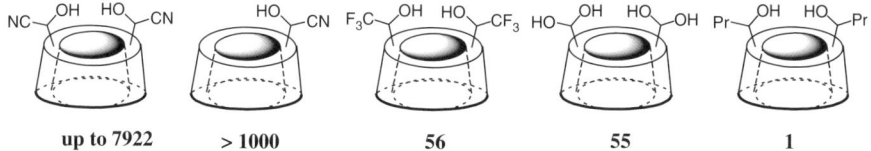

| up to 7922 | > 1000 | 56 | 55 | 1 |

Scheme 4.95 k_{cat}/k_{uncat} values compared for the hydrolysis of *p*-nitrophenyl β-D-glucoside catalyzed by various hydroxyalkyl substituted β-cyclodextrins at 59–60 °C in the presence of phosphate buffer.[236] The parent CH_2OH compound shows no catalysis.

for the general acid catalyzed hydrolysis of the model substrate 2-(p-nitrophenoxy)tetrahydropyran (**4.119**, Scheme 4.77) in 50% aqueous dioxan.[207]

The mechanism (**4.158**) suggested in Scheme 4.94 is the best available explanation of the evidence, but the evidence is not conclusive. Problems include the exact role of the phosphate dianion and particularly the general acid catalysis. Known general acid catalysts for the hydrolysis of the *p*-nitrophenyl acetal **4.119** are carboxylic acids, comfortably stronger acids than the *p*-nitrophenol leaving group (pK_a close to 7). But proton transfer from a cyanohydrin OH group, with a pK_a close to 11, never becomes thermodynamically favourable, so that the mechanism as drawn does not conform with the libido rule (Section 1.3.2).[4]

4.4 Hydrogen Transfer

Introduction

The transfer of the smallest atomic particle from a donor **X** to an acceptor **Y** (Scheme 4.96) has been described, with feeling, as a "deceptively simple process." (A recent 4-volume multiauthored monograph deals with the subject, rather comprehensively, in 49 chapters).[237] Mechanistic complexity arises because of the exceptional versatility of this smallest atomic particle, which can be transferred between molecules under biological conditions as a hydrogen atom, a proton or a hydride anion.

Proton transfer is the commonest of all biological reactions, which generally involve heterolytic (two-electron) processes taking place in an aqueous medium at controlled pH. The great majority of these involve one or more proton transfers between the electronegative centres of Brönsted acids and bases. These reactions are typically diffusion-controlled processes, and thus in principle fast enough (thermodynamics permitting) to support enzyme reaction rates without catalytic assistance; at least in active sites with reasonable access to the solvent. Proton transfers slow enough to need catalysis include those concerted with the making or breaking of other bonds, discussed above under

$$X\text{—}H + Y \rightarrow X + H\text{—}Y$$

Scheme 4.96 Hydrogen transfer: "a deceptively simple process" (see the text).

the heading of General Acid-Base Catalysis in Section 1.3; and transfers to and from carbon, which are intrinsically slow. The most familiar of these is enolization, discussed below in section 4.4.1.

Hydrogen atom and hydride transfers of biological interest also involve C–H bonds, often in key steps in oxidation-reduction (redox) processes. Most of the reactions concerned are beyond the normal capabilities of the modest collection of functional groups available on the side-chains of the common amino-acids (Section 1.1.3). The enzymes involved generally rely on assistance from coenzymes (Section 1.7). This economical arrangement, whereby a specialist auxiliary molecule may serve in the active sites of hundreds of different enzymes, expands enormously the range of available chemistry.[238] But not the principles of catalysis: once the specialist functional group is installed in the active site, by standard covalent or noncovalent binding interactions, these are the same as for all enzymes; depending as usual on productive binding of the substrate, selective stabilization of the transition state and efficient release of the product. We discuss here a single high-profile example of coenzyme chemistry involving hydride transfer (Section 4.4.2), chosen to illustrate this powerful way of extending the capabilities of protein enzymes.

4.4.1 Enolization: Proton Transfer from Carbon

The crucial step in organic synthesis is the formation of the C–C bond. This is achieved generally in the laboratory and almost exclusively in biological systems by heterolytic reactions between nucleophilic and electrophilic carbon centres. Chemically, the most difficult step in such reactions is the generation of nucleophilic carbon, often in the shape of a carbanion. The synthetic organic chemist can use highly reactive organometallic reagents in aprotic solvents, but for biological reactions, taking place in a predominantly aqueous medium, a carbanionic reagent must be efficiently stabilized or protected, or both.

The problems are both kinetic and thermodynamic, with a special kinetic dimension in the case of stabilized carbanions like enolates. The thermodynamic problem is simply expressed in terms of the high pK_as of typical carbon acids (Scheme 4.97).[239] Enzymes like aldolases, enolases and racemases must use the usual active-site bases, with pK_as close to 7, to remove α-hydrogens from these compounds under physiological conditions, in reactions that are clearly thermodynamically highly unfavourable. Furthermore, proton transfer in systems like those of Scheme 4.97 is not a simple Eigen process,[240] for two fundamental and related reasons:

1. C–H cleavage rather than a diffusion process is rate determining: that is to say, protonation on carbon of (all but the most) strongly basic carbanions is not diffusion controlled.
2. Cleavage of the C–H bond may not be a simple synchronous process. In model systems removal of H appears to run ahead of the delocalization necessary for full stabilization of the developing negative charge (principle of nonperfect synchronization[241]).

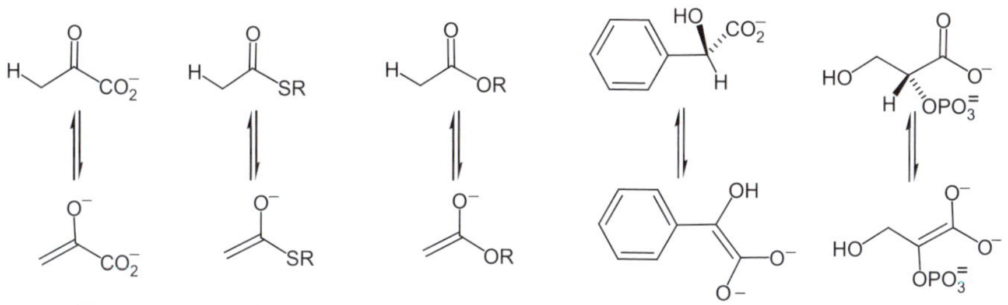

| pK_a | 17 | 21 | 26 | 23 | ~ 34 |

Scheme 4.97   pKas of typical carbon acids involved in enzyme-catalyzed enolization reactions. (Enolate oxygens are much less basic, with pKas typically between 10–11.)

Likely explanations are that viable carbanions generally depend on delocalization of their negative charge, which involves both reorganization of solvation at sites remote from the nominal C^-, and, in particular, significant changes in geometry as sp^3-hybridized centres become planar sp^2. An "exception that proves the rule" is HCN, which behaves as an Eigen acid[242] even though it is as weak an acid as a β-diketone (pK_a = 9). The linear CN^- carbanion undergoes no change of hybridization or geometry when HCN ionizes, and consequently its protonation is diffusion controlled.

The mechanism of enolization (Scheme 4.98) is in principle simple: a general base removes a proton from a carbon α to a carbonyl group, at a useful rate if the resulting enolate is sufficiently stable: as is the case under mild conditions for aldol-type reactions of simple aldehydes and ketones. Esters are weaker acids (Scheme 4.97) and need more vigorous conditions, but thiolesters are closer in carbon acidity to ketones (and are involved extensively in biosynthesis).

The most effective mode of stabilization of enolate anions is the introduction of a positive charge, especially in combination with delocalization (Scheme 4.99). Thus, the ammonium form of glycine ethyl ester is already as strong an acid as a thiolester, and the formation of the iminium derivative with acetone reduces the pK_a further, by a massive 7 units. While replacing the carbonyl oxygen of a ketone by the more electronegative iminium nitrogen brings neutralization to form the enamine (the final example in Scheme 4.99) within range of active-site group pK_as. This is of course the basis of the mechanism of action of pyridoxal coenzyme, and of various enzymes that use active-site lysines to activate ketones.[22]

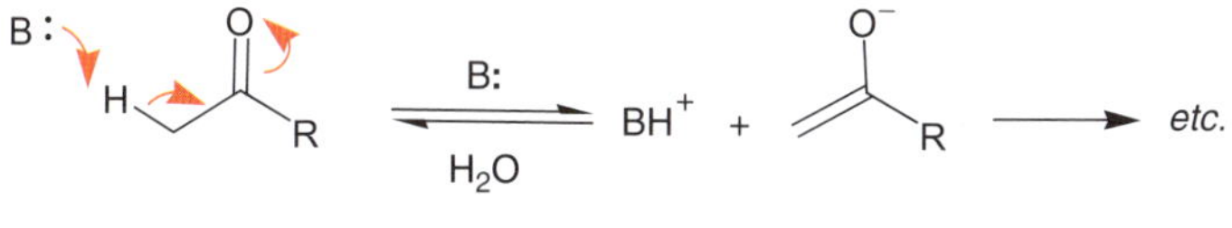

Scheme 4.98   The enolization mechanism (see also Section 1.3.1).

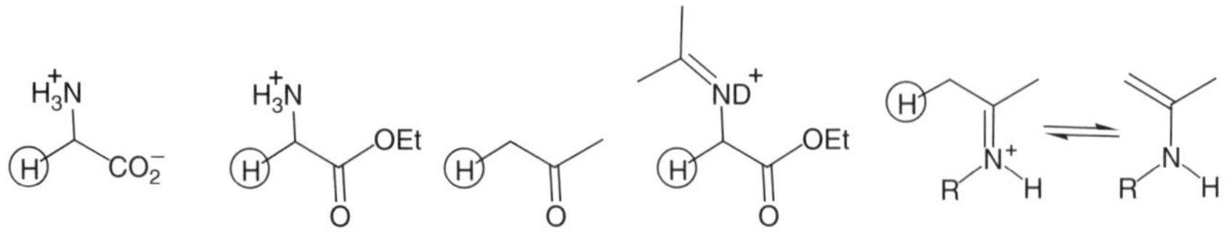

pK$_a$ 29   21   19   14   11

Scheme 4.99   Stabilization of carbanions by delocalization and local positive charges, as measured by the pK$_a$s of their conjugate acids. The pK$_a$s listed refer to the removal of the marked (circled) proton.

4.4.1.1   Simple Models

General base catalysis of enolization by small molecule bases is a well-behaved, rather inefficient reaction in water. Second-order rate constants depend on the basicity of the general base according to the Brönsted equation (see Section 1.3.1). The Brönsted coefficient β, which contains information about the degree of proton transfer in the transition state, increases with the pK$_a$ of the carbonyl compound, as shown in Scheme 4.100.[243]

As expected, the transition state for C–H cleavage is reached earlier in the process for the more reactive system: while the limiting value of β of unity (actually slightly greater than 1.0, as explained in Scheme 4.101) indicates that in the extreme case the reaction of an ester enolate with a general acid of pK$_{BH}$ near 7 (typical of those in the active sites of enzymes) is diffusion controlled.[244] This means that the ester enolate carbanion is so strongly basic that its diffusion apart from BH$^+$ has become rate determining (Scheme 4.101).

Catalysis of enolization by general acids is also weak, with second-order rate constants some 2–10 times smaller than for general base catalysis (representative rate data appear in Scheme 4.106). It is also weak compared with catalysis by H$_3$O$^+$, though observed for relatively strong acids (pK$_a$<5) reacting with aliphatic aldehydes and for acetone: a Brönsted $\alpha = 0.56$ has been

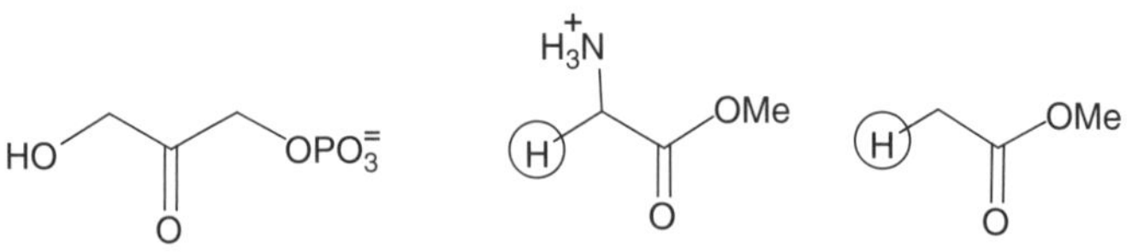

Brönsted β   0.5   0.92   1.1

Scheme 4.100   The sensitivity of the second-order rate constants for enolization catalyzed by tertiary (quinuclidine) bases (measured by the Brönsted coefficient β) increases for less acidic carbonyl compounds (for the relevant pK$_a$s see Schemes 4.97 and 4.99). Catalysis is more readily observed for methyl acetate than for methyl glycinate because the background hydroxide-catalyzed reaction is faster for the cationic ester.[239]

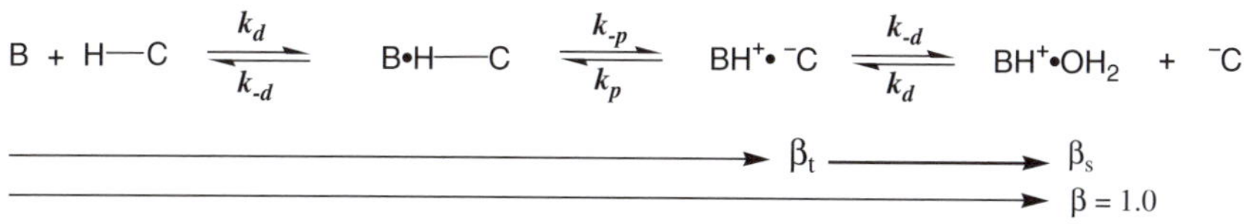

Scheme 4.101   The Brönsted β for the removal of an α-proton from an ester can be greater than unity because the final, diffusional separation step is rate determining. The overall $\beta_{equilibrium}$ is unity by definition, but the observed β_t takes no account of the contribution β_s from solvation of BH^+ as it is released from the intimate ion pair.[244]

measured.[245,246] A contribution from a third-order reaction may be observed, involving also the conjugate base of the general acid.[245] This is interpreted in terms of the familiar (see Section 1.3.2, Scheme 1.4) concerted mechanism shown in Scheme 4.102. A Brönsted plot for the third-order terms has a positive slope ($\beta = 0.2$), indicating that reaction is dominated by the contribution of the general base.

As might be expected, a metal cation can take the place of the general acid of Scheme 4.102, but the contribution to catalysis is not significantly greater in simple systems. Metal-ion catalysis can become more important where the substrate has functional groups that can support chelation, as discussed in Section 4.4.1.3.

4.4.1.2   Intramolecular Models

Effective molarities for enolization, as for the great majority of general acid and general base-catalyzed reactions are typically small (Scheme 4.103), but exceptions to this generalization have emerged.

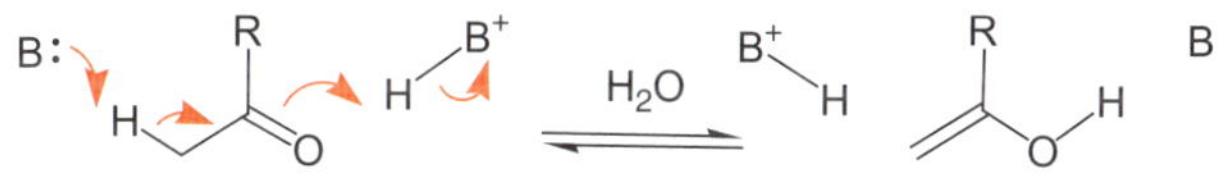

Scheme 4.102   Concerted mechanism for the general acid–base catalyzed enolization of a ketone or aldehyde.

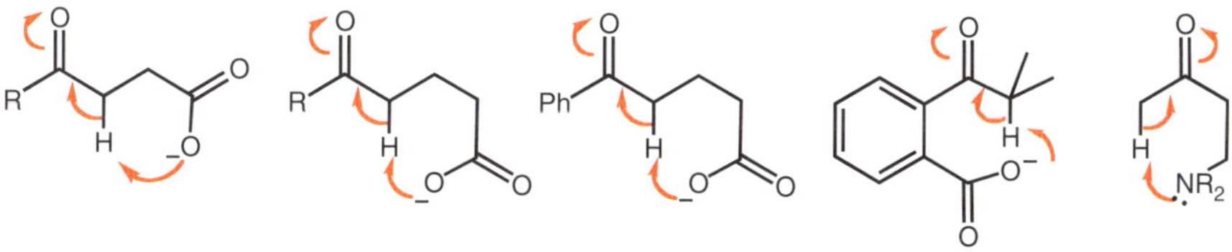

| EM | 0.1-0.5 | 9 | 50 | 56 | < 0.5 |

Scheme 4.103   Effective molarities (EM) for intramolecular enolization reactions.[8]

The key to increased efficiency in general acid–base catalysis (as discussed in Section 4.3.3) is stabilization of the transition state for proton transfer by a strong intramolecular hydrogen bond. Proton transfer to carbon shows similar increases in efficiency in systems **4.160** and **4.161** (Scheme 4.104), with geometries designed to support strong intramolecular hydrogen bonds in the immediate products of reaction. The reaction of **4.160** involves rate-determining proton transfer to carbon with a substantially increased efficiency (compare the EMs in Schemes 4.103 and 4.104), but the reaction of **4.161** is more complicated – and more interesting. The evidence suggests that the proton transfer step is not rate determining, and that the reverse reaction (**4.162**, arrows) is faster than the conformational reorganization. This makes the high estimated EM only a lower limit: and the reverse reaction a candidate for the most efficient known intramolecular general base-catalyzed reaction involving proton transfer from carbon.

This remains the state of play for this reaction. No example of efficient intramolecular general base catalysis of enolization has been reported. A possible explanation is that the reverse reaction, for example that defined by the curved arrows in **4.165** (Scheme 4.105), is too fast to allow any intermolecular reaction of the enolate intermediate – even deuterium exchange with the solvent. An efficient ketonization process will always be very fast in the

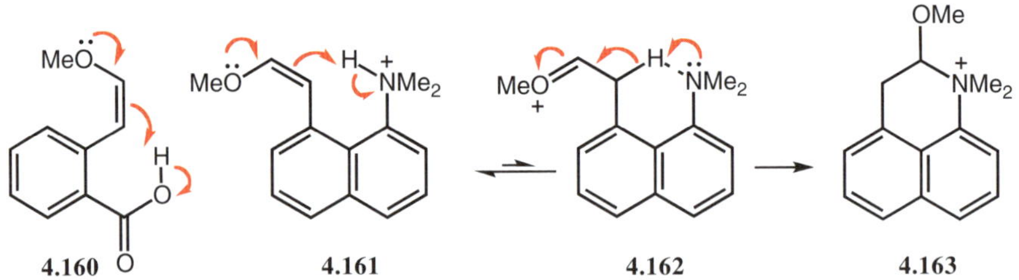

EM   2000   EM   > 60,000

Scheme 4.104   Effective molarities (EM) for intramolecular general acid catalysis of enol ether ketonization.[225,247]

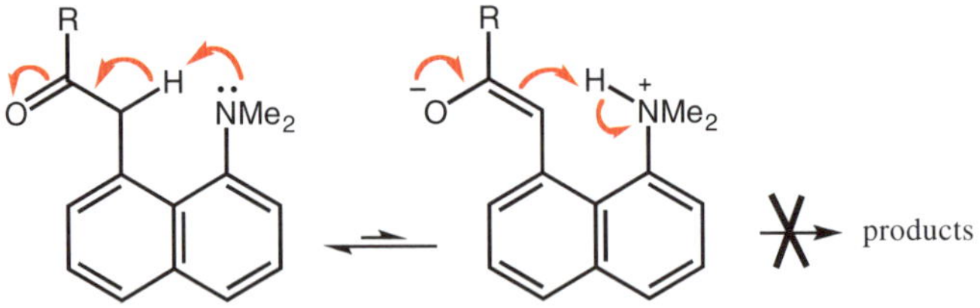

Scheme 4.105   Further reactions of an enolate anion intermediate may be too slow to compete with the rapid reverse-ketonization reaction.

thermodynamically favourable direction and the arrangements in enzyme-active sites have evolved to take account of this.

4.4.1.3   Catalysis by Metal Ions

Recent work has shown that general base catalysis of the enolization of simple ketones is modestly enhanced in the presence of metal ions (top row of Scheme 4.106), but significantly increased by the presence on a substrate ketone of functional groups that support chelation (Scheme 4.106).[243] For standard states of 1 M water and 1 M Zn^{2+} the metal stabilizes the transition state for proton transfer from the α-CH$_3$ and α-CH$_2$OH groups of hydroxyacetone by 4.4 and 6.3 kcal (18.2 and 26 kJ) mol^{-1}, respectively. The lesson seems clear: electrophilic catalysis by a properly positioned metal cation can make a significant, but not overwhelming contribution to catalysis of enolization.

Probably the largest effect in a simple model was reported by Kimura, who showed that a zinc cation bound to the cyclen **4.166** coordinates also the carbonyl oxygen of the pendant phenacyl group (**4.167** in Scheme 4.107).[248]

Titration of an aqueous solution of **4.167** with base gave a pK_a of 8.41 and a 3:1 mixture of the Zn(OH) complex and **4.168**. A pK_a of 8.41 would represent a massive reduction, of the order of 10 units, in the pK_a of a C–H proton of a phenacyl group (note that it is also close to that expected for the enol OH). The rate of H–D exchange of the CH$_2$ group of **4.167** at pD $=$ 7 in D$_2$O is only 100

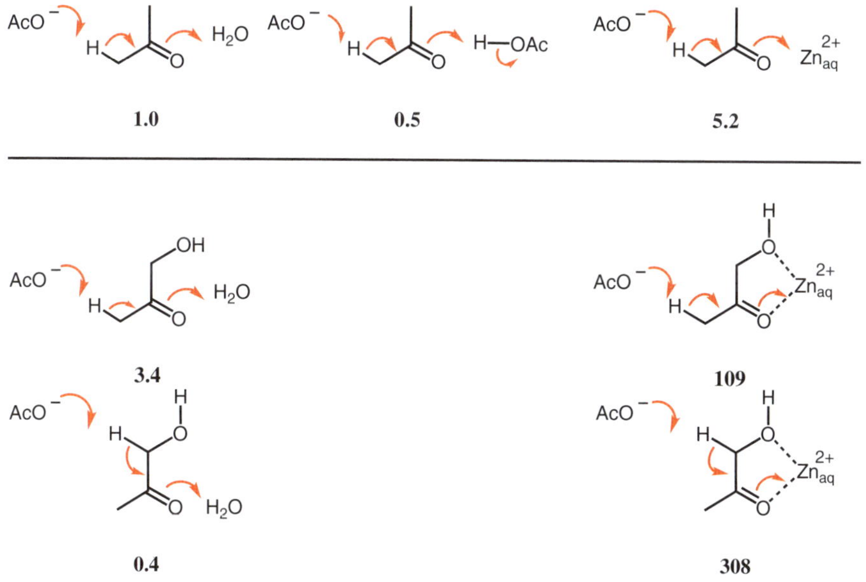

Scheme 4.106   Relative rates of enolization for protons of the methyl groups of acetone and hydroxyacetone in acetate buffers in the presence and absence of zinc. Comparisons are based on second- (left-hand column) and third-order rate constants (columns 2 and 3).

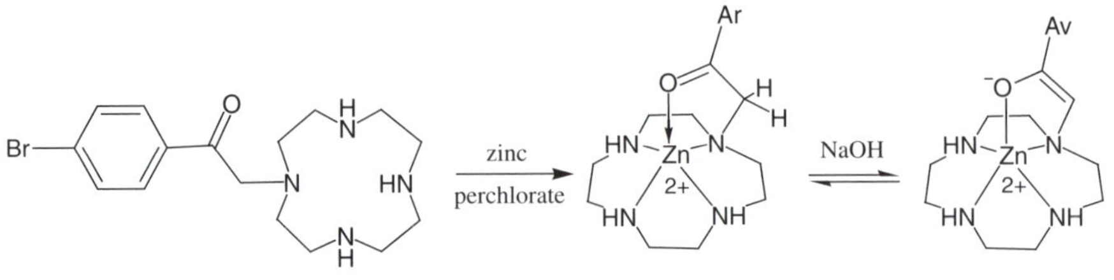

Scheme 4.107 Cyclen **4.166** binds Zn^{2+}, which coordinates the C=O group and thus catalyzes the enolization of the pendant ketone. Complex **4.167** was characterized by a crystal structure, and the enolate **4.168** could be isolated from its reaction with methoxide in acetonitrile.[248]

times faster than that of the parent ligand **4.166**, so transition-state stabilization is relatively modest.

4.4.1.4 Enzymes Catalyzing Enolization

Since general base catalysis of enolization by small molecule general bases, and the consequent aldol condensation, can be observed with reactive carbonyl compounds in water[249] it is not surprising that aldolases can catalyze reactions of aldehydes and ketones without the need for cofactors. Proton transfer from carbon is conveniently studied independently using isomerase and racemase enzymes, and has been the subject of a series of classical investigations.

4.4.1.4.1 Triose Phosphate Isomerase.   Triose phosphate isomerase (affectionately known as TIM) catalyzes the reversible interconversion of dihydroxyacetone phosphate (DHAP, **4.169** in Scheme 4.108) and glyceraldehyde-3-phosphate (GAP, **4.170**). Possible mechanisms include a hydride shift involving the alkoxide anions (**4.171** in Scheme 4.98), but – with few exceptions (of which the xylose isomerase reaction is thought to be one[250]) – enolization (proton transfer) pathways seem generally to be preferred.

The accepted mechanism is outlined in Scheme 4.109: removal of a proton α-to the carbonyl group of either isomer by Glu-165 gives the enediol(ate) **4.172**, which can be reprotonated on either carbon of the double bond.

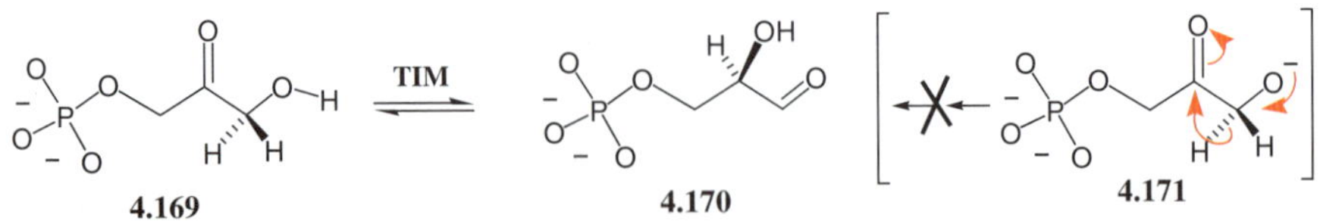

Scheme 4.108   The triose phosphate isomerase reaction (**4.169** ⇌ **4.170**) involves enolization, rather than a hydride-transfer mechanism (**4.171**)

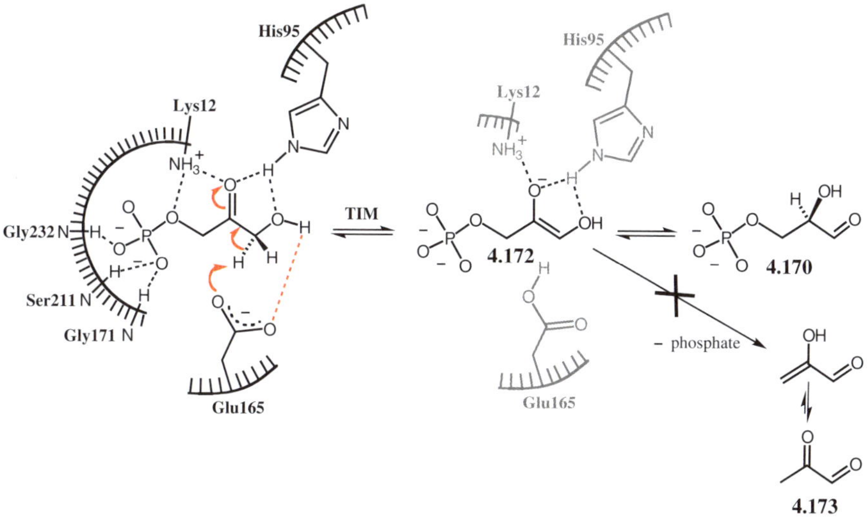

Scheme 4.109 Proposed mechanism for the conversion of dihydroxyacetone phosphate to glyceraldehyde phosphate (**4.170**) catalyzed by triose phosphate isomerase (TIM). The active-site general base Glu-165 removes the pro-R proton from C(1) to generate a bound enediol(ate) intermediate that is stabilized by H-bonding to neutral His-95, as well as by Lys-12 and Asn-10 (not shown).

The mechanism of this relatively simple reaction, which requires no cosubstrate, metal ion or coenzyme, is revealing in a number of ways:

(i) The substrates are small molecules, with both functional groups (C=O and OH) involved directly in the reaction. A major contribution to binding the substrate, and particularly, it is suggested,[251] the transition state, is a phosphate binding motif (found also in other enzymes handling short-chain C, H and O compounds as their phosphate esters). This involves hydrogen bonding to the NH_3^+ group of Lys-12, and to three main-chain amide NH protons.

(ii) In the absence of enzyme the enediolate resulting from the base-catalyzed enolization of **4.169** or **4.170** eliminates phosphate to give methylglyoxal **4.173**. A different enzyme, methylglyoxal synthase, is known to catalyze this elimination of phosphate. The specificity of TIM for proton-transfer is thought to be achieved by binding the substrate in a conformation like **4.172a** (Scheme 4.110), with the C–OPO$_3^{2-}$ bond in the plane of the enediolate (Scheme 4.110).[252] This minimizes the $\pi-\sigma^*_{C-O}$ overlap necessary for cleavage of the C–OPO$_3^{2-}$ bond. No doubt methylglyoxal synthase binds the enediolate in an alternative conformation close to **4.172b** (Scheme 4.110) that favours overlap.

(iii) It was originally assumed that His-95, the general acid in the initial enolization step, would be involved as the more reactive conjugate acid.

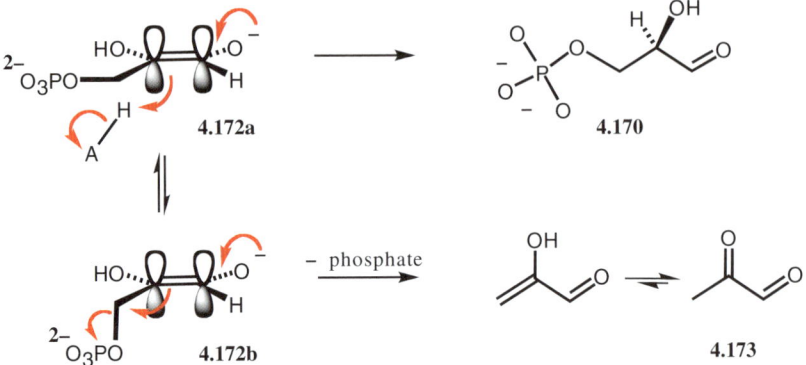

Scheme 4.110   Different conformations **a** and **b** of the enediolate **4.172** have different
intrinsic reactivities. Both can be easily ketonized in the presence of a
general acid, to give either DHAP (**4.169**) or GAP (**4.170**), but only
4.172b, with the C–OP bond parallel to, and thus overlapping with, the
π system, can eliminate phosphate.

Careful NMR investigations showed that the neutral imidazole is the
active form, so that the key O–H–N hydrogen bond stabilizing the
enediolate (pK_a ~10) involves a weak proton donor (pK_a ~14 rather
than pK_a~7 for the histidine conjugate acid). (These pK_as refer to water
as solvent, and will be modified in a more hydrophobic environment, to
favour neutral forms.) The bifurcated hydrogen bond involved in stabilizing the enediolate is still not fully understood: it needs to be strong
enough to stabilize the enediolate, enough to support its brief existence
but not so strong as to hinder subsequent reactions. This stabilization of
an enolate anion by a neutral histidine imidazole is not unique to the
triose phosphate isomerase reaction (see the following section, Section
4.4.1.4.2).

(iv) Triose phosphate isomerase is of special importance in the history of
    enzyme mechanisms because the evidence suggests that it has achieved
    catalytic perfection.[253] The enolization step would be rate determining
    for a reaction of this sort in the absence of the enzyme, but no primary
    deuterium isotope effect was observed for the TIM-catalyzed reaction of
    [2-^{2}H]GAP. This evidence that C–H cleavage is not rate determining for
    the enzyme-catalyzed reaction, combined with the high value of k_{cat}/K_M
    (greater than $10^8 \, M^{-1} \, s^{-1}$) suggested that the reaction is diffusion-con-
    trolled: with binding or release of the less-stable isomer, GPA (**4.170**)
    primarily rate determining. This conclusion was confirmed by experi-
    ments that showed that k_{cat}/K_M is reduced when the viscosity of the
    aqueous medium is increased (by adding sucrose or glycerol).[254] The
    profound implication is that for this enzyme (and for many other
    examples identified since) all the chemical steps are catalyzed efficiently
    enough that they are not significantly rate limiting: more efficient

catalysis would not make the reaction any faster because the rate of substrate binding or product release is the rate-limiting step.

4.4.1.4.2 Citrate Synthase. Citrate synthases catalyze the synthesis of citrate at the entry point to the tricarboxylic acid cycle, using acetyl coenzyme A as the enol(ate) and oxaloacetate as the acceptor electrophile (Scheme 4.111). The remainder of the coenzyme A structure (not shown) provides ample binding recognition, thus playing a role similar to that of the phosphate group in the triose phosphate isomerase reaction (Scheme 4.109), and the two mechanisms appear to be closely similar: with Asp-375 acting as the general base and His-274 as the lone electrophilic side-chain group, acting here also in the neutral form. (The ionic form of His-274, and thus the details of the enolization process, are still under discussion.[255]) A bound water provides additional hydrogen-bonding stabilization for the enolate oxygen.

Enolization is only observed in the presence of the second substrate oxaloacetate, or other four-carbon dicarboxylic acid, indicating that the productive geometry for catalysis of the overall reaction is only established when both substrates are bound (the oxaloacetate by a network of three arginines).

4.4.1.4.3 The Enolase Superfamily[256]. Enolase is a central enzyme in the glycolytic pathway, catalyzing the reversible dehydration of 2-phospho-D-glycerate (2-PGA) to phosphoenolpyruvate (PEP) (Scheme 4.112). The *anti*-elimination of water from 2-PGA involves an initial enolization α- to the carboxylate group, and thus nominally the generation of the dianion of a carboxylic acid. This is thermodynamically far less favourable than the reactions of ketones and thiolesters discussed so far in this section, with the pK_a for the α-CH likely to be at least as high as that (ca. 34) of the acetate anion.[239,257] So it is no surprise that metal cations are involved in catalysis:

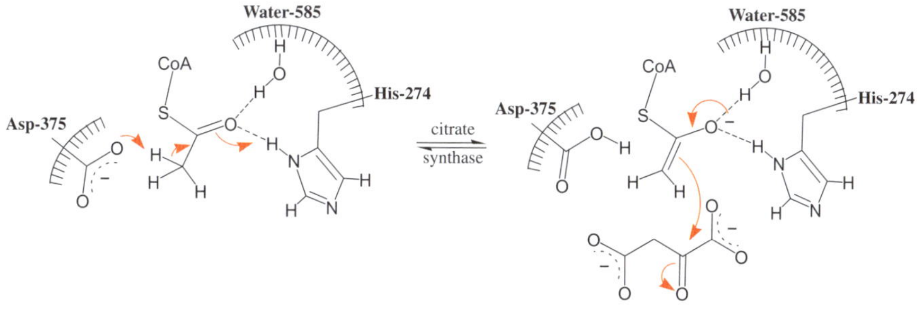

Scheme 4.111   The mechanism of enolization of acetyl coenzyme A catalyzed by citrate synthase shows obvious similarities to that used by triose phosphate isomerase (Scheme 4.109: see the text). The arrows delineating the second, the C–C bond formation step shows how the citryl CoA, the initial product of the reaction, is formed.

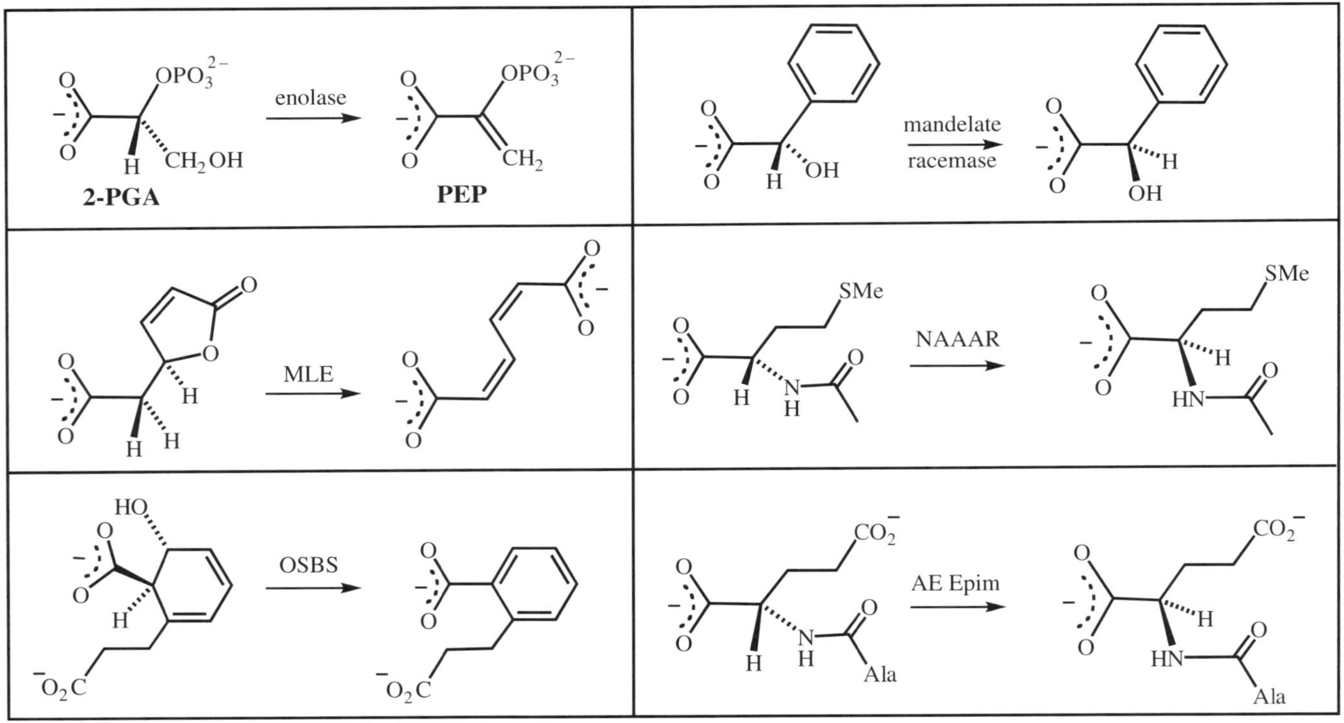

Scheme 4.112 Reactions catalyzed by enzymes of the enolase superfamily.[258] MLE is muconate lactonizing enzyme, OSBS o-succinylbenzoate synthase, NAAAR N-acylamino-acid racemase and AE Epim is L-Ala-D/L-Glu epimerase. Active site features of these and other members of the superfamily are shown in Figure 5.27.

nor, perhaps, that the successful mechanistic solution to this basic problem is used by many different enzymes.

Enzymes of the enolase superfamily show well-defined structural and sequence similarities and share an initial step in which an active-site base removes the α-CH of a carboxylate anion, to generate an enolate intermediate stabilized by coordination to a Mg^{2+} cation.[256,258] A strongly conserved cluster of carboxyl groups makes up the binding site for the divalent cation essential for catalysis, but the identity and position of the general base that abstracts the proton varies. The superfamily is named for the "senior" enzyme in the classification, though it was identified through detailed investigations of several other enzymes, notably mandelate racemase: which, like triose phosphate isomerase, catalyzes specifically the enolization step. (The enolase subgroup amounts to over 600 different proteins specific for the conversion of 2-PGA to PEP, and uses two active-site magnesiums. One of these, dubbed the "conformational" magnesium, induces a conformational change in the enzyme and is involved in the binding of the phosphate group: and also in stabilizing the enolate anion, though only after the second, "catalytic" metal ion binds.) We discuss the enolization reaction in the context of mandelate racemase, which uses a single magnesium.

4.4.1.4.4 Mandelate Racemase. Mandelate racemase interconverts R and S-mandelate (Scheme 4.113).[259] As a benzylic α-hydroxy-acid mandelate has a pK_a for the α-CH proton of around 23 (Scheme 4.97), and the primary role of the Mg^{2+} is to lower this pK_a, by helping to neutralize the negative charge that must be accommodated on the carboxylate group (Scheme 4.113). Note that the developing dianion would be an enediolate, like that in the triose-phosphate isomerase reaction (**4.172** in Scheme 4.109, above); but different in that it is electronically highly unsymmetrical. The second negative charge is effectively neutralized in the course of general acid catalysis by the carboxyl group of Glu-317. Racemization is achieved very simply: the planar enediolate intermediate is formed, by general base-catalyzed removal of the α-CH proton, with a general acid on each face (Scheme 4.113). Protonation by His-297 on the *re*-face gives R-mandelate, by the Lys-166 NH_3^+ group on the *si*-face gives S-mandelate.

Proof that a highly reactive intermediate is generated in the active site of mandelate racemase derives from the reaction with *p*-bromomethyl mandelate (**4.176**, Scheme 4.114). The enzyme accepts the modified substrate, but this is not racemized but rather converted to *p*-methylbenzoyl formate **4.179**. The *p*-$BrCH_2$ group conjugated with the enediolate carbanion can stabilize itself by eliminating bromide (**4.177**) to generate the quinone methide **4.178** as a short-lived intermediate.[260] Both enantiomers of **4.176** react, because MR has active-site bases on both faces of the potential enediolate; though if either is mutated to a nonbasic group enantiospecificity is restored.

4.4.1.4.5 Models. Enolization is a well-understood process, and fast enough to be studied in natural, inactivated systems in the absence of enzymes. The

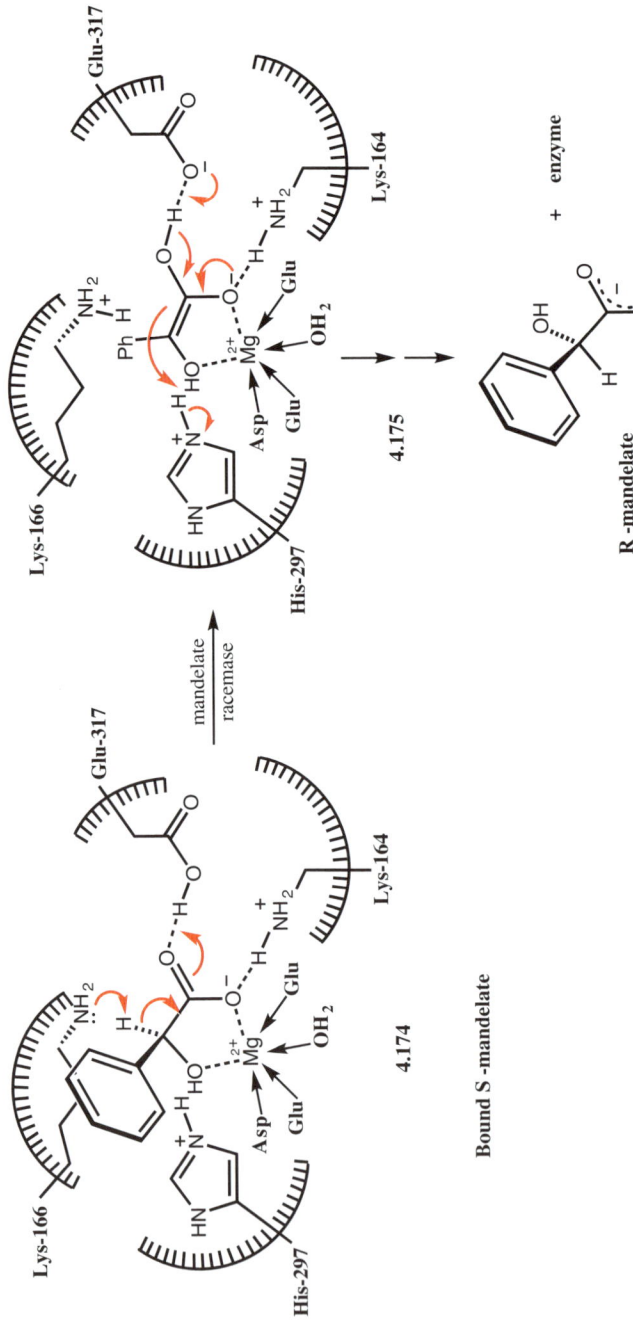

Scheme 4.113   Outline mechanism for the mandelate racemase reaction. The planar enediolate, stabilized by multiple interactions with general acids and Mg²⁺, is sandwiched (**4.175**) between two general acids, one above and one below the plane, on the *si*- and *re*-faces, respectively.

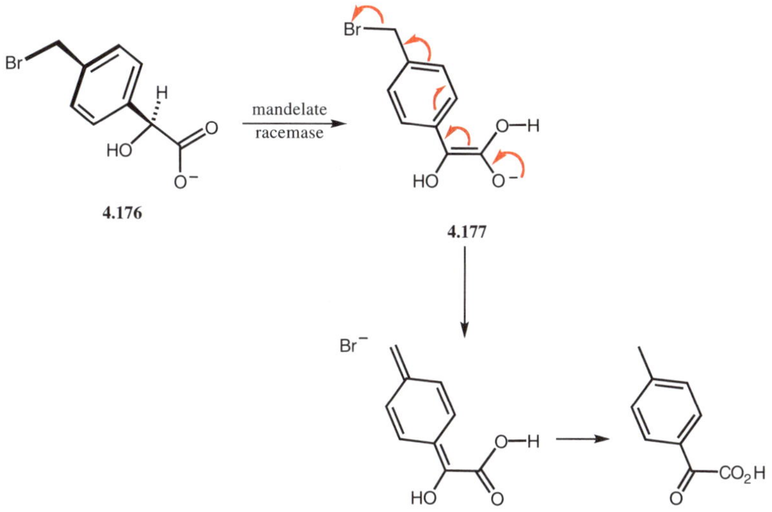

Scheme 4.114 Mandelate racemase converts *p*-bromomethyl mandelate **4.176** to the expected enediolate **4.177**, which collapses with the rapid elimination of bromide.[260]

reaction catalyzed by mandelate racemase, for example, is catalyzed by methylamine, imidazole and acetate general bases in very slow but measurable processes that allow estimates of the effective molarities of the active site Glu-317, His-297 and Lys-166 NH_3^+ of $> 10^5$ M, $> 10^3$ M and > 600 M, respectively;[261] consistent with the expectation that general acid–base catalysis can be highly efficient in enzyme-active sites.

The ketonization of the enediolate **4.175** (Scheme 4.113, generated *in situ* by flash photolysis of phenyldiazoacetic acid, $Ph(C=N_2)COOH$) has been studied directly, and the equilibrium constant for enolization of the mandelate anion estimated as 4.55×10^{-20}.[262] This provides a measure of the thermodynamic barrier the enzyme has to overcome of over 27 kcal/mol (the intermediate must be lower in energy than the transition state leading to it).

4.4.2 Hydride Transfer

Coenzymes extend the capabilities of protein enzymes by providing functionality not available from the catalytic groups of the naturally occurring aminoacids. Chemically, the single most important missing functionality is oxidation and reduction. This is provided by various coenzymes and metal complexes, and we discuss briefly how one of these, the $NAD^+ \rightleftharpoons NADH$ coenzyme system (Scheme 4.115), is integrated into the active sites of some of the many enzymes that use it. Though some coenzymes are bound covalently to active-site functional groups, the nicotinamide reductants are bound noncovalently.

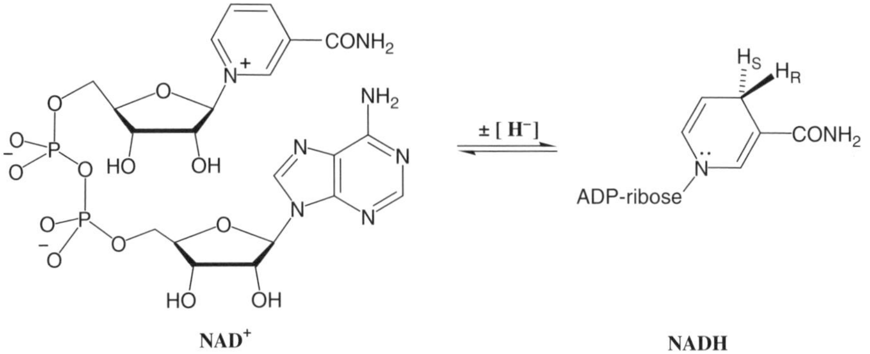

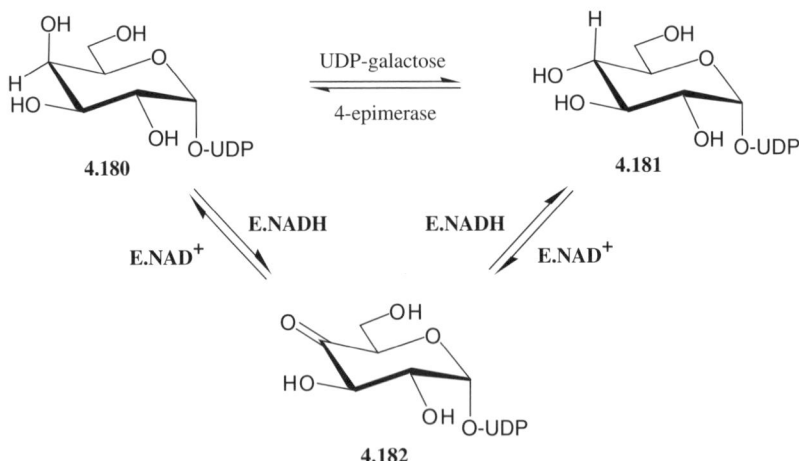

NAD⁺   **NADH**

Scheme 4.115 The redox functionality of the coenzyme nicotinamide adenine diphosphate (NAD⁺) resides in the pyridinium ring. This can act as an oxidizing agent by accepting a hydrogen atom, together with a pair of electrons: to give the reducing agent NADH, which is the strongest of the biological reducing agents.

4.4.2.1   Uridine Diphosphate-galactose-4-epimerase

This enzyme uses a tightly bound molecule of NAD⁺ to catalyze the epimerization of the galactosyl residue attached to the nucleoside diphosphate (**4.180**, Scheme 4.116), to give the epimer, UDP-glucose (**4.181**, the activated glycosyl donor in the biosynthesis of glycogen). The first step of the reaction involves oxidation of the secondary alcohol at C(4) to the ketone (**4.182**, Scheme 4.116). The C=O group could allow epimerization at C(3) or C(5) of **4.182** (and this strategy is used by some enzymes), but UDP-galactose-4-epimerase finds a

Scheme 4.116 Uridine diphosphate-galactose-4-epimerase catalyzes the reversible oxidation of the substrate to the keto-sugar (**4.182**). Reduction on the opposite face gives the gluco-epimer.

chemically simpler solution, reducing $C=O$ back to the epimeric 4-CHOH by transferring H to the opposite face.

This is an unusual result. Reactions involving NADH are almost invariably stereospecific at both donor and acceptor centres, in consequence of their strictly controlled geometry of approach in the active site. In this case the pro-S hydrogen (H_S in Scheme 4.115) is transferred selectively, as expected, but to either face of the $C=O$ group. (Scheme 4.116).

This is possible because while the NADH is tightly bound (behaving effectively as part of the enzyme), the keto-sugar ring of the substrate is not: it is firmly anchored by the bound nucleoside diphosphate group, but retains sufficient conformational flexibility to offer either face of the planar (4)$C=O$ group to the H-donor group, as sketched out in Scheme 4.117.[263]

4.4.2.2 Dehydrogenases

The coenzyme is not always bound so tightly to dehydrogenases, but it always forms part of the active site as encountered by the substrate, by virtue of the strictly imposed ordered binding sequence; in which NAD^+ is first to bind and NADH last to diffuse away. Coenzyme binding involves a specialized "nucleotide binding fold" that is also found in other nucleotide binding proteins, which are functionally complete only when the coenzyme is bound. The reduction of a ketone or aldehyde is not a particularly difficult reaction, but the oxidation of a primary or secondary alcohol to $C=O$ is more energetically demanding, and needs as a minimum the removal of the OH-group proton. Lactate dehydrogenase uses the imidazole group of His-195 as the general base needed to remove this proton (Scheme 4.118).

For the oxidation of simple, short-chain alcohols generally stronger activation is needed. In some cases the phenolate O^- group of a tyrosine ($pK_a \sim 9$) is used as the general base, but the widespread class of alcohol dehydrogenases uses an active-site zinc (Scheme 4.119). The combination of metal coordination and the electrostatic effect of the positively charged pyridinium system bound close by reduce the pK_as of bound alcohols by up to 9 units.[264]

4.4.2.3 Models

Coenzymes are involved in most enzyme-catalyzed reactions involving serious organic chemistry, and the mechanisms of their reactions have always proved fascinating and often instructive to the synthetic organic chemist. As a result, models and mimics for coenzyme reactions, and NADH models in particular, have provided a fertile field of investigation. Simple N-alkyl dihydropyridines will reduce carbonyl compounds *in vitro*, but only if the $C=O$ group is activated: by electron-withdrawing substituents, by metal-complex formation, or by both. And they will do so stereospecifically if a chiral group is built in. The 4-CH_2 protons of NADH itself are diastereotopic (Scheme 4.115: H_R and H_S have different NMR chemical shifts), but the stereospecificity of hydride transfer in enzyme-active sites is determined primarily by the controlled

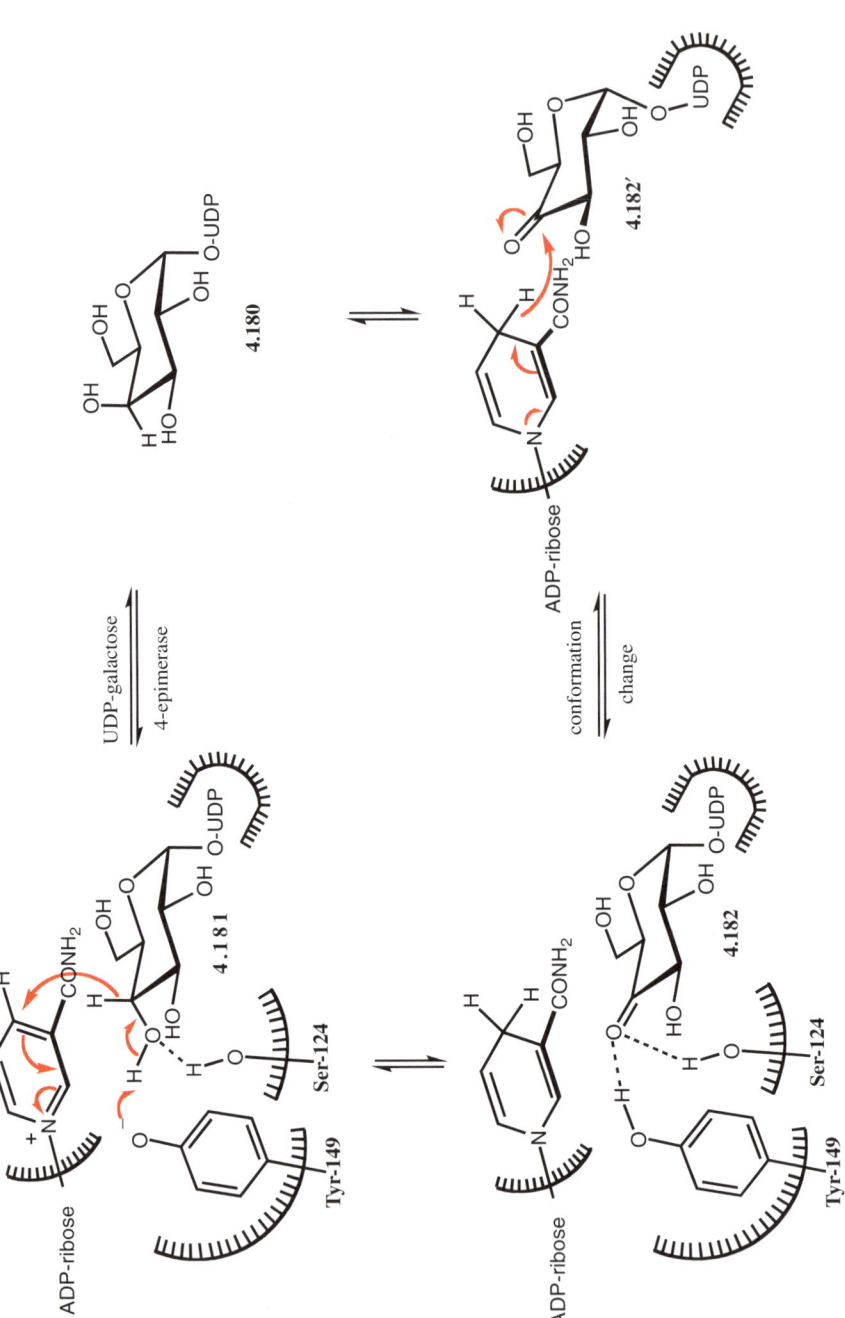

Scheme 4.117 The keto-sugar (**4.182**) attached to uridine diphosphate has the conformational flexibility to accept hydride, from NADH bound in the active site of uridine diphosphate-galactose-4-epimerase, on either face of the C=O group. Note that the general base involved in the oxidation step is the phenolate anion of Tyr-149.

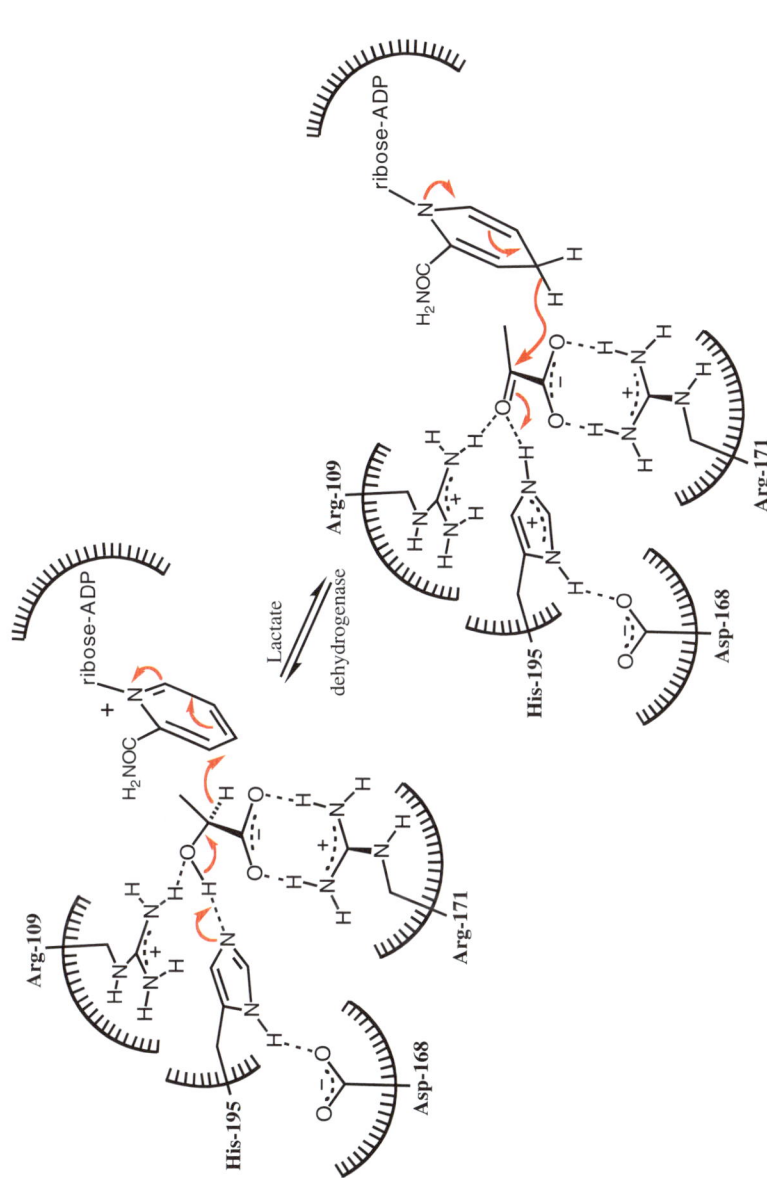

Scheme 4.118 Lactate dehydrogenase uses the imidazole group of His-195 as a general base to activate the CHOH group of lactate.

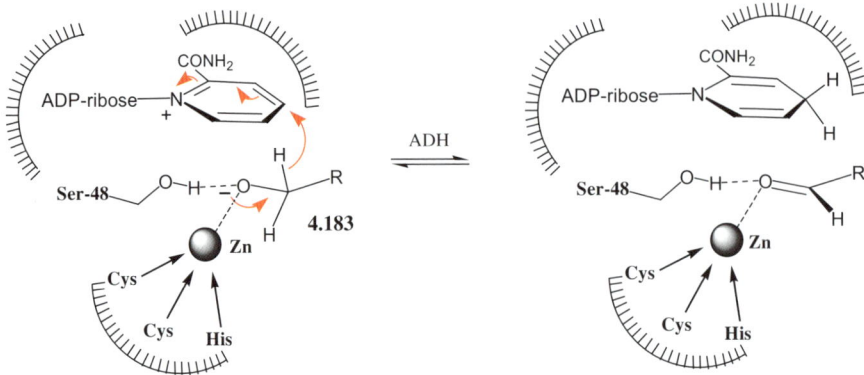

Scheme 4.119 Liver alcohol dehydrogenase uses an active-site zinc to activate the OH
group of simple alcohols. The pK_a of the hydroxyl group of bound
ethanol (**4.183**, R = CH$_3$) coordinated to Zn is about 6.4,[264] and it is
present as the alkoxide in the presence of bound NAD$^+$.

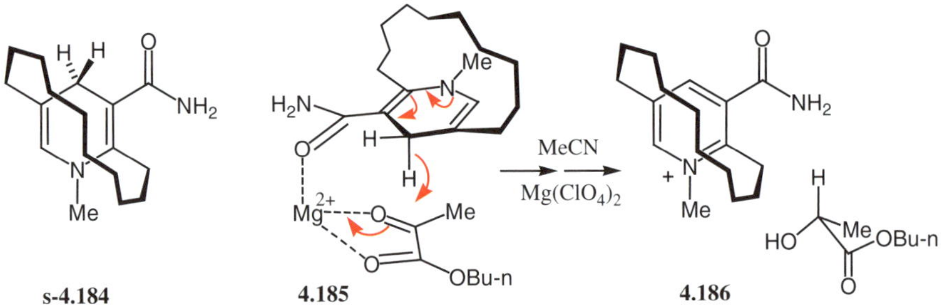

Scheme 4.120 Suggested mechanism for the enantiospecific reduction of a pyruvate
ester by S-**4.184**.[265] The reaction gives 83% of R-lactate, with 97%
e.e., in 5 days at room temperature.

geometry of approach of the hydride acceptor to the exposed face of the
dihydropyridine. A model that nicely illustrates many of these properties is the
system S-**4.184** (Scheme 4.120). Approach to one face is hindered effectively by
the ansa-bridge, and S-**4.184** reduces n-butyl pyruvate (**4.185**) to R-lactate in
the presence of magnesium perchlorate with 97% enantiomeric excess (rather
slowly, in acetonitrile).[265]

The basic hydride-transfer reaction from a dihydropyridine to a C=O group
generates two charged centres, and is as expected faster in more polar solvents,
especially in water.[266] However, hydride transfer in water must compete with
other reactions, and in particular with acid-catalyzed hydrolysis of the dihydropyridine ring. Simple dihydropyridines are very sensitive to pH, and rapidly
hydrolyzed via general acid-catalyzed attack on the ring.[267] Electron-withdrawing groups, in particular the amide group of dihydronicotinamide, are
stabilizing, but the hydride-transfer reactions in aqueous solvents still have to
be studied at relatively high pH.

Most model reactions involving hydrogen transfer, and the vast majority of enzyme-catalyzed reactions involving NADH, are known, thought or simply assumed to involve hydride transfer. A single-electron transfer mechanism is in principle possible, and has been shown to account for a number of model reactions involving the reactions of dihydropyridines with known single-electron oxidants like ferricyanide.[268,269]

4.4.2.4 *Intramolecular Models*

The little evidence available suggests that intramolecular hydride transfer is a relatively inefficient process, not a great deal more efficient than typical intramolecular proton transfers (see above, Section 1.4.1). No hydride-transfer reaction was observed for (relatively unreactive) **4.187** (Scheme 4.121) or for several homologues;[270] or for the ketone **4.188**,[271] though both offer 6-membered cyclic transition states for the hydride transfer process. However, the reaction is observed for the α-ketoester **4.189**. Triester **4.189** is stable in ether at $-17\,°C$, but undergoes rapid intramolecular hydride transfer ($t_{1/2}$ 6–7 min at $39\,°C$) in aqueous solution between pH 3–6.[271] The reaction is pH-independent (no general acid catalysis was observed), and almost 100 times faster than the acid-catalyzed hydrolysis of the dihydropyridine ring under these conditions. (The rate falls off at higher pH as the β-ketoester group ionizes to the enolate anion.)

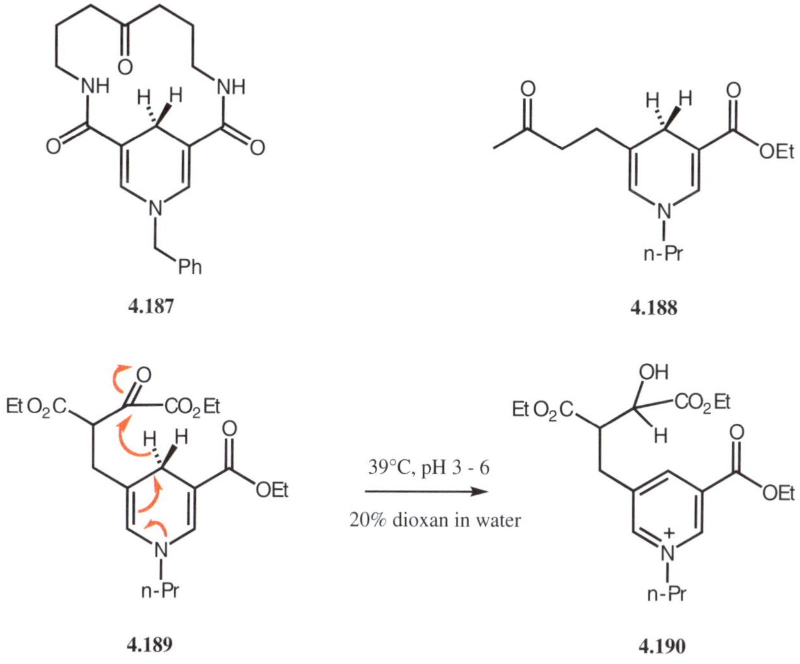

4.187

4.188

4.189

39°C, pH 3 - 6

20% dioxan in water

4.190

Scheme 4.121   Intramolecular hydride transfer to a keto-group also needs the $C=O$ to be activated. Only **4.189** shows the reaction, in aqueous solution and in acetonitrile in the presence of $Mg(ClO_4)_2$.

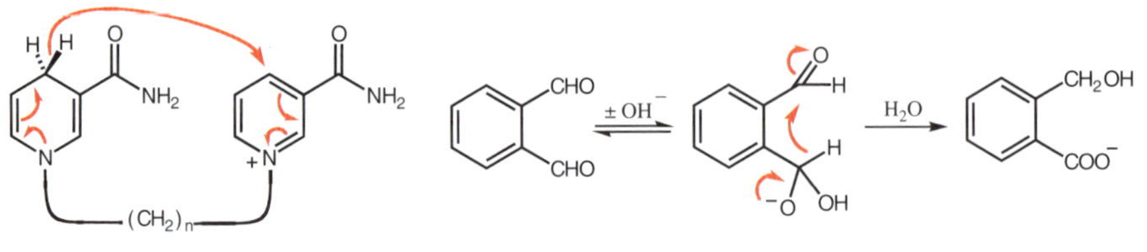

4.191   **4.192**

Scheme 4.122   Intramolecular hydride-transfer reactions for which effective molarities have been estimated. The long-range arrow in **4.191** should not
be taken at face value: hydride transfer will only happen when the
reacting centres are close: probably brought together in this system by
stacking of the donor and acceptor rings.

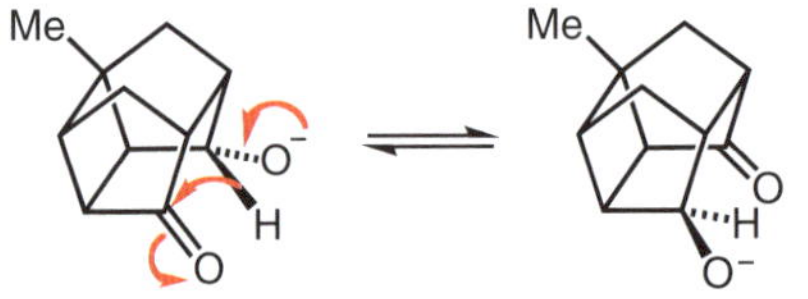

4.193

Scheme 4.123   The rapid base-catalyzed interconversion of the hydroxyketone **4.193**
and related systems[275] is very sensitive to the geometry of the interaction of the reacting centres.

An effective molarity for the reaction of **4.189** could not be obtained in
water, because the (inevitably slower) comparison intermolecular processes were
complicated by side reactions; but an estimate of 6 M was obtained for the
reaction in acetonitrile catalyzed by $Mg(ClO_4)_2$.[272] This figure is consistent with a
handful of available comparable values involving rate-determining hydride
transfer (Scheme 4.122). Hydride transfer between the dihydropyridine and
pyridinium groups of **4.191** in DMSO at 25 °C, followed by spin-saturation
transfer NMR, allowed estimates of 4 and 210 M for the reactions of **4.191**
($n = 3$ and 4), respectively: in the same range as the figure of 14 M for the
hydride transfer involved in the intramolecular Cannizzaro reaction of phthal-
dehyde (**4.192**).[273]

These results indicate that EMs of the order of 10–100 are likely to be typical
for intramolecular hydride-transfer reactions. As discussed above for proton-
transfer reactions, it seems certain that enzyme-catalyzed hydride transfers
from NADH, *etc.*, will be more efficient than this. Results for just one system
suggest that the figure can be raised substantially, if the geometry is optimized
and (or?) sufficient strain is built into a reactant keto-alcohol. The base-catalyzed interconversion of the hydroxyketone **4.193** (Scheme 4.123) proceeds at
NMR rates, and an EM of over 10^5 can be estimated, using for comparison the
intermolecular reduction of cyclopentanone by isopropoxide.[274]

4.4.3 Hydrogen-Atom Transfer

Proton and hydride transfers account for the great majority of hydrogen-transfer reactions in enzymes. Hydrogen-atom transfers (Scheme 4.124) were until recently looked on as exotic exceptions, necessarily involving as intermediates highly reactive, notoriously nonselective radicals, evidently inappropriate guests within the delicate architectural arrangements of enzyme-active sites.

We should never underestimate enzymes. Radicals typically react indiscriminately because they are reactive enough to make bonding interactions with either bonding or antibonding orbitals of any molecule they encounter. To control their reactions it is therefore essential to limit their freedom of movement, to permit contacts only along the desired reaction coordinate. We should not be surprised that enzymes can achieve this degree of control, and that mechanisms involving radical intermediates are being convincingly proposed for more and more enzyme-catalyzed processes.[276] Since not all these reactions involve hydrogen-atom transfers they are discussed in the following section under the general heading of radical reactions. However, one topic specific to hydrogen-atom transfer (HAT) reactions should be mentioned.

These reactions fall into the more general class of proton-coupled electron-transfer (PCET) reactions.[277] Put simply, the proton and the electron do not necessarily move together. Movement of the proton, as the heavier particle, is fundamentally limited to short distances, whereas transfer of the much lighter electron is possible over a much longer range. The situation is summarized in Scheme 4.125.

In the typical hydrogen-atom transfers familiar from radical chemistry, the proton and the electron come from the same atom of the donor molecule X–H, in close contact with the acceptor functionality in an encounter complex, and their movements are strongly coupled. However, electron transfers are possible

$$X{-}H \;+\; {\cdot}Y \rightleftharpoons \; {\cdot}X \;+\; H{-}Y$$

Scheme 4.124 Hydrogen-atom transfer (HAT) involves radical centres in both reactants and products.

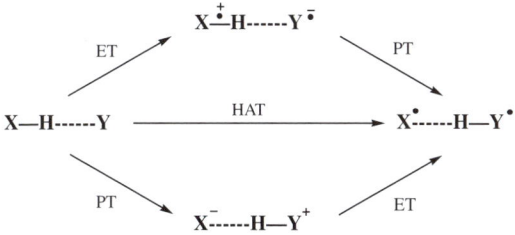

Scheme 4.125 Hydrogen-atom transfer (Scheme 4.124) is a special case of the generalized proton-coupled electron-transfer (PCET) reaction.

over much longer distances, which protons can cover only in a series of steps. Enzymes can also handle such uncoupled processes: by providing functional group "staging posts" for multiple short-distance proton hops, or pathways via structured water channels. Sophisticated models for these uncoupled processes are becoming available,[277] but we will be concerned primarily with short-range, closely coupled proton transfers.

4.5   Radical Reactions

Radicals are typically highly reactive and thus short-lived species, and their reactions therefore subject to special constraints. The initiation step, the generation of the radical in the first place (Scheme 4.126), is a crucial, enabling event; and of particular importance for homolysis reactions (eqn (i)) where recombination is usually faster than diffusion apart.

Neither mechanism is available to the usual functional groups available in enzyme-active sites, and enzyme-catalyzed radical reactions depend for their initiation steps on coenzymes. (Exceptions are a few reactions involving molecular oxygen as a substrate.[278]) We are concerned here specifically with the radical chemistry going on post-initiation, rather than with the fascinating chemistry of the coenzymes involved.[279] However, initiation is the imperative origin of almost all radical chemistry, so we introduce the subject with a brief discussion of the process as implemented by two important, related coenzyme systems.

4.5.1   Coenzyme Initiators Based on Adenosylcobalamin

The most familiar biological source of radicals from a thermal homolysis is the vitamin B_{12} coenzyme adenosylcobalamin (Scheme 4.127). The bond-dissociation energy of the Co–C(5$'$) bond is some $30 \, kcal \, (125 \, kJ) \, mol^{-1}$, making it reasonably stable under physiological conditions but susceptible to ready activation by host enzymes when substrates are bound. Dissociation gives the reactive 5$'$-deoxyadenosyl radical, and the stable, paramagnetic cob(II)alamin: in which the unpaired electron resides at the metal centre, stabilized and protected by the corrin ring system and the dimethylbenzimidazole ligand.

The 5$'$-deoxyadenosyl radical can also be generated by a single-electron transfer (eqn (ii) in Scheme 4.126) to S-adenosylmethionine (**SAM**, Scheme 4.127). (Recent results suggest that **SAM** – dubbed the biological methyl group

$$R{-}X \; \rightleftharpoons \; R{\cdot} \; {\cdot}X \; \xrightarrow{\text{diffusion apart}} \; R{\cdot} \; + \; X{\cdot} \quad \text{(i)}$$

$$R{-}X \; + \; e^{-} \; \longrightarrow \; R{\cdot} \; X^{-} \; \xrightarrow{\text{diffusion apart}} \; R{\cdot} \; + \; X^{-} \quad \text{(ii)}$$

Scheme 4.126   Initiation steps for radical reactions, involving (i) homolysis of a covalent bond, and (ii) single-electron transfer to the $\sigma^{*}_{R{-}X}$ antibonding orbital.

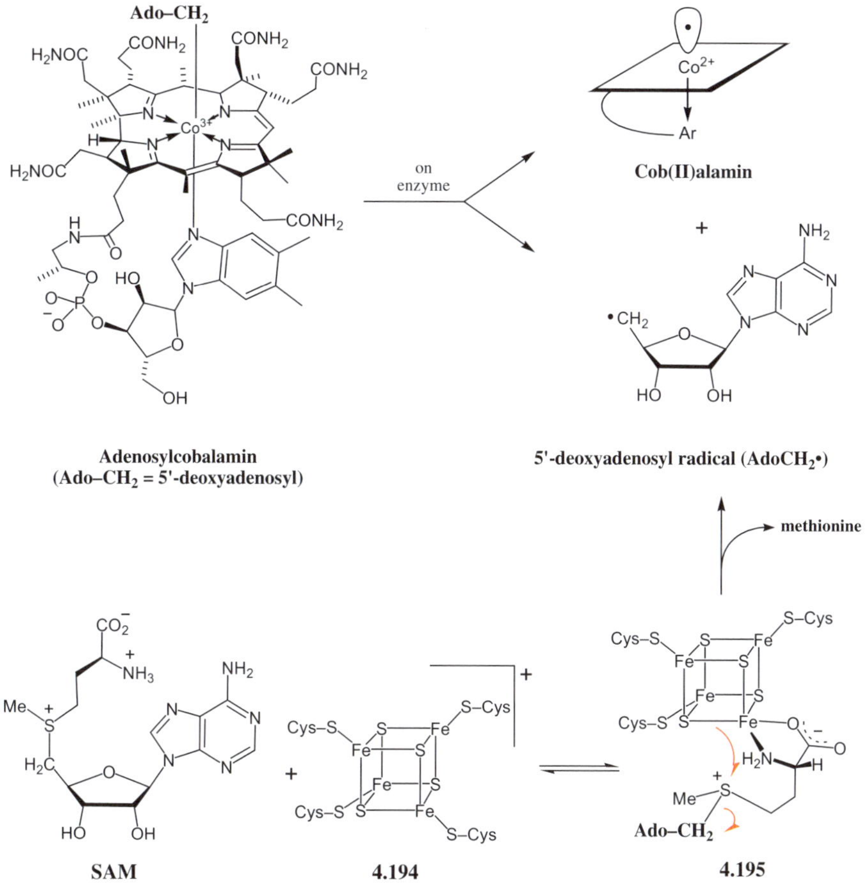

Scheme 4.127 The initiation step of the many reactions involving the B_{12} coenzyme involves the homolysis of the Co–C bond, to give the 5'-deoxyadenosyl radical and cob(II)alamin. The same radical is also commonly generated by a single-electron transfer from the iron–sulfur cluster [4Fe–4S] (**4.194**), probably via an initial ligand exchange (giving **4.195**; see the text).

donor – initiates more radical reactions than adenosylcobalamin.[280,281]) The source of the single electron is another coenzyme, the iron–sulfur cluster (**4.194**). The mechanism of the process is still under investigation, but seems likely to involve an initial ligand exchange, making the electron transfer intramolecular, as suggested (**4.195**) in Scheme 4.127.

For turnover, it is necessary that an enzyme be restored to its active form when reaction is complete. Radical reactions generally continue until one radical meets another and both are destroyed by the electron-pairing interaction. Turnover is achieved in reactions initiated by the 5'-deoxyadenosyl radical **Ado–CH$_2$·** when the product, itself a newly formed radical, removes a hydrogen atom from the methyl group of the **Ado–CH₃** formed in the initiation step (Scheme 4.128). This step reverses the original initiation process.

Some enzymes do not recycle **Ado–CH₂·** in this way, but use it as a stoichiometric oxidizing agent, releasing deoxyadenosine as a byproduct, and binding a new molecule of SAM for each cycle. Two cases illustrate possible fates for the radical centre under this regime. The active form of pyruvate-formate lyase (PFL, see Section 5.2.1) has a relatively stable main-chain radical centre at Gly 734, produced post-translationally by the reaction with an activase enzyme: which itself generates **Ado–CH₂·** by the radical SAM reaction.[281] The PFL activase reaction is thus complete with the formation of the first radical **4.196** (Scheme 4.129). The case of biotin synthase (the second example

Substrate H—S + Enzyme

H —S ⇌ AdoCH₃ + •S

•AdoCH₂

H —P ⇌ AdoCH₂ —H •P

Enzyme + Product H—P

Scheme 4.128 Catalytic cycle supporting turnover in enzymes using the 5′-deoxyadenosyl radical **Ado–CH₂·** to initiate radical reactions. The radical is generated as described in Scheme 4.127. The reaction catalyzed by the enzyme is the conversion of substrate to product radical.

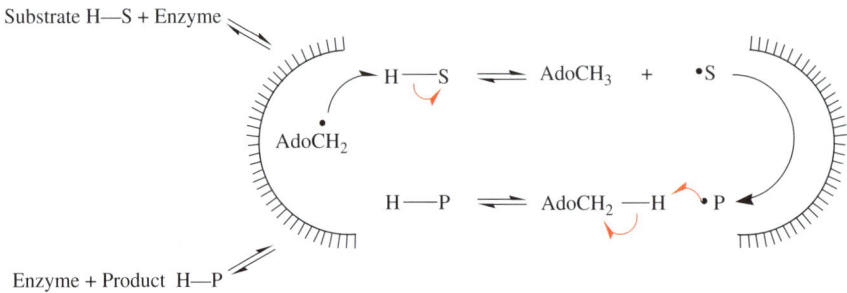

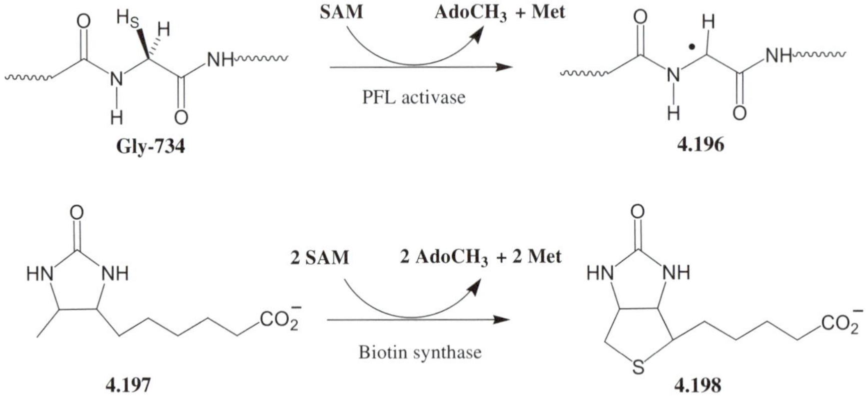

Scheme 4.129 PFL activase generates a main chain radical **4.196** at Gly 734 of pyruvate formate lyase; while biotin synthase removes hydrogen atoms from unactivated CH_3 and CH_2 groups of **4.197** in the biosynthesis of the vitamin biotin **4.198**. In neither case is a product radical available for recycling AdoCH₃.

in Scheme 4.129) is equally logical, but leaves no radical at the end of the process. This enzyme uses the unique capability of alkyl radicals to remove hydrogen from sp^3-hybridized carbon: illustrating graphically the extraordinary control an enzyme can exert over such a reactive centre. In this case two radical or potential radical centres are generated in close proximity: evidently these combine with each other to form the second, tetrahydrothiophen ring of the coenzyme. (The details of the actual sulfur insertion process are not known at the time of writing.)

4.5.2 Radicals in Enzyme-Active Sites

The radical centres generated in coenzyme structures are typically well protected, and approach to them well controlled. But reactions of radicals post-initiation, involving ordinary substrate and active-site groups, evidently need special attention. Control of reactivity, rather than reactivity itself, becomes the main problem to be solved. Because radicals are uncharged, their reactivity is not sensitive to the polarity of the environment, so proximity is the key: radicals must come into contact only with their proper target groups. We discuss two different enzyme types, which catalyze important metabolic and biosynthetic reactions.

4.5.2.1 Pyruvate-formate Lyase

Pyruvate-formate lyase, the enzyme (Scheme 4.129) with the main-chain radical centre at Gly 734, catalyzes the cleavage of pyruvate, to give acetyl-coenzyme A and formate (Scheme 4.130).

The enzyme handles the conversion as two consecutive reactions, the pyruvate-cleavage process and the transfer of the acetyl group produced from the resulting acetyl enzyme intermediate to coenzyme A. The first reaction (step 1 in Scheme 4.131) involves two consecutive hydrogen-atom transfers initiated by the Gly 734 radical, via the SH groups of two adjacent cysteines, Cys-418 and 419. The Cys-418 thiyl radical produced then adds to the neighbouring C=O group of bound pyruvate to form a tetrahedral addition intermediate radical **4.199**. This breaks down (Step 2 of Scheme 4.131) with the elimination of the intermediate CO$_2$ anion radical **4.200**: which completes the cycle by abstracting H from the main-chain Gly 734 (Scheme 4.131, Step 3).[278,282]

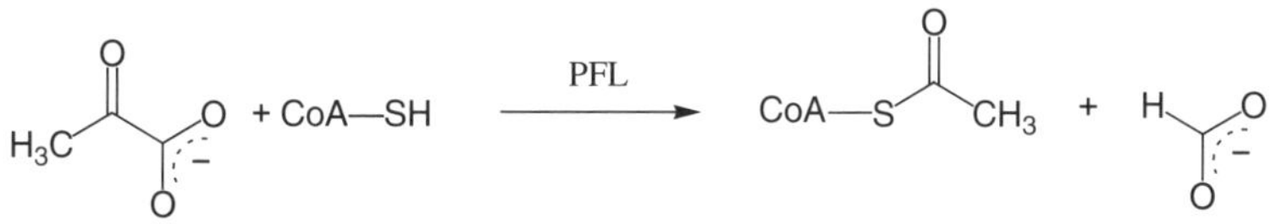

Scheme 4.130 The reaction catalyzed by pyruvate-formate lyase (PFL).

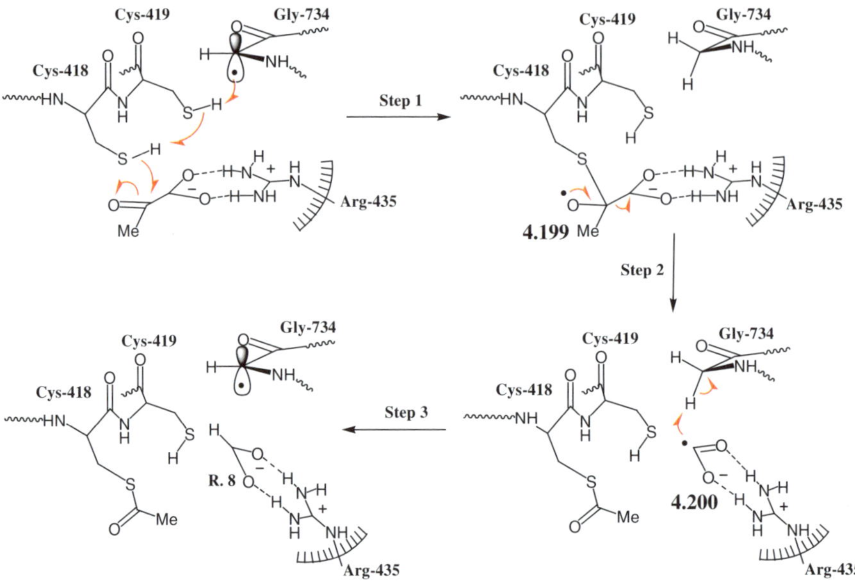

Scheme 4.131   Suggested outline mechanism for the first half-reaction catalyzed by pyruvate-formate lyase.[282] See the text. Step 1 summarizes what are probably three separate reactions initiated by the Gly 734 radical: which is regenerated, to complete the cycle, in Step 3.

4.5.2.2   Ribonucleotide Reductases

One of the most important groups of enzymes to use radical mechanisms of this sort comprises the various ribonucleotide reductases, which convert ribo- to deoxyribonucleotides (Scheme 4.132), and provide the only route for the biosynthesis of these essential building blocks for DNA.

The common first step of the conversion (Scheme 4.133) is the abstraction of a hydrogen atom from the 3'-position of a ribonucleotide di- or triphosphate by an active-site thiyl radical: the various classes of enzyme differ in the radical initiation step, and thus also in how the cycle is completed, but all use radical-transfer sequences to generate the active-site thiol.

Class I ribonucleotide reductases use a binuclear iron centre to generate a relatively stable tyrosyl radical. This initiates a series of hydrogen-atom transfers, similar to those in Scheme 4.131, to convert a cysteine in the distant

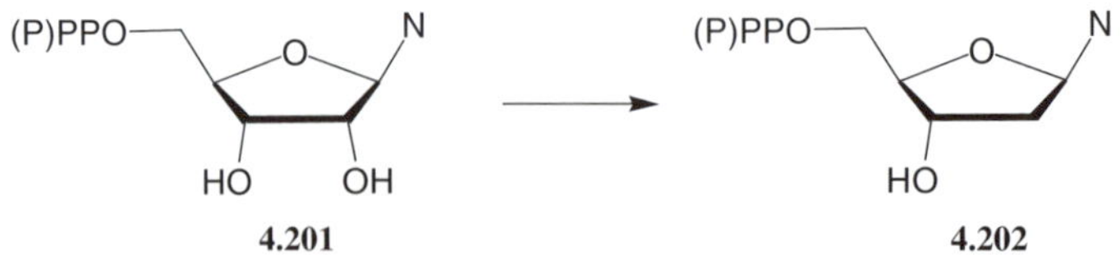

Scheme 4.132   The overall reaction catalyzed by ribonucleotide reductases.

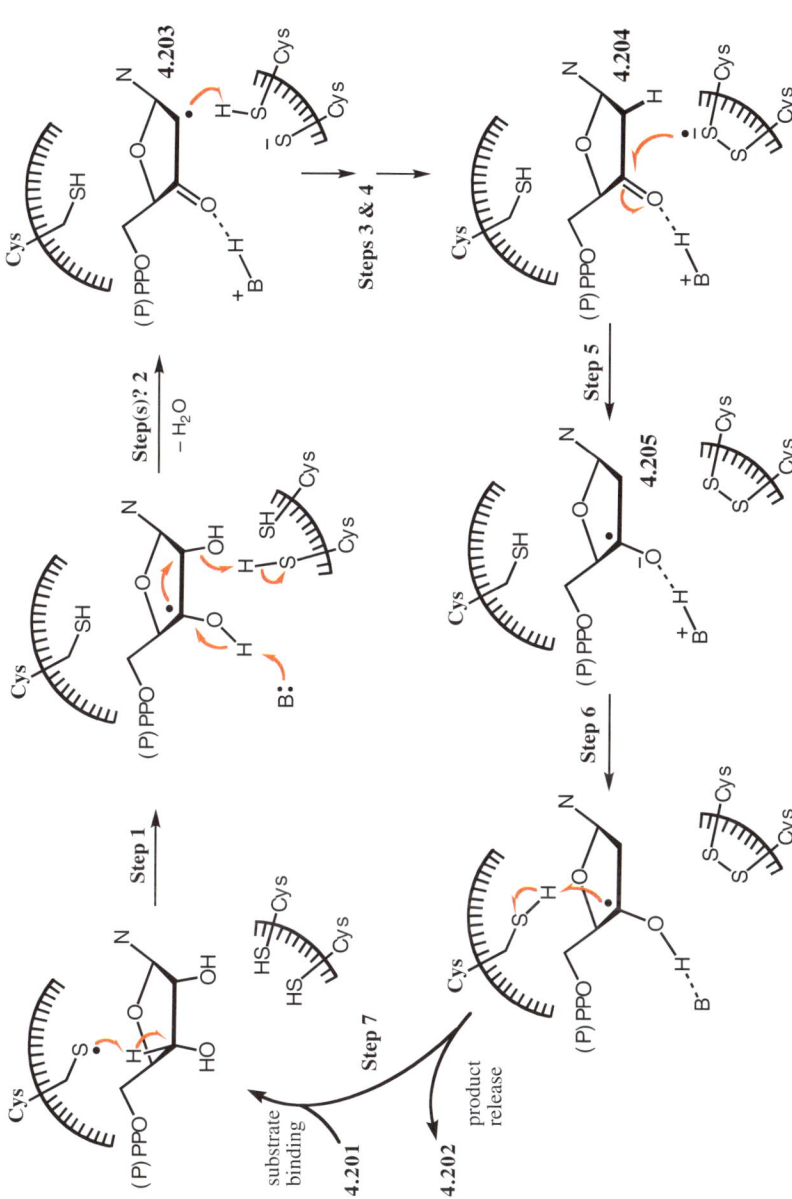

Scheme 4.133 The conversion of a ribo- to a deoxyribonucleoside involves a complex series of reactions, as suggested here for the Class I and II enzymes. (Scheme based on Licht and Stubbe.[284]) Three cysteines are involved in the sequence, one at the point of entry into the cycle, and a contiguous pair involved in the reduction process. N is one of the four nucleic acid bases.

active site to the active thiyl radical.[283] Class II enzymes generate the thiyl radical directly from adenosylcobalamin, no doubt via formation of a deoxyadenosyl radical. While the active forms of class III reductases (anaerobic enzymes) rely on a stable (but oxygen-sensitive) glycyl radical, in this case at the C-terminus, produced originally by an activase-catalyzed radical SAM reaction. Like the similar glycyl radical in PFL this continuously generates the thiyl radical required for the activation of the substrate.

The central reaction, the reductive dehydration of the ribonucleotide, starts with the abstraction of the C(3′) hydrogen atom (Step 1 of Scheme 4.133). The 3′-radical formed breaks down with loss of the 2′-hydroxyl group (Step 2, written here as a 2-electron process with general acid catalysis of the departure of OH as water). For further evidence concerning this process see Scheme 4.138.

This would produce a radical cation, which is neutralized by a general base, acting on the 3′-OH group, to give a ketone radical **4.203**. (Hint for Step 2: removing the 3′-OH proton gives the ketyl radical anion, which can be written in either form, $^-C{-}O^{\bullet}$ or $^{\bullet}C{-}O^-$.) Abstraction of H from a cysteine SH (Step 3) gives the ketone **4.204** and a thiyl radical, which is thought to interact with the neighbouring thiolate anion (Step 4) to give a disulfide radical anion, ready to reduce the ketone by a single-electron transfer (Step 5) to produce the ketyl **4.205**. The conversion is completed by a final H-transfer from the cysteine SH generated in Step 1, diffusion away of the product, and reduction of the disulfide (by external thioredoxin or related reagent). This regenerates the active form of the enzyme, allowing the cycle to begin again when a new molecule of ribonucleotide binds.

The (anaerobic) class III enzymes follow a similar sequence as far as the ketone **4.204** (steps 1–4), but use a single cysteine plus formate as the reducing agent rather than the disulfide system (Scheme 4.134). The key reduction step is a hydrogen-atom transfer from formate to an active-site Cys–S· radical, followed by a single-electron transfer to the $C{=}O$ group from the H-bonded CO_2 radical anion **4.200** thus formed (Step 6′). Structure **4.200** figured as an intermediate in the first half-reaction catalyzed by pyruvate-formate lyase.

Although the three classes of the ribonucleotide reductases are not evolutionarily conserved in terms of primary sequence, quaternary structure or cofactor preference, the close similarities in the chemistry and the active-site arrangements, as summarized in Schemes 4.133 and 4.134 clearly suggest a common evolutionary origin for these important enzymes.[286–288]

4.5.3 Models

The radicals involved as intermediates in the reactions described in Schemes 4.131, 4.133 and 4.134 are short lived, present in very low concentration and generally show no convenient chromophore. The first direct evidence for a protein-based, cysteinyl radical at the active site of any enzyme came from EPR identification of a coupled $Co(II) \cdots S{-}C408$ radical pair in a class II ribonucleotide reductase active site.[289] But detection, usually by EPR (the same as

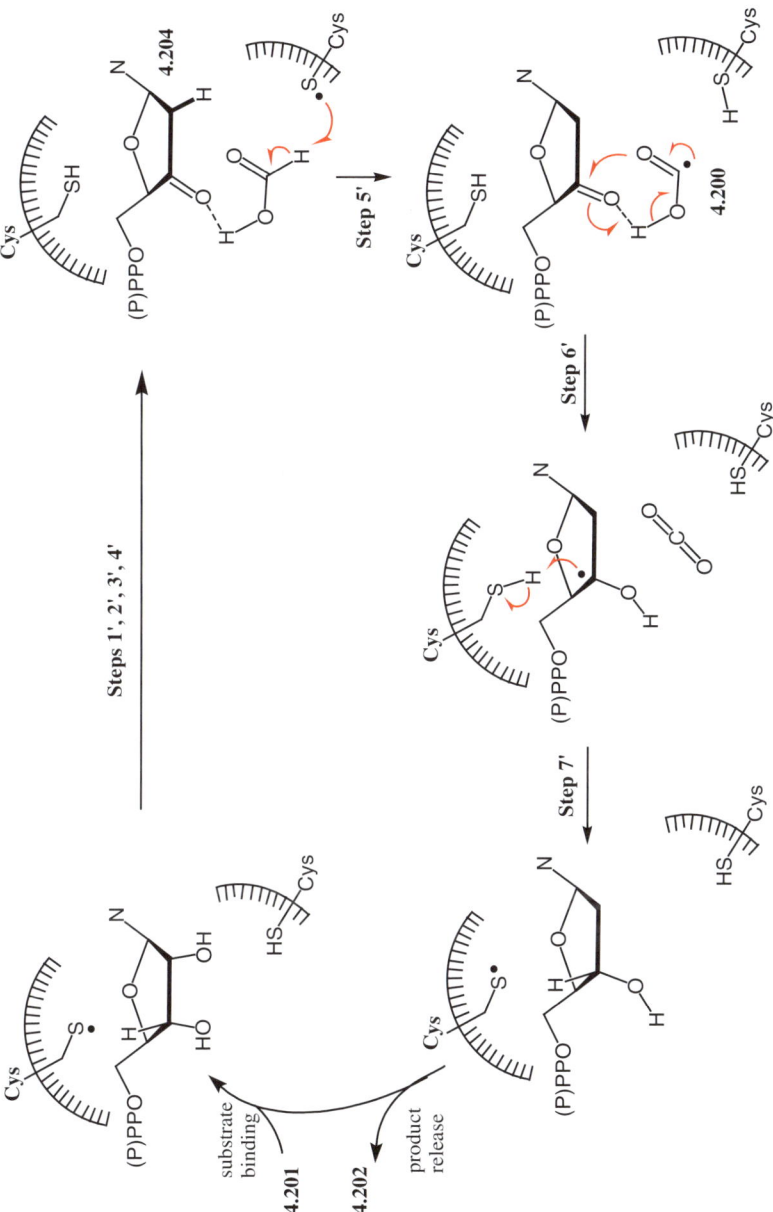

Scheme 4.134 Proposed mechanism for the reduction of the ketone **4.204** by class III ribonucleotide reductases. Steps 1'–4' in the formation of **4.204** from the ribonucleotide are closely similar to those (Scheme 4.133) in the reactions catalyzed by class I and II enzymes (with formate probably acting as the general base in Step 2).[285]

Table 4.4   Selected R–H bond dissociation energies (BDE).

Bond-dissociation reaction	BDE $(kJ\,mol^{-1})$
EtO–H → EtO$^{\bullet}$	435
RCH$_2$–H → RCH$_2$$^{\bullet}$	410
RS–H → RS$^{\bullet}$	380
ROCH$_2$–H → ROCH$_2$$^{\bullet}$ ArO–H → ArO$^{\bullet}$	370
RCOCH2–H → RCOCH$_2$$^{\bullet}$ ArCH$_2$–H → ArCH$_2$	360

ESR: electron paramagnetic (or spin) resonance) and unambiguous assignment are not always possible. So the detailed chemistry of the various species involved is often most conveniently investigated either computationally, or experimentally by the use of simple models: based always on the best available X-ray structural information concerning the relevant enzymes. This approach has been notably productive and informative in the study of mutase and isomerase reactions initiated by the B$_{12}$ coenzyme system,[276] but there is relatively little such work available on the more recent chemistry involving active protein radicals.

Relative reactivities of radicals R$^{\bullet}$ in hydrogen-atom transfer reactions can be predicted in broad terms by comparing R–H bond dissociation energies (Table 4.4). Reactions that are exothermic or close to it are expected to go at rates close to diffusion controlled, and certainly fast enough to account for the observed overall rates of most of the enzymes concerned. Note that the thiyl radical RS· is well placed to remove hydrogen, if necessary reversibly, from C–H bonds activated by adjacent electron donors or acceptors: but not from an unactivated methyl group as in Ado–CH$_3$.

4.5.3.1   Initiation Stages

Heating adenosylcobalamin (10^{-4} M) under anaerobic conditions in water at pH 7 (phosphate buffer) in the absence of any enzyme gives the cyclonucleoside **4.206**, evidently produced via intramolecular attack of the **AdoCH$_2$**$^{\bullet}$ radical on the 8-position of adenine (Scheme 4.135, reaction **A**). This must be a relatively inefficient process, because under the same conditions in the presence of an excess of 2-mercaptoethanol the almost quantitative products are cobalamin, 5'-deoxyadenosine (**AdoCH$_3$**) and the disulfide **4.207**, derived from a competing hydrogen-atom transfer (reaction **B**).[290] Cystine (CysS–SCys) is formed in the same way, by thiyl radical recombination, when cysteine is present. The reaction is zero-order in thiol, as expected if the original homolytic cleavage of adenosylcobalamin (some 10^{10} times slower than in the enzyme-catalyzed reaction) is rate determining.

4.5.3.2   Hydrogen-Atom Transfers

Hydrogen-atom transfers are well understood at the simple level, and can be expected to be exothermic if the new H–X bond is stronger (Table 4.4). More

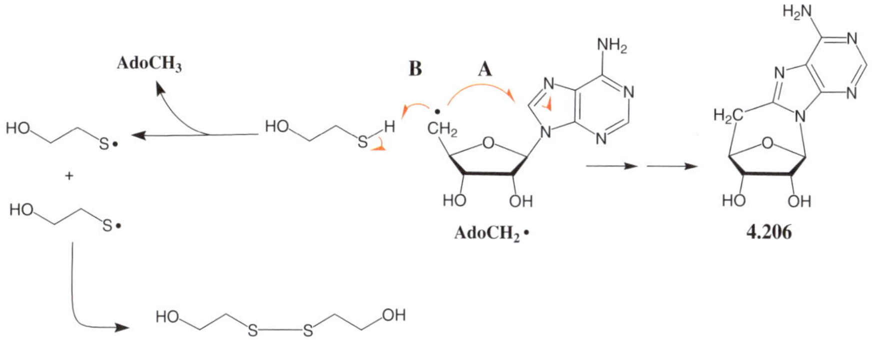

Scheme 4.135 In the presence of a good hydrogen-atom donor RSH the 5′-adenosyl radical is rapidly converted to **AdoCH₃·** and the RS· radicals formed recombine to give the disulfide (*e.g.* **4.207**, Reaction **B**). In the absence of such a donor **AdoCH₂·** cyclizes to **4.206** (Reaction **A** goes via a tetrahedral addition intermediate that requires the services of another radical to remove the hydrogen atom from the adenine C(8).)

complex processes, like the series of hydrogen-atom transfers that ferry the radical centre from the initiating tyrosyl in the class I ribonucleotide reductase mechanism (Scheme 4.134) to a specific active-site thiol are beyond the scope of this discussion. The complete control of the PCET process achieved by the enzymes (Section 4.4.3) is remarkable, seeing that the two functionalities are in two different proteins, and some 40 Å apart.

The generation of the main-chain glycyl radicals in the activation of PFL and the class III ribonucleotide reductases described above involve hydrogen-atom transfer to AdoCH₂· produced via the radical SAM reaction. A simple intramolecular model (Scheme 4.136)[291] illustrates the chemical feasibility of this reaction. Deuterium atom transfer to the primary alkyl radical **4.208**, generated conventionally in solution, to give product **4.210**, competes with the intramolecular 1,5-hydrogen-atom transfer to give the more stable glycyl radical **4.209**. Deuterium atom transfer to the glycyl radical so formed gives the second, isomeric product **4.211**.

The removal of the 3′-H of the nucleotide (Step 1 of Scheme 4.133) is a type of reaction familiar in small-molecule radical chemistry. The new radical is stabilized by the adjacent OH group and the reaction likely to be at least close to exothermic, and certainly so if the electron-donor properties of the OH group are enhanced by full or partial removal of the proton by an active-site general base.[292] A simple intramolecular model (Scheme 4.137) shows that the basic reaction is possible without the assistance of a general base.

The presence of the thiyl radical **4.213**, generated by standard radical chemistry in the presence of the good H-atom donor Bu₃SnH, is confirmed by the formation of **4.214**. Under high dilution conditions intramolecular H-abstraction to give **4.215** competes, giving up to 21% of **4.216** by elimination of the stable phenylsulfinyl radical.

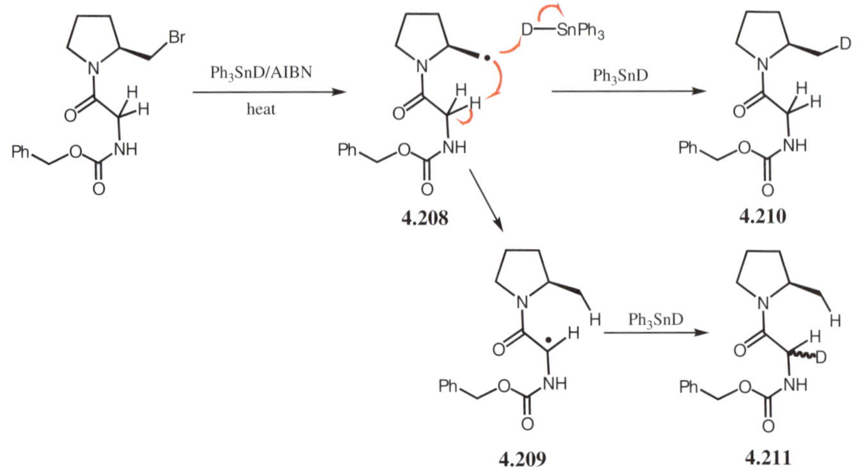

Scheme 4.136 Simple model for the generation of a glycyl radical, by 1,5-hydrogen transfer to a primary alkyl radical.

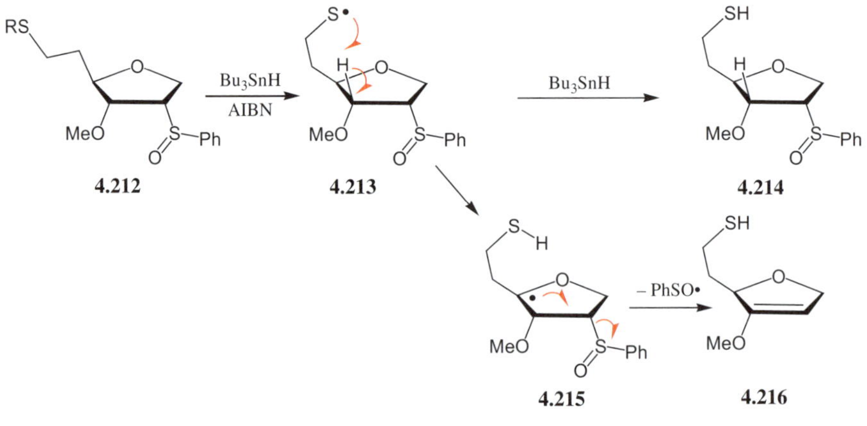

Scheme 4.137 Simple intramolecular model for Step 1 of Scheme 4.133. Thiyl radical **4.213**, generated from **4.212** (R = PhOCS.O(CH$_2$)$_2$) undergoes the reaction of interest, to form **4.216** via the 3′-radical **4.215**, in competition with direct H-transfer to give **4.214**.[293]

The loss of the 2′-OH group (step 2 of Scheme 4.133) is less straightforward. Direct elimination of HO · is at least formally possible, as shown by the rapid elimination of the phenylsulfinyl radical from **4.215** in Scheme 4.137. However, the hydroxyl radical is a particularly high-energy species, and the currently favoured mechanism, shown in Scheme 4.133, involves loss of the 2′-OH group as water. This is consistent with the results of model systems involving simple diols:[284] it has been shown, for example, that the dehydration of the 1,2-dihydroxyethyl radical is acid catalyzed. The most specific model involves the selective generation of the 3'-adenosyl radical **4.218** by photolysis of the phenylselenyl ester **4.217** (Scheme 4.138).[294]

Irradiation of **4.217** in the presence of a good hydrogen-atom donor gave a mixture of xylo- and ribonucleosides **4.220**, products of the direct reduction of the 3'-adenosyl radical (Reaction **A** in Scheme 4.138). The competing elimination of the 2'-OH (Reaction **B**) to give the ketone radical **4.219** is general base catalyzed, and predominates in buffered aqueous solution. Hydroxide is generally a poor leaving group (acetate is eliminated faster in a related system), and is likely to also need some assistance from a general acid, as it is believed to do in the enzymes (Scheme 4.133, above). The final product from this pathway is the ketone **4.222**, produced by rapid elimination of adenine from the enolate of ketone **4.221**.

One of the less-familiar steps in the mechanism suggested (Scheme 4.133) to account for the reactions of the class I and II ribonuclease reductases is the formation of a cystine disulfide radical anion, formed (Step 4) from a thiyl radical and a thiolate anion, and supposed to reduce the carbonyl group of the ketone **4.204** (Scheme 4.133) to the ketyl by a single-electron transfer. This is a known process, nicely exemplified by the reaction of dithioerythritol **4.223**

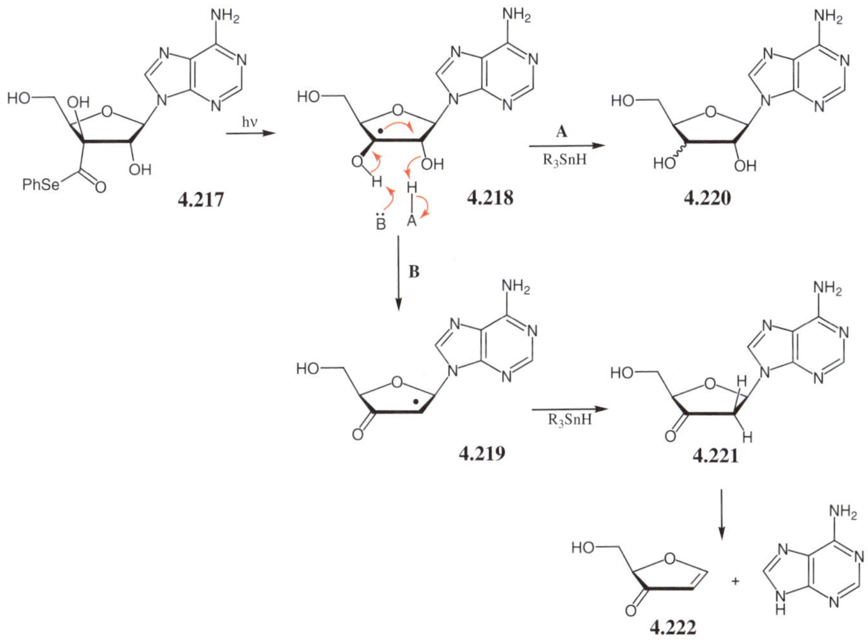

Scheme 4.138   The selective formation of the 3'-adenosyl radical **4.218** is achieved by photolysis of the phenylselenyl ester **4.217**. Radicals formed are trapped by R₃SnH, to give **4.220** and ketone **4.221** (which rapidly breaks down under the conditions to give adenine and the unsaturated ketone **4.222**).[294]

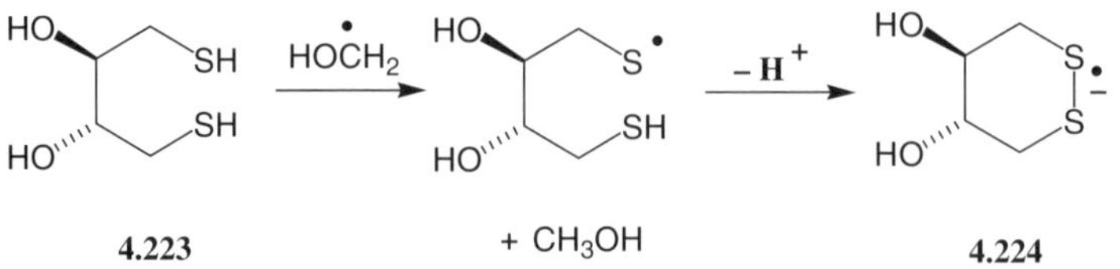

4.223   + CH$_3$OH   **4.224**

Scheme 4.139   Model reaction for Steps 3 and 4 of Scheme 4.133 (above), observed in water at pH 7.5. The initial product radical cyclizes spontaneously, with the loss of a proton.

(Scheme 4.139) with α-hydroxyalkyl radicals:[295,296] this gives the radical anion **4.224**, thus modeling Steps 3 and 4 of Scheme 4.133.

The single-electron transfer from the disulfide radical anion to the C=O group (Step 5 of Scheme 4.133) is a well-precedented type of reaction, and expected to be rapid; though likely to need assistance from a general acid (*i.e.* steps 5 and 6 are likely to be concerted).[284] And the final H-abstraction from Cys-SH by the secondary alcohol radical (Step 7) to regenerate the initial active-site configuration is expected to be rapid in solution.

The new reaction involved in the mechanism of action of the class III ribonucleotide reductases is the use of formate as the reducing agent for the reduction of the intermediate ketone **4.204** (Scheme 4.134). The evidence that the formate is stoichiometrically oxidized to CO$_2$, and that the isotope from [^{3}H]formate appears in water, is consistent with the suggested H-atom transfer to the thiyl radical (Step 5 of Scheme 4.134), followed by a single-electron transfer from the CO$_2$ radical anion produced.[297] The mechanism of the reaction is unknown (though the participation of the iron–sulfur complex centre during electron transfer from formate to substrate must be considered), and clear-cut evidence from a model system would be welcome.

4.6   Pericyclic Reactions

Pericyclic reactions are rare in biological chemistry, catalyzed by only a handful of properly characterized enzymes. The best-known example is chorismate mutase, which catalyzes a 3,3-sigmatropic rearrangement on the main biosynthetic route to the essential aromatic amino-acids. Stereochemically reliable C–C bond forming reactions are invaluable in synthesis, making them popular targets for "artificial enzymes" designed to catalyze key steps in the synthesis of complex homochiral systems. The Diels–Alder is the highest profile such reaction, and natural Diels–Alderases have been tentatively identified: to universal interest, but not universal acceptance, mostly because stepwise mechanisms have not been conclusively ruled out.[298,299]

These reactions are also of special mechanistic interest in the context of enzyme catalysis. They were at one point dubbed no-mechanism reactions[300] because they "show no response to free-radical initiators or inhibitors, and are relatively insensitive to acid–base catalysis, structural changes and solvation effects," and no intermediates are involved: which seems to rule out most of the catalytic methodology available to enzymes! We now understand the orbital-symmetry considerations that control reactivity in pericyclic processes, and can design receptors for computationally well-defined transition states. So this has become an area unique in enzyme model chemistry, where the models outnumber the (known) enzymes. We will discuss the chorismate mutase reaction, and some important models for the Diels–Alder reaction.

4.6.1 Chorismate Mutase

Chorismate mutase catalyzes the conversion of chorismate to its isomer prephenate by a Claisen rearrangement, as shown in Scheme 4.140. The mechanism is well established as the 3,3-sigmatropic rearrangement **4.225**. Heavy-atom isotope effects are consistent with a concerted but asynchronous transition state **4.226**, in which C–O bond breaking runs ahead of C–C bond formation. Unlike practically every example we have discussed so far the enzyme-catalyzed reaction is fast enough to be followed in the absence of the enzyme, so that direct comparisons are possible between the enzymic and uncatalyzed reactions.

Closely related mechanistically is isochorismate pyruvate lyase, which catalyzes the cleavage of isochorismate by a [1,5]-sigmatropic pathway (Scheme 4.141, Reaction **A**). This enzyme also shows significant (though lower by 2 orders of magnitude) chorismate mutase activity.[301]

There is general agreement that these are genuine pericyclic reactions,[302] concerted but asynchronous as described above, with transition states not very different in degree of bond making and breaking in the presence or absence of the enzyme.[303] So the origin of the enzymic acceleration is of much interest. Since the reactions concerned can be followed in the absence of enzyme it is evident that large accelerations are not involved: in terms of k_{cat}/k_{uncat} (for this first-order reaction) these are of the order of 10^6.

Explanations fall into two main classes. Clearly the enzyme must bind the substrate in a productive conformation. The required pseudo-diaxial conformation (**4.225** in Scheme 4.140) is not the ground-state conformation of chorismate, but it does account for 10–20% of the conformational distribution in aqueous solution.[304] So, selective binding can only be a minor factor: unless

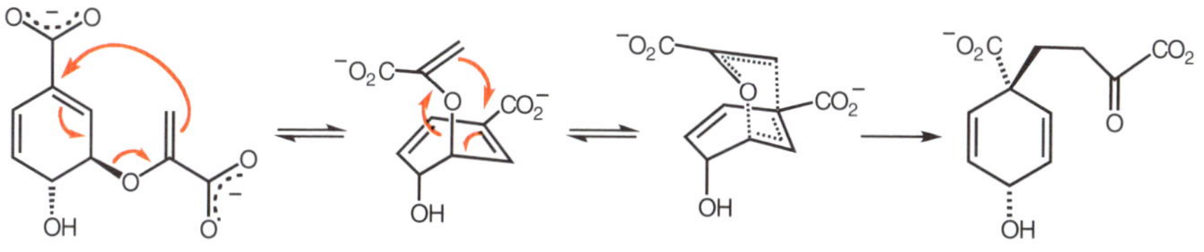

| Chorismate | **4.225** | **4.226** | Prephenate |

Scheme 4.140 The reaction catalyzed by the chorismate mutases.

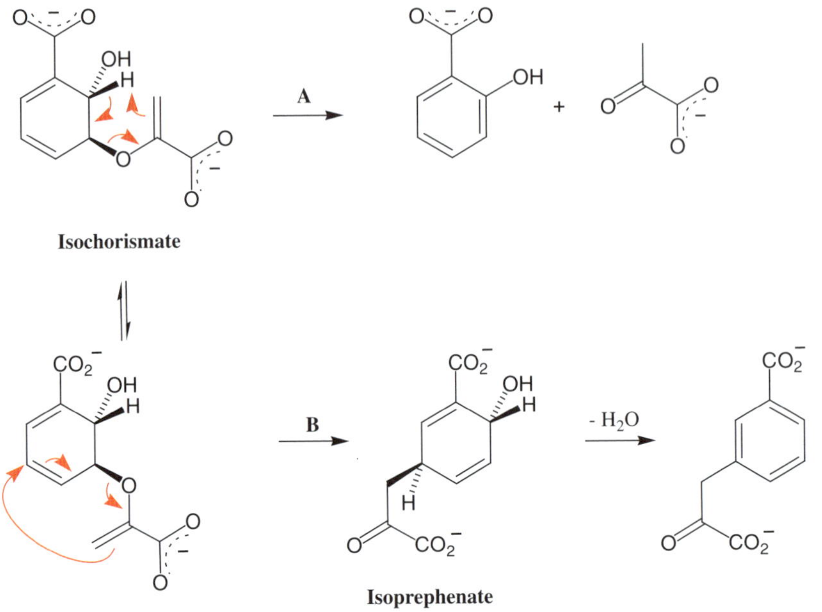

Isochorismate

Isoprephenate

Scheme 4.141 Isochorismate pyruvate lyase catalyzes the 1,5-sigmatropic rearrangement of isochorismate to salicylate and pyruvate (Reaction **A**). In
the absence of the enzyme the 3,3-sigmatropic rearrangement to isoprephenate (Reaction **B**) is some 8 times faster.

binding exerts some degree of steric strain, for example to bring the reacting
sp^2-C centres into closer than van der Waals contact. Strain of this sort is never
likely to be a large factor because proteins are (necessarily) flexible hosts.
Bearing in mind the important general caveat (Section 1.5) that no single factor
is likely to be uniquely responsible for the acceleration observed in any particular case, the most important seems to be modest stabilization of the transition
state by electrostatic effects (including H-bonding) in the active site.[305,306]
Calculations generally agree that C–O bond breaking runs ahead of C–C bond
formation; and the Claisen rearrangement is typically several hundred times
faster in water than in an apolar medium, even though secondary isotope effects
detect little difference in transition-state character.[307] It seems eminently reasonable that reactions that are faster still in chorismate mutases should benefit
from the highly polar active-site environments (*e.g.* Scheme 4.142) that these
enzymes have evolved.

4.6.1.1 Models: Catalysis by Antibodies

There is little need for models for the *reactions* catalyzed by the chorismate
mutases and related enzymes, other than to study structural or medium effects,
simply because these unimolecular processes can themselves be studied in detail

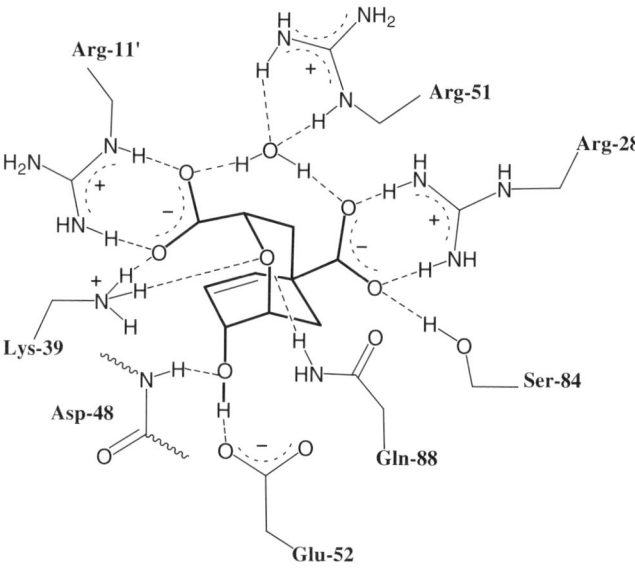

Scheme 4.142 Binding at the active site of *E. Coli* chorismate mutase.[308] The bound molecule (in bold) is a transition-state analogue (see **4.226** in Scheme 4.140), which acts as a competitive inhibitor for the (dimeric) enzyme. (Arg-11′ is from the second monomer chain.)

in the absence of enzyme. It is, however, of interest to design host-catalysts for pericyclic reactions such as Claisen rearrangements and Diels–Alder reactions, and some successes with catalytic antibodies[vi] have proved particularly instructive.

A handful of antibodies that catalyze the chorismate mutase reaction have been raised against haptens designed as transition-state analogues. The bound molecule shown in bold in Scheme 4.142 is a typical hapten, known to inhibit the natural enzyme and thus presumed – and in this case shown, by a crystal structure – to be complementary to the active site. Antibodies raised against this molecule will bind it well, perhaps in different conformations: and since binding the substrate in the correct conformation for reaction is the one catalytic function certainly involved in the enzyme reaction, some may be expected to stabilize the transition state selectively, and thus catalyze the isomerization of chorismate to prephenate. Antibodies 1F7 and 11F1-2E11 catalyze the rearrangement moderately well, only a few hundred times more slowly (Table 4.5) than the enzyme which they can match in either enthalpy or entropy of activation: but not in both. More detailed interpretation would be speculation: thermodynamic data are always difficult to interpret, especially when a protein is involved in the reaction: and in this context the antibodies' prime function is not catalysis but to bind the hapten.[309]

[vi] Catalytic antibodies are discussed in more detail in Section 5.2.

Table 4.5 Kinetic and thermodynamic data compared for the rearrangement of chorismate to prephenate catalyzed by antibodies 1F7 and 11F1-2E11 and *E. coli* chorismate mutase.[310]

Catalyst	k_{cat} (s⁻¹)	K_M (μM)	k_{cat}/k_{uncat}	$(k_{cat}/k_{uncat})/K_M$	ΔG_{TS} (kcal/mol)	$\Delta G^{\ddagger}$ (kcal/mol)	$\Delta H^{\ddagger}$ (kcal/mol)	$T\Delta S^{\ddagger}$ (kcal/mol, 298 K)
Antibody 1F7	1.2×10^{-3}	51	250	4.9×10^{6}	9.13	21.3	15.1	-6.55
Antibody 11F1-2E11	0.045	260	1×10^{4}	3.4×10^{7}	10.3	18.7	18.3	-0.34
Chorismate mutase	13.5	290	3×10^{6}	1.0×10^{10}	13.6	15.9	15.9	0
Uncatalyzed						24.2	20.5	-3.85

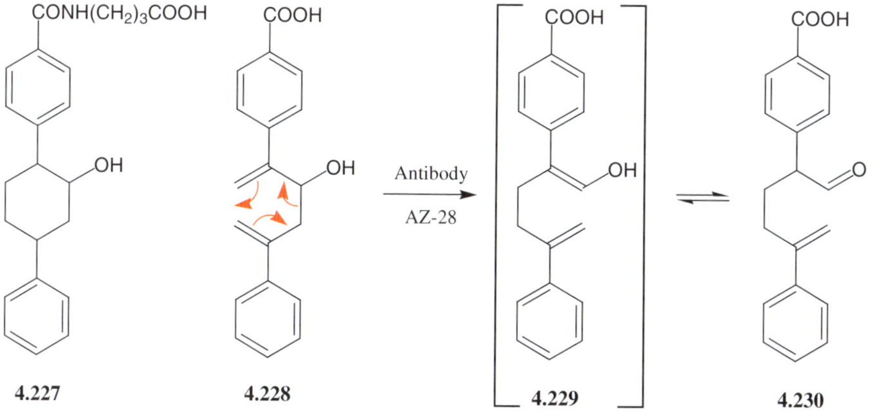

Scheme 4.143 Antibody AZ-28, raised against hapten **4.227**, catalyzes the oxa-Cope rearrangement of **4.228** to **4.230** (via the enol **4.229**). Product inhibition of the antibody was prevented by *in-situ* conversion to the oxime.

One other catalytic antibody is of interest here because it catalyzes a non-natural pericyclic reaction. Antibody AZ28 was raised against hapten **4.227** (Scheme 4.143), designed to mimic the cyclic (chair-like) transition state (see **4.226** in Scheme 4.140) for the [3,3]-sigmatropic rearrangement of the diene **4.228**.[311] Catalytic efficiency is comparable with similar antibodies, with $k_{cat}/k_{uncat} = 5300$. Interestingly the germ-line precursor antibody of AZ-28 is over 30 times more efficient ($k_{cat}/k_{uncat} = 1.6 \times 10^5$), even though its affinity toward the hapten is much lower.[311] (One more piece of evidence for the important generalization (Section 1.5) that tight substrate binding can be a catalytic disadvantage.)

It is clear from these, and many similar results, that antibody catalysts do not come close in efficiency to enzymes catalyzing similar reactions of natural substrates. Antibody 11F1-2E11 (Table 4.5) comes closer than most, primarily because chorismate mutase is one of the least efficient enzymes. Houk *et al.*[16] found that transition states for antibody-catalyzed reactions are typically bound some 10^3 times more strongly than the substrates. This compares with ratios of 10^{18} or more for efficient enzymes (Section 2.4).

4.6.2 Antibodies Catalyzing the Diels–Alder Reaction

A generic problem for antibody catalysts raised against transition-state analogues is that reaction products are always likely to be structurally closely similar to the haptens used in immunization. This makes product inhibition a major factor in the low catalytic efficiency of hydrolytic antibodies,[311] and it is of special relevance for antibodies designed to catalyze pericyclic reactions. A transition-state analogue for an isomerization, like **4.226** for the Claisen rearrangement (Scheme 4.140), necessarily has the same basic structure as the product, so it will often be bound at least as well.

Product inhibition is a particular problem for catalysis of cycloaddditions, like the Diels–Alder reaction, where the transition state (*e.g.* **4.232** in Scheme 4.144) differs from the product only in fine geometrical detail, in this case principally in the lengths of the C–C bonds being formed: so that a transition-state analogue will inevitably be close in structure to the product also.

A neat, though not general, solution to the problem is to arrange that the product as formed is short lived. Thus, the most successful antibody catalyst for the Diels–Alder reaction is Hilvert's 1E9, raised against a hapten **4.234** (Scheme 4.145) which models the Diels–Alder adduct **4.236**.[312] **4.236** rapidly loses SO_2 to give (after rapid aerial oxidation) a planar final product **4.237**, very different in geometry to the hapten, and thus rapidly released from the catalytic site, opening the way to multiple turnovers. The effective molarity for the reaction of the bound reactants was estimated as > 100 M.

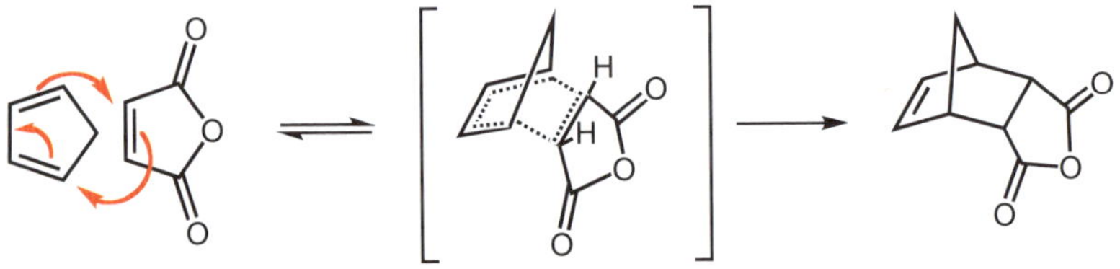

| **4.231** | **4.232** | **4.233** |

Scheme 4.144 The transition state for the Diels–Alder reaction has very specific, product-like geometry, imposed by the orbital interactions involved in bond formation.

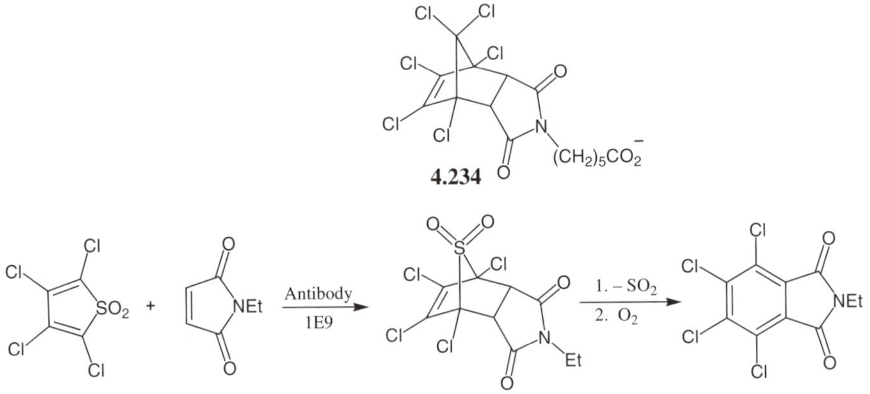

4.234

| **4.235** | **4.236** | **4.237** |

Scheme 4.145 The Diels–Alder reaction of tetrachlorothiophen (**4.235**) and N-ethylpyrrole gives the unstable adduct **4.236**. This rapidly loses SO_2 to give a dihydrobenzene derivative, which is converted under the conditions to the planar **4.237**.

4.6.2.1 *Supramolecular Catalysis of the Diels–Alder Reaction*

In antibody catalysis the design principles are expressed exclusively in the structure of the hapten, with implementation left to the immune system and appropriate selection. Recent progress in our understanding of the principles of molecular recognition has reached the point where it is possible to design and synthesize hosts capable of binding specific guests at least as strongly as enzymes bind substrates, and in favourable cases in specific orientations. But the apparently simple next step, of introducing functional groups in the correct position for catalysis, has not so far produced notably efficient enzyme mimics. However, pericyclic reactions typically need no catalysis from functional groups: binding the substrate in the correct conformation appears to be the basis of catalysis by chorismate mutase: and favourable binding to a host with a cavity big enough to bring together diene and dienophile might be expected to overcome the substantial negative entropy of activation involved in bimolecular cycloadditions like the Diels–Alder reaction.

This expectation was encouraged by early results with Mock's cucurbituril host, still one of the most successful examples of a host–guest catalyst for a pericyclic reaction. Cucurbituril (Scheme 4.146: named for its resemblance to a pumpkin (species *cucurbita*)) is a synthetic cage compound that binds ammonium cations RNH_3^+ particularly strongly, through hydrogen bonds to the six carbonyl dipoles symmetrically arrayed around each face.[313] Appropriate R groups are bound within the cavity, and the rigidity of the host structure makes binding particularly selective. As usual, the release of solvent molecules provides favourable entropic and hydrophobic contributions to binding.

A catalytic amount of cucurbituril accelerates the 1,3-dipolar cycloaddition of the alkyl azide and alkyne shown in Scheme 4.146. Reaction follows Michaelis–Menten kinetics, with rate-determining product release; and is regiospecific, giving **4.238** as the sole product. The regiospecificity is simply

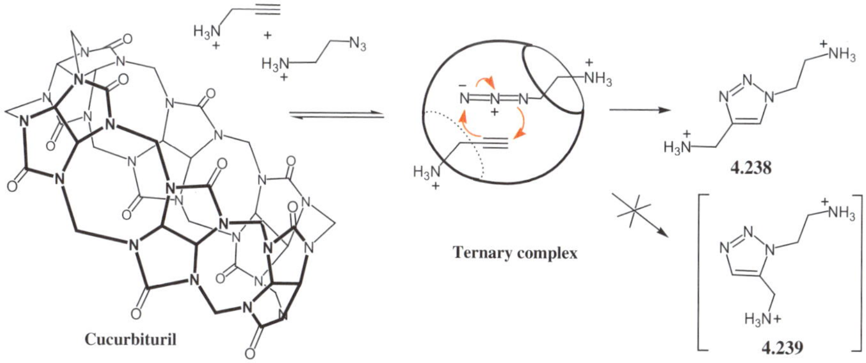

Scheme 4.146   In the presence of catalytic amounts of cucurbituril the 1,3-dipolar cycloaddition of the alkyne and alkyl azide shown gives the single product **4.238** (see the text),[314] whereas in solution (in 88% aqueous formic acid) both isomers, **4.238** and **4.239**, are formed.

explained if the ammonium groups are bound on opposite faces of the host with the substituents extending into the interior in the ternary complex (Scheme 4.146), thus aligning the reacting groups to favour the formation of **4.238**.[314] Supramolecular catalysis is remarkably efficient, with a half-time for the reaction of the bound reactants of just over half a minute at 40 °C, corresponding to an EM estimated at 1.6×10^4 M.[315]

This is still the most efficient supramolecular catalysis observed for a pericyclic reaction: or indeed for any bimolecular reaction. For efficient catalysis it is not enough simply to bring reactants together in the same cavity: the bound reactants must be bound in, or have easy access to, a productive – *i.e.* transition-state-like – conformation, and transition-state binding must be stronger still. Strong non-covalent binding requires shape complementarity between host and guest, and this is likely to be possible only for a specific arrangement of two guest molecules in a given cavity. This arrangement may or may not be productive, whereas an encounter complex of two molecules in solution has the flexibility to sample different approach geometries. So, not surprisingly, some hosts tested as catalysts for Diels–Alder reactions bind the reactants more strongly than the transition state.[316] Diels–Alder reactions in solution have strongly negative entropies of activation, reflecting the very specific geometries imposed by the orbital interactions involved in bond formation (Scheme 4.144). While this is a measure of the potential advantage of bringing the reactants together in exactly the right relative configuration, it is equally an indication that the probability of it happening is not high. It is no coincidence that the exceptionally efficient catalysis found for Mock's cucurbituril involves a cycloaddition of two linear reactants, which need only to be more or less parallel for reaction.[315]

Supramolecular catalysis of the Diels–Alder reaction has been expertly reviewed, in the context of two-substrate reactions,[315] and only a few high-profile examples will be discussed here. Cyclodextrins are poor catalysts, for several reasons. They do not have the capacity to accommodate two large reactants, and even if they do bind donor and acceptor **D + A** in preference to **D, A, D + D** or **A + A**, the favourable effect of bringing them together is offset by the less favourable hydrophobic environment of the cavity compared with water (cyclodextrins are conveniently soluble only in aqueous media). β-cyclodextrin does catalyze the cycloaddition of diethyl fumarate and cyclopentadiene with Michaelis–Menten kinetics (Scheme 4.147),[317] with apparent K_Ms, in the milli-molar range, normal for small molecules interacting with a cyclodextrin cavity,

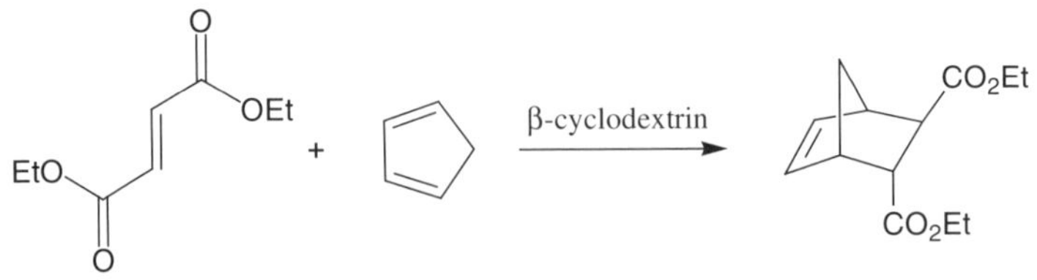

Scheme 4.147   The Diels–Alder reaction of cyclopentadiene with diethyl fumarate in aqueous solution is catalyzed by β-cyclodextrin.

and k_{cat}/k_{uncat} about 100.[316,317] It also catalyzes the reaction with acrylonitrile ($H_2C=CH.CN$), but it *inhibits* the reaction with ethyl acrylate ($H_2C=CH.CO_2Et$);[40] no doubt because of different binding preferences.

The construction of larger cavities leaves more room for binding – including of course unproductive binding. Sanders' solution to this problem was to organize the binding of the reactants, using substrates that are also ligands for metal-ion centres in the host. Free in solution the reaction between diene **4.240** and maleimide **4.241** is reversible, giving initially the kinetically favoured *endo*-adduct, seen only in the early stages of the reaction: the thermodynamically favoured *exo*-adduct (Scheme 4.148) is the only product at equilibrium.[318]

In the presence of stoichiometric amounts of the porphyrin trimer **4.242** (Scheme 4.149) the formation of the *exo*-adduct is catalyzed specifically, while that of the *endo*-isomer is inhibited. The mechanism of this reaction has been studied in some detail, and the results are instructive.

Unlike most examples we have discussed, the reaction is carried out in an organic solvent ($C_2H_2Cl_4$), so that solubility problems for organic substrates are reduced (and hydrophobic effects on binding eliminated). In the absence of the host the reaction (Scheme 4.148) between 9 mM **4.240** and **4.241** is preparatively useful, with a half-time of about an hour at 60 °C for 1 M reactants. Using 0.9 mM substrates the rate is over 10^6 times slower, and equilibrium favours starting materials: but in the presence of one equivalent of porphyrin trimer **4.242** a 65% yield of *exo*-adduct is obtained after two days, corresponding to a 200-fold acceleration. This substantial if not spectacular catalysis results from more efficient binding, by 13 kJ (3.1 kcal) mol^{-1}, of the transition

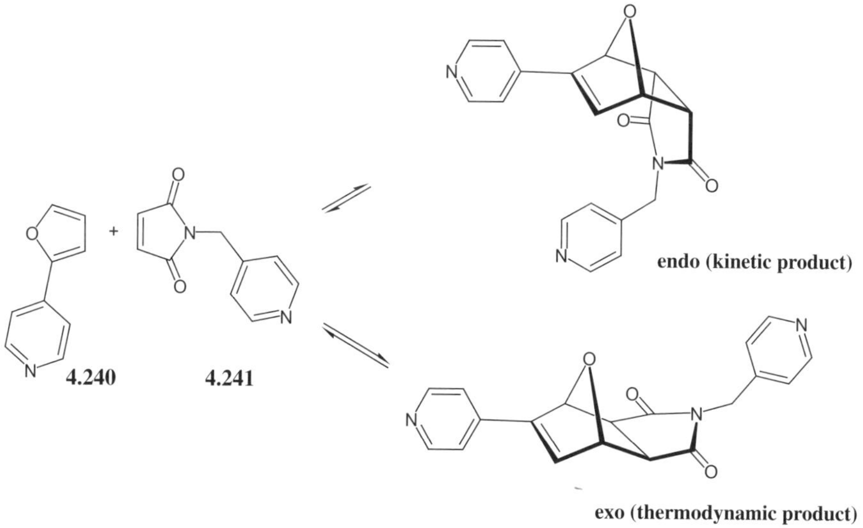

endo (kinetic product)

exo (thermodynamic product)

4.240   **4.241**

Scheme 4.148 Kinetically and thermodynamically favoured products of the Diels–Alder reaction between **4.240** and **4.241**. The thermodynamically favoured *exo*-adduct is formed in 50% yield at equilibrium under preparative conditions (at 60 °C in $C_2H_2Cl_4$).[319]

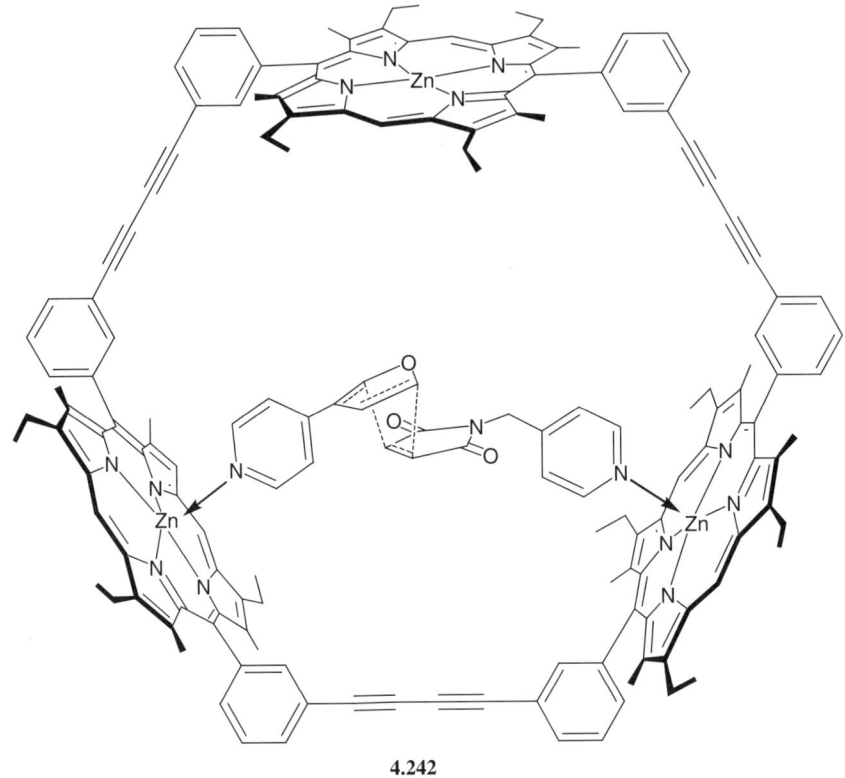

4.242

Scheme 4.149   Proposed transition state for the Diels–Alder reaction between **4.240**
and **4.241** in C$_2$H$_2$Cl$_4$ catalyzed by porphyrin trimer **4.242**, which gives
only the *exo*-adduct (Scheme 4.148).[319]

state leading to the *exo*-adduct (Scheme 4.149), corresponding to an EM of at
least 4 M.

Extensive work on related large synthetic systems had little success in the
search for more efficient catalysis, and has led to the exploration of self-
assembly as a way of generating combinatorial libraries of catalysts complementary to transition-state analogues (TSAs, Section 3.5) for reactions of
interest. Dynamic combinatorial chemistry generates mixtures of compounds
from a set of small building blocks, each equipped with two (or more) functional groups capable of the reversible formation of covalent bonds. When
thermodynamic equilibrium is set up in the presence of a fitting template
molecule strong binders can be amplified, and subsequently isolated after
equilibrium is frozen by quenching the reaction.[46] Thus, the principle of affinity
for a stable transition-state analogue (TSA), on which the development of
catalytically active imprinted polymers and catalytic antibodies is based, could
be applied to the selection of a catalyst from a dynamic combinatorial library
for the Diels–Alder reaction shown in Scheme 4.150.[320]

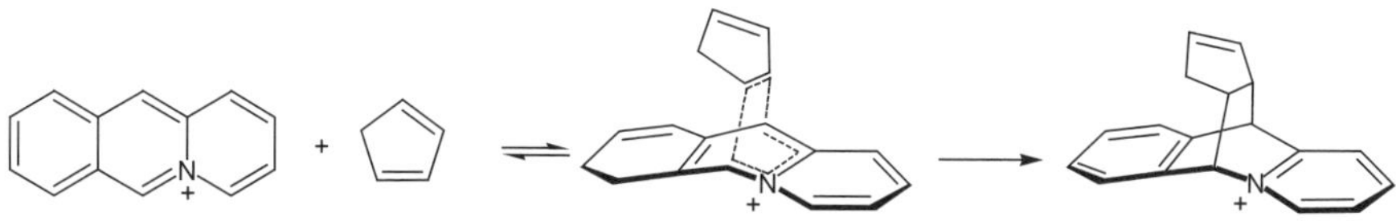

Scheme 4.150 The Diels–Alder reaction of cyclopentadiene and acridizinium cation **4.243** produces the adduct **4.245** by way of a transition state **4.244** of similar geometry.[320]

The template molecule used was the Diels–Alder adduct **4.245** itself (not exactly a true transition-state analogue!) and the dynamic combinatorial library generated by disulfide exchange from the mixture of dithiols **4.246, 247** and **248** shown in Scheme 4.151. In the presence of **4.245** the two macrocyclic hosts **4.249** and **4.250** were amplified. **4.249**, with the smaller cavity, showed a greater affinity for the reactant (acridizinium cation **4.243**) than for the product **4.245**, or presumably the transition state, so was catalytically inactive. But **4.250** bound the product more strongly, and was evidently able to bind both reactants in its larger cavity: because in the presence of an equivalent amount of the macrocycle cycloaddition was some 10 times faster.[320] The catalysis is modest, equivalent to an EM of 0.08 M for the reaction within the cavity; partly no doubt because **4.245** is not a true transition-state analogue but the product of the reaction, so bound efficiently by **4.250**. Nevertheless, product inhibition is not strong enough to prevent some turnover.

4.6.3 Catalysis by RNA

Our understanding of catalysis by "ribozymes" – RNA molecules with catalytic activity, despite their lack of obvious catalytic functionality – has made significant advances in recent years.[321] Efficient *in vitro* evolutionary screening based on catalytic activity (see Section 5.3) allows informed searches for activity beyond the range of the mostly phosphate-transfer reactions catalyzed by ribozymes *in vivo*. An early success involved the isolation from a combinatorial RNA library of a set of ribozymes that accelerate the Diels–Alder reaction between an anthracene, covalently tethered to the ribozyme, and biotinylated maleimide (Scheme 4.152) by factors of up to 18500 under single-turnover conditions. The majority of the active sequences share a common short structural motif, and a synthetic oligoribonucleotide 49-mer based on this motif was shown to be catalytically almost as active. In contrast to most known RNA aptamers this system has a preformed tertiary structure, containing a well-defined hydrophobic pocket that accommodates both diene and dienophile, and shows no major changes when the (single enantiomer) product is bound. Evidently, the ribozyme binds the transition state for the cycloaddition more strongly than the reactants; but not as strongly as the product, so that turnover is very slow.

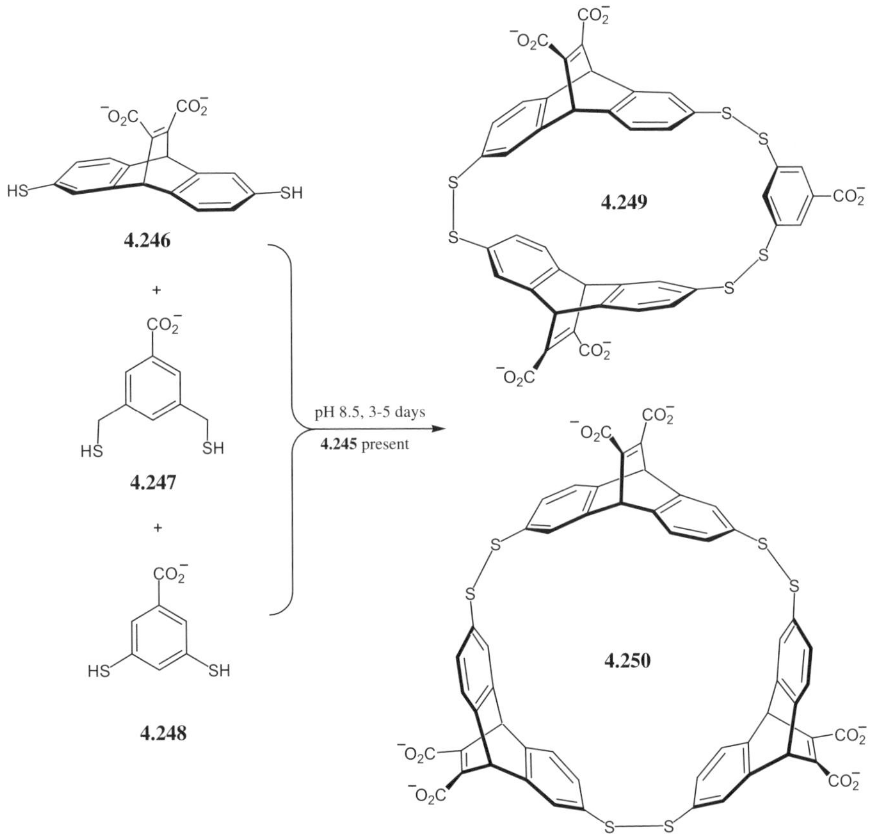

Scheme 4.151   A dynamic combinatorial library was generated by disulfide exchange from the mixture of dithiols **4.246**, **4.247** and **4.248**. Disulfide formation occurs readily by oxidation of thiols upon exposure to air. Exchange takes place under mild conditions (pH 7–9) in the presence of thiolate, and is quenched under acidic conditions or after this is removed. After equilibration in the presence of **4.245** (Scheme 4.150) the combined yield of the two macrocyclic hosts **4.249** and **4.250** was 79%.[320]

The same ribozyme (49-mer with the 5′-terminal OH free) was found to catalyze the Diels–Alder reactions of free 9-hydroxymethylanthracene with various N-alkyl maleimides. Catalytic efficiency was measured using initial rates and the more soluble derivatives **4.253** and **4.254** (Scheme 4.153). Their reaction followed Michaelis–Menten kinetics, with K_M values of 0.37 and 8 mM for the diene and dienophile, respectively, and a k_{cat} of 0.35 s^{-1}; corresponding to a >1100-fold acceleration, and an EM for the dienophile of 6.6.M.[322] Turnover is improved with the removal of the tether, to some 6 transformations per catalyst molecule per minute.

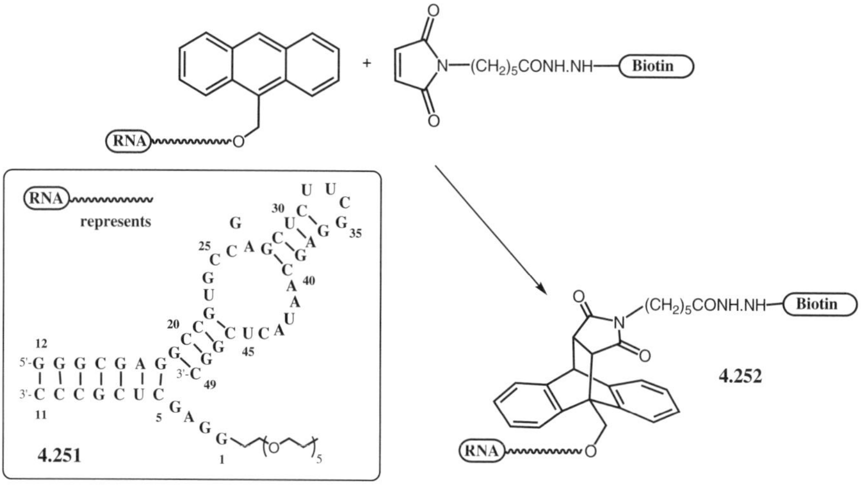

Scheme 4.152 The ribozyme (**4.251**) is made up of 38-mer (G12–C49) and 11-mer (G1–C11) chains: the anthracene is linked to the 5′-terminus by a hexaethyleneglycol tether, and sits in a preformed hydrophobic pocket. This can also accommodate the biotin-linked maleimide dienophile, and the product (**4.252**) of the reaction.

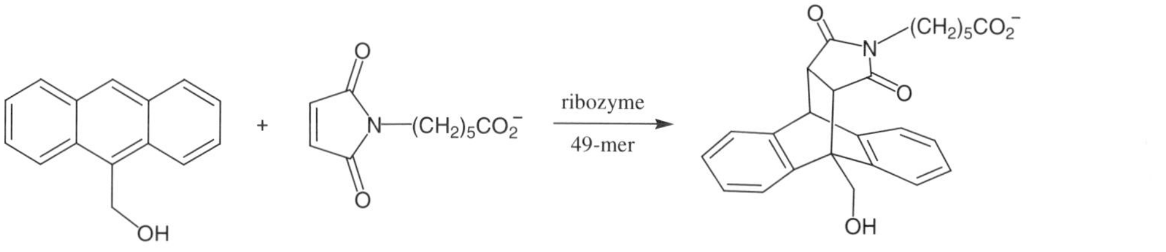

Scheme 4.153 Ribozyme **4.251** (Scheme 4.152) is an authentic model enzyme, catalyzing the Diels–Alder reactions of unattached 9-hydroxy-methyl-anthracene and N-alkyl maleimides with Michaelis–Menten kinetics.

Like many ribozymes (for more detailed discussion see Section 5.3), this system is a genuine model enzyme, binding two substrates – not too strongly and with a certain amount of selectivity – and catalyzing their enantioselective ($>95\%$ e.e.) reaction with a reasonable rate of turnover.

There are broad similarities in binding and catalytic efficiency between this and related ribozymes and (the much larger) antibodies catalyzing Diels–Alder reactions, and no doubt similar mechanisms are involved.[323]

Recommended Further Reading

H. Dugas, *Bioorganic Chemistry*. 3rd. edn.; Springer-Verlag, New York, 1996.

Breslow, R.; Dong, S. D., Biomimetic reactions catalyzed by cyclodextrins and their derivatives. *Chemical Reviews* **1998**, *98*, 1997–2011.

Davies, G.; Sinnott, M. L.; Withers, S. G., Glycosyl Transfer. in *Comprehensive Biological Catalysis*; Academic Press, London, 1998

Cleland, W. W.; Hengge, A. C., Enzymatic mechanisms of phosphate and sulfate transfer. *Chemical Reviews* **2006**, *106*, 3252–3278.

Parkin, G., Synthetic analogues relevant to the structure and function of zinc enzymes. *Chem. Rev.* **2004**, *104*, 699–767.

Design vs. Iterative Methods – Mimicking the Way Nature Generates Catalysts

Introduction

The objective of most synthetic chemistry in the enzyme-model area has been an understanding of why a molecule has a particular activity, to support the rational design of an improved or novel molecule to fulfil a given function. In a number of areas of biological chemistry this paradigm is being challenged, and even superseded.

Despite the enormous advances in our understanding of complex biological systems and in specific receptor characterization, random screening of potential drug candidates remains a cost-effective way of identifying promising leads for small-molecule enzyme inhibitors: the activity of successful hits to be improved only later by cycles of rational redesign. This is a typical, practical approach to a problem too complicated for our current understanding. In the case of drug leads, interactions with relatively floppy protein targets, and the multiple interactions with other molecules in complex natural systems, typically render this understanding exceedingly difficult. And as we have seen in Chapter 4, the design of model enzymes efficient enough to rival their natural counterparts is another area where the complexity of the challenge has not so far been convincingly met by predictive design.

A consequence of imperfect understanding is imperfect design: and where a first attempt cannot be expected to be totally successful, a procedure for systematic, *iterative* improvement becomes essential. The idea of iterative improvement is familiar from the natural evolution of functional molecules, and has become familiar to the bioorganic chemist. Directed evolution under

From Enzyme Models to Model Enzymes
By Anthony J. Kirby and Florian Hollfelder
© Anthony J. Kirby and Florian Hollfelder 2009
Published by the Royal Society of Chemistry, www.rsc.org

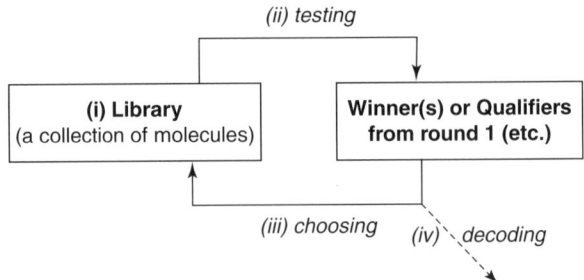

Scheme 5.1 Outline of the basic iterative approach to the creation of functional molecules. (i) The starting library can be a collection of small molecules, polymers, DNA, RNA or proteins. (ii) The test system needs only to permit the efficient assay of members of the library in large numbers. (iii) Evolutionary improvement measured by the assay is evaluated as the basis for choosing which part of the current library will be carried forward into the next cycle. (iv) Decoding of the selected molecules is the final step of an iterative approach: for biomolecules such as proteins and nucleic acids this usually involves a PCR amplification followed by sequencing; in the case of other methods, this often means analyzing the selected survivors directly.

laboratory conditions uses techniques from molecular biology that, though not trivial, are simple enough to be implemented in a chemistry laboratory. The approach effectively extends the biomimetic theme, from learning from the defining features of natural molecular recognition, to imitating the procedure by which these features were brought about. Scheme 5.1 shows a typical basic cycle of iterative improvement. The simple sequence involves (i) constructing a library, (ii) characterizing the ability of each member to carry out a given task and then (iii) using the results to choose the most promising candidate catalyst. The more of these cycles one can go through, the bigger the potential improvement.

A wide variety of systems for such iterative improvement is available. There are two different approaches to evaluating the results of a high-throughput assay: screening and selection. In Nature molecules are *selected*: if a new function allows an organism to survive, it is selected: this is essentially a digital yes/no decision determined by the selection threshold. *Screening* by contrast can generate more data, which can be of high quality: for example by the quantification of product yields or the measurement of rate constants. A decision still has to be made (= *selection*!) about which molecules are carried on into the next iterative round, but the process provides more information: for example, how many members of the library are functional, and how large is the difference between the best and the worst library member (reporting on the sensitivity to diversification). It also offers the possibility to steer iterative cycles, *e.g.* by increasing the stringency of successive iterative rounds of improvement.

A second practical consideration, relevant specifically to the directed evolution of functional biomolecules, is the linkage of genotype and phenotype. "Phenotype" defines a functional trait, such as binding or catalytic activity: "genotype" the relevant nucleic acid "blueprint," which can be replicated. It can be argued that selections from nucleic-acid libraries for aptamers, ribozymes or DNAzymes have advanced much faster because nucleic acids themselves may have binding or catalytic activities, making them selectable phenotypes, so that genotype and phenotype are identical. Proteins, Nature's preferred biocatalysts, on the other hand, cannot be amplified directly. The evolution of proteins requires linkage between the phenotypes exhibited by the proteins and the genes (nucleic acids) that encode them. Nature links genotype and phenotype by compartmentalizing genes in cells. Creating systems with such a linkage suitable for directed evolution has been the technical bottleneck for research in this area and a number of practical solutions are described in Section 5.4

Synthetic molecules by contrast are not encoded by a DNA "blueprint" that can be amplified by PCR and read out. This has the consequence that many more copies of a successful hit are needed to identify the species of interest: enough for spectroscopic identification being the absolute minimum (typically micromoles). For experiments carried out in multiwell plates, for example in the synthesis of an enzyme model using different combinations of reagents, the spatial address of the well in which a successful functional molecule is made, can be used to decode the "genotype" of a synthetic molecule.

Apart from the very different technological requirements for these iterative methods, the approaches are distinguished by their potential for exploration of possible solutions to a given catalytic problem. The following factors determine the ease of traversing chemical and biological space: (i) the type of library: chemical libraries can in principle incorporate more diversity as they are not constrained by naturally accepted building blocks. (ii) The actual size of a library: this can vary between hundreds (in the case of typical chemical libraries) to 10^{15} members (in the case of RNA libraries), determining how much of "diversity space" can be explored. (iii) The speed of carrying out an assay and its volume will be important as the fraction of library members that can be assessed will matter. (iv) Finally, each library member will have to be "decoded". The time and effort for sequencing (after PCR amplification) or the analysis of the chemical composition of the catalyst will impose another limit.

Table 5.1 summarizes the various systems where an iterative approach has been applied to the quest for more efficient enzyme models. The systems are discussed in this chapter in the order shown, with the iterative component increasing as the synthetic design element diminishes (until it eventually disappears).

5.1 Catalytic Polymers

Enzymes are polypeptides, and typically large molecules. This may be for various reasons (see discussion in Section 1.1.2), but supporting catalytic efficiency is certainly one. So it is logical that large synthetic molecules, which allow an almost infinite range of possible structures, should attract interest as potential enzyme models. Many types of non-natural polymer are readily accessible by simple polymerization processes, and many more by

Table 5.1 Systems using iterative methods for the development of artificial enzyme models.

Section/catalyst system	Library synthesis	Type of library	Typical library size	Copies per library member	Testing for activity	Selection for . . .	Decoding
5.1 Catalytic Polymers 5.1.1 Synzymes	Chemical modification of a polymer	Constrained by chemical linkage, but in principle diverse	10^2–10^3	many	in multiwell plates	Multiple turnover catalysis	Spatial address (see the text)
5.1.2 Dendrimers and Molecular imprinting	Chemical synthesis		10^2–10^3	many	in multiwell plates or on beads	Multiple turnover catalysis	Sequencing of peptide dendrimers off bead
5.2 Catalytic antibodies	Antibody repertoire (immune system or *in vitro* generated antibody library)	Constrained by peptide linkage, ribosomal synthesis and (usually) natural amino-acid repertoire	Selection for TSA binding: 10^8 Screening for catalysis: 10^2–10^3	1 many	• "panning" for TSA binders • in multiwell plates	TSA binding Multipe turnover catalysis	amplification by PCR, sequencing
5.3 RNA or DNA	DNA manipulation • error-prone PCR	Constrained by nucleotides that can be incorporated	Up to 10^{15}	1	• self-modification	Single turnover catalysis	Genotype = phenotype: Amplification by PCR, sequencing
5.4 Proteins	DNA manipulation • error-prone PCR • DNA shuffling • saturation mutagenesis • nonhomologous recombination	Constrained by peptide linkage, ribosomal synthesis and usually natural amino-acid repertoire	Multiwell plates: $<10^4$ Cells: $<10^8$ *In vitro*: $<10^{12}$	1	• in multiwell plates	Multiple turnover catalysis	Genotype and phenotype must be linked (scheme 5.26) or co-compartmentalized in cells, microwells or droplets→amplification by PCR and sequencing

subsequent modification. They include both linear and branched systems, with key factors the number of catalytic groups and – in multiple-group systems – their distribution. Enzymes are typically constructed around a single active site identified with the catalytic functionality. A polymer-based enzyme model may similarly carry a single functional group (or a single cluster of them), but it can equally well be generated from a functionalized monomer, thus deploying a large number of potential catalytic groups. These will be similar in reactivity but not identical – depending on their position along the chain and its overall conformation. (The result may no longer qualify as a true enzyme model, but if the objective is simply a catalytic system capable of carrying out a simple reaction efficiently, multiple functionality can make good sense.)

An early example came from the work of Overberger on poly 4(5)-vinylimi-dazole (**5.1**, Scheme 5.2).[324] The imidazoles in the polymer are on average less effective catalysts than imidazole itself for the hydrolysis of *p*-nitrophenyl acetate below about 80% free base, but become increasingly effective with the changing microenvironment at higher proportions of the free base form at high pH.

Catalysis by these and similar multifunctional polymers is not particularly efficient, but the realization that it can be enhanced in simple model systems by modification ("tuning") of the local microenvironment has inspired a good deal of work on various polymer forms.

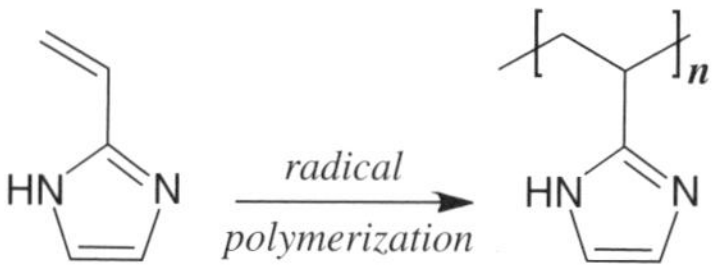

5.1

Scheme 5.2   Poly 4(5)-vinylimidazole (**5.1**) catalyzes the hydrolysis of activated esters (see the text). For a useful review of this and related systems see Dugas' *Bioorganic Chemistry*.[325]

5.1.1   Synzymes

Prominent among these is polyethylenimine (PEI), commercially available as a highly branched structure (Scheme 5.3) containing a mixture of primary, secondary and tertiary amine centres, which are protonated to various extents under physiological conditions.

In aqueous solvents near pH 7 the amine and ammonium groups of the parent structure offer an array of potential nucleophiles, general bases and general acids, built in to a polycationic backbone capable of binding anionic substrates and transition states. This recurring functionality offers multiple opportunities for modification and for further functionalization, and Klotz showed that alkylating or acylating PEIs with long-chain alkyl groups can add hydrophobic binding capability. Thus, PEI itself shows substantial rate accelerations for the aminolysis of activated esters, but PEI with 10 residue per cent

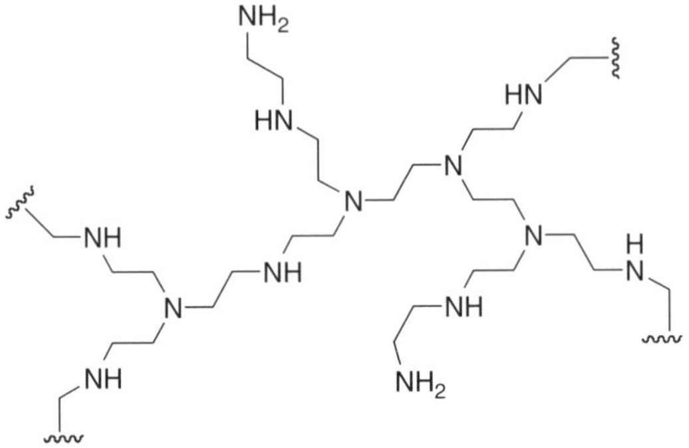

Scheme 5.3   Polyethylenimine (PEI) presents a heterogeneous macromolecular fra-
             mework based on multiple primary, secondary and tertiary amino-
             groups, available for catalysis or for further modification or functiona-
             lization. Several polymeric forms are commercially available, with
             molecular weights ranging from 600 to 60 000.

of lauryl groups is acylated faster by *p*-nitrophenyl laurate > caproate >
acetate.[326]

Catalysis by a modified PEI [327] of the Kemp decarboxylation (see Scheme
3.3), a reaction known to be particularly sensitive to medium effects,[33]
emphasizes that such environmental effects can be uniquely easily accessed
using these larger enzyme models. A wide range of work with such "synzymes"
is summarized in an authoritative review by Suh and Klotz.[328]

This approach has been expanded more recently by introducing different
combinations of several different alkyl substituents, to generate synzyme
libraries comprising several hundred different modified PEIs. The convenient
test reaction, allowing a simple colorimetric assay in 96-well plates, was the
eliminative ring-opening of benzisoxazoles like **5.2**, involving proton transfer
from carbon (Scheme 5.4):[329] which has the further advantage of being
mechanistically an uncomplicated, single transition-state process.

The working microenvironment of the polymer's amino-group general bases
(B in Scheme 5.4) was systematically modified by alkylation with various
combinations of dodecyl iodide, benzyl bromide, and methyl iodide. This was
designed to generate (i) a range of hydrophobic cavities or regions to drive
substrate binding, in close proximity to (ii) a range of amine groups to serve as
catalytic bases, embedded in (iii) the positively charged polymer framework,
expected to stabilize the delocalized negatively charged transition state (**TS** in
Scheme 5.4). The simple synthesis, combined with the simple and efficient
screening protocol, made possible the generation and assay of hundreds of
modified polymer forms, in the quest for the high catalytic activity that would
signify optimal transition-state stabilization.

These synzymes turn out to be rather good enzyme models, albeit for reactions of selected, non-natural substrates. Reactions follow Michaelis–Menten
kinetics, with rate accelerations k_{cat}/k_{uncat} as high as 10^6 and at least 1000
turnovers per basic site. Estimated EMs of up to 5000 M for the reactive base in

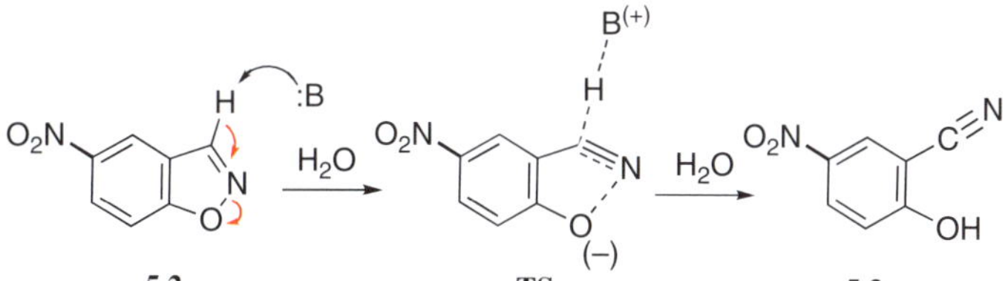

Scheme 5.4   The Kemp elimination, which can be particularly sensitive to the polarity of the medium. Note that the product phenol **5.3** (or its anion) differs significantly in geometry from the transition state (**TS**) leading to it. (See Table 5.8 for an overview of quantitative data for this and other models catalyzing the Kemp elimination.)

the reaction of Scheme 5.4 are remarkably high for general base catalysis: and thought to result from efficient transition-state stabilization by specific, localized medium effects rather than the precise positioning of catalytic amino-groups.[329,330] Product inhibition is insignificant, but the reaction is inhibited specifically by what can be loosely described as transition-state analogues, small molecules like SDS (sodium dodecyl sulfate) or triiodobenzoate, which have substantial hydrophobic regions attached to a negatively charged group. In the case of catalysis by an enzyme, or a protein like BSA, this could be cited as evidence for a specific active-site configuration; however, synzymes do not have defined 3-dimensional structures, but rather dynamic arrangements of amine general bases and positively charged hydrogen-bond donors (protonated amines) placed in productive proximity to dynamic apolar, hydrophobic regions. Specific combinations of alkyl substituents give – reproducibly – the most efficient catalysis, but this must result from the combined contributions of something like a dozen, constantly varying catalytic sites per macromolecule, some of which are presumably more and some less efficient at a given moment.

Recent work has extended the scope of synzymes to catalysis of phospho-diester cleavage, using the cyclization of HPNP (Scheme 5.5) as (another relatively easy) test reaction (see Section 4.2.2.3.1).[331] As befits a model system designed to mimic the active sites of enzymes that catalyze phosphate transfer the polymer catalyst **5.4** is equipped with guanidinium groups.

PEI itself shows little catalytic activity for the cleavage of HPNP, but screening 200 synzymes, with different combinations of alkyl substituents and varying degrees of guanidinylation, identified a variant capable of catalyzing the cleavage of HPNP with Michaelis–Menten kinetics, in water in the absence of metal ions, with a rate acceleration k_{cat}/k_{uncat} of 4.6×10^4. This variant shows over 2000 turnovers per catalyst molecule, and catalysis can be specifically and competitively inhibited by anionic and hydrophobic small molecules. The doubly charged phosphate monoester naphthyl-1-phosphate, similar in charge to the transition state, is the strongest inhibitor, binding more tightly ($K_i = 26$ mM) than either substrate or product. This increased binding of a molecule that mimics the TS of the reaction suggests that the combinatorial approach has indeed achieved some degree of selectivity for transition-state

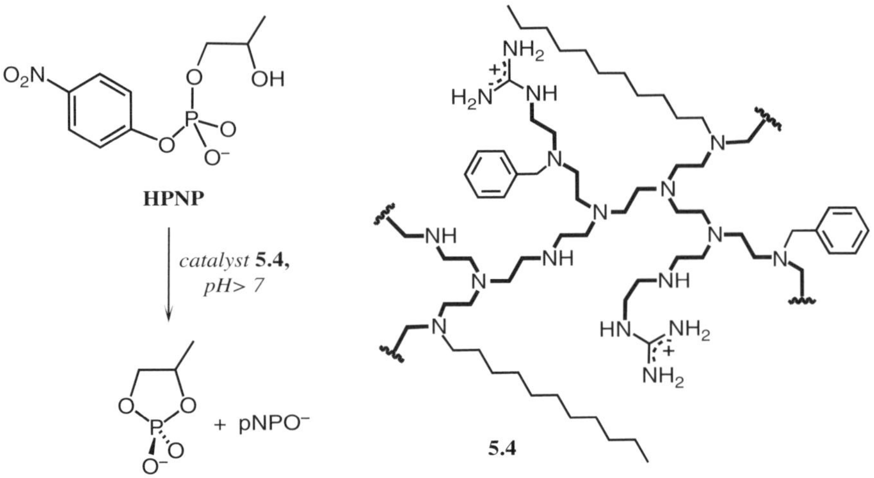

Scheme 5.5 Part structure **5.4** of PEI-based catalysts used for the cleavage of HPNP.[331] The guanidinylation reaction is specific for the primary amino-groups of PEI. The alkyl chain length was varied from C_2 to the C_{12} derivative shown.

binding based on charge interactions. Increasing the length of the alkyl chain results in rate increases of up to about 100-fold, and increasing guanidylation in rate increases of up to about 200: but these effects are almost entirely interdependent. Increasing the chain length to C_{12} has no effect in the absence of guanidinium groups, and increasing the guanidinium count has only a very small effect (2-fold) in the absence of alkyl groups. The most efficient catalyst showed a catalytic proficiency $(k_{cat}/k_M)/k_2$ estimated as $1.8 \times 10^8 \, M^{-1}$, and a rate acceleration k_{cat}/k_{uncat} per active site of ~ 3600, corresponding to an average K_{TS} of 71 nM:[331] making this variant of synzyme **5.4** (Scheme 5.5) the most effective synthetic phosphate-transfer catalyst not to use a metal.

5.1.2 Dendrimers

Dendrimers are regular tree-like macromolecules, obtained by standard chemical synthesis, which typically adopt a globular shape comparable to many protein structures. They offer technical advantages as catalysts because of their easy separation, by nanofiltration or precipitation, from small-molecule products of reaction.[332]

Dendrimers come – like PEI – in a range of molecular weights, but differ in having well-defined[i] structures, assembled one step at a time.[333] The multivalency of the resulting structures, the possibility of cooperative interactions and the creation of a special microenvironment that envelopes small molecules

---

[i]This will of course depend on the yields of individual steps if generations are not separately purified.

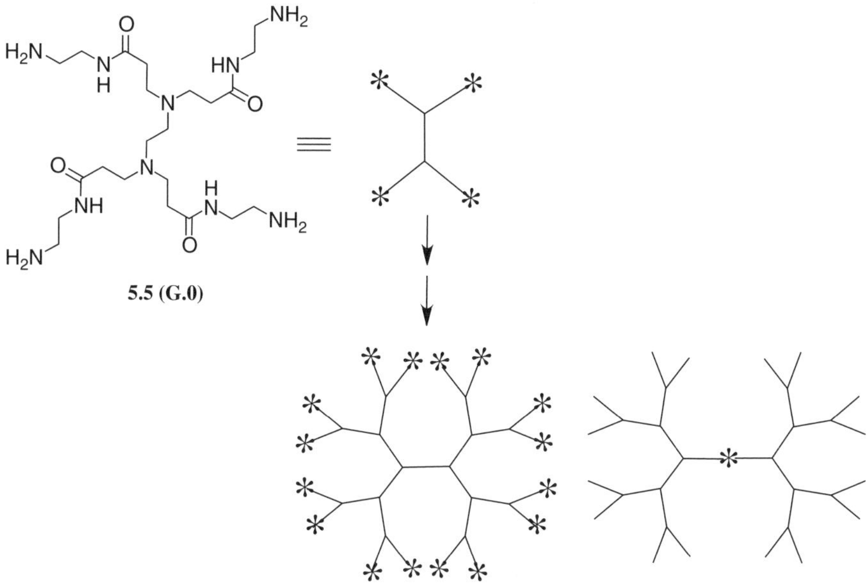

Scheme 5.6 Dendrimer synthesis and typical representation, based on the PAMAM system. The branching unit is added in two steps: double Michael addition of methyl acrylate to free terminal primary amines, followed by the generation of new terminal amino-groups using ethylenediamine. The asterisk * represents catalytic functionality (here the NH_2 groups), **G.n** is the generation number.

are potential sources of catalytic rate enhancements.[334] For example, the commonest class, the polyamidoamine (PAMAM) dendrimers, is prepared (Scheme 5.6) by the repeated addition of branching units to an amine core (typically ammonia or ethylene diamine). Each generation of growth doubles the number of terminal groups, and the molecular weight thus increases exponentially.

This protocol can produce "peripherally modified" dendrimers like **5.6(G.2)**, with multiple functionality nominally on the surface of what eventually becomes a globular polymer: or alternatively systems represented by **5.7(G.2)**, with a single, central "core" functionality (Scheme 5.6). The centre typically becomes progressively more sterically hindered, so that the most effective catalysts are generally peripherally modified.[333]

Instructive core-modified examples are Breslow's transaminase mimics, which have the coenzyme pyridoxamine or pyridoxal at the dendrimer centre. Pyridoxal-centred PAMAM dendrimers, **5.8(G.n)**, Scheme 5.7, with *n* increasing from 1 to 6, show positive dendrimer effects in the racemization of α-amino-acids in aqueous buffer,[335] with modest increases in rate for higher generations. However, while NHAc-terminated dendrimers **5.8(G.6)** racemize alanine only 3–5 times faster than does pyridoxal itself, the corresponding

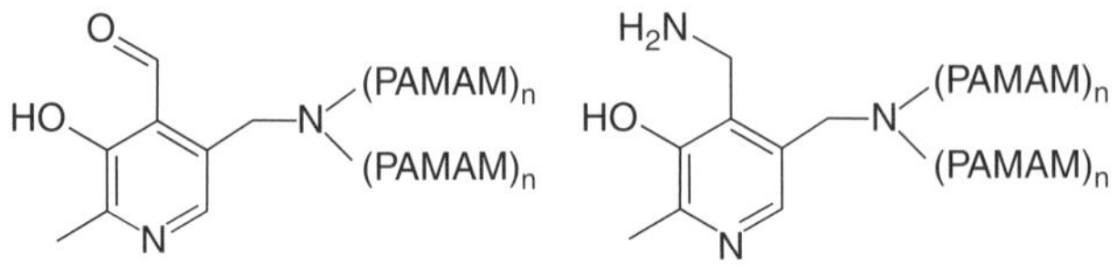

5.8 (G.n)   **5.9 (G.n)**

Scheme 5.7   PAMAM dendrimers built on the modified hydroxymethyl group of the coenzyme pyridoxal/pyridoxamine catalyze the racemization of α-amino-acids and the transamination of α-keto acids in aqueous buffer.[335,336]

NMe_2-terminated pyridoxal dendrimers are 50–100 times faster. Evidently 64 peripheral dimethylamino groups offer more efficient general acid–base catalysts for the racemization than the many tertiary amino groups in the body of the polymer: presumably indicating that they can fold back into the interior of the dendrimer structure.

The more complex transamination reaction of pyruvate (using phenylglycine as a "sacrificial" amino-acid) is similarly catalyzed by the corresponding pyridoxamine-centred dendrimers **5.9(G.n)**, with Michaelis–Menten kinetics and positive dendrimer effects, particularly on substrate binding, but with limited turnover.[336] Here too, the peripherally modified derivatives proved more efficient.

5.1.2.1 *Peptide Dendrimers*

Some of the most instructive – and convincing – enzyme models are those based not just on the principles that underlie enzyme catalysis, but also on the same building blocks. Polypeptides can offer the same physicochemical advantages as enzyme proteins, and developing an "artificial" polypeptide model enzyme from first principles, *i.e.* starting from the amino-acids, is an enormously attractive – if enormously challenging – target. Higher generations of peptide dendrimers can be expected to develop naturally into globular structures,[333] and – in contrast to the folding-based approximation of functionality involved in setting up an active site from a linear polypeptide – their stepwise synthesis allows the controlled incorporation of catalytic function at specific positions. Combinatorial libraries of peptide dendrimers can be prepared, and screened for catalytic activity.

The first catalytic peptide dendrimers were reported by Reymond and coworkers for the hydrolysis of quinolinium esters **5.12** (Scheme 5.8), which give a fluorescent product. Small libraries of dendrimers **5.10** and **5.11** were constructed, using solid-phase synthesis, from the three amino-acids Asp, His and Ser, known to be involved in the active site of the serine proteases; with diamino-acids as branching points and cysteine as the C-terminus. Oxidizing the six possible second-generation dendrimer thiols **5.10** to the disulfides gave a further 21 dimers **5.11**, which were also tested for activity against the ester substrate. The dendrimers with 8 surface histidines (at positions A^3 in **5.10**)

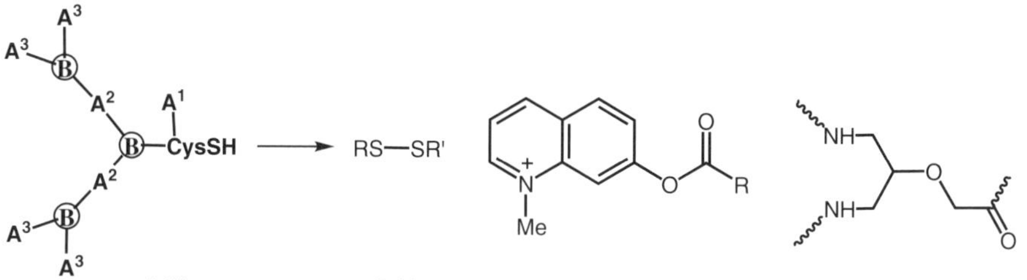

Scheme 5.8 Second-generation dendrimers **5.10**, based on all six possible combinations of the three amino-acids (A^1, A^2 and A^3), = aspartate, histidine and serine, and their 21 possible disulfide dimers **5.11**, were tested (as $5\,\mu M$ solutions) for hydrolase activity against fluorogenic ester substrates **5.12**. The branching unit **B** was **5.13**, based on the symmetrical 1,3-diaminoisopropyloxyacetic acid.

were catalytically active, showing Michaelis–Menten kinetics with useful substrate binding ($K_M \sim 0.1\,mM$) and rate accelerations k_{cat}/k_{uncat} of the order of 1000.[332,337,338]

Compared with the second-order rate constant k_2 for catalysis by 4-methyl imidazole this represents a 350-fold acceleration [$(k_{cat}/k_M)/k_2$], corresponding to a factor of about 40 per histidine imidazole. This is similar to figures obtained by the Baltzer group, who attached clusters of histidines to conformationally well-defined 4-helix bundle peptides.[339] These results suggest some cooperativity between imidazole and perhaps imidazolium groups, but without significant assistance from serine or aspartate residues.

The almost exclusive involvement of surface residues in catalysis suggested that these systems form compact structures with sterically limited access to core functionality, and led to the development of a second group of dendrimers with the general structures **5.10** and **5.11**, with the same three catalytic amino-acids but using 3,5-diaminobenzoate as a more rigid branching unit (**B**), expected to support a more open structure.[340] These compounds gave some improvement in catalytic efficiency, with rate enhancements k_{cat}/k_{uncat} of up to 4000-fold: and delivered in addition rather well-defined substrate selectivities. Two dendrimer dimers, one of them compound **5.14** (Scheme 5.9), now with surface aspartates nominally shielding the histidines, catalyzed the hydrolysis of the same cationic ester **5.12** as before (Scheme 5.8): two others, *with* surface histidines (*e.g.* **5.16**), did not; but were instead specific for the hydrolysis of anionic esters **5.15**.[340] This specificity is consistent with substrate binding favoured by electrostatic attraction: to the surface aspartate carboxylate groups in the case of **5.14** and to surface histidine imidazolium groups in the case of **5.16**. Catalysis is still exclusively histidine driven in both cases, but in the case of cationic ester **5.12** by histidines that are less immediately accessible.

More extensive further developments of these enzyme model dendrimers take advantage of the ready accessibility of polypeptides by solid-state synthesis. A simple extension of the original structure **5.10** → **5.11** (Scheme 5.8)

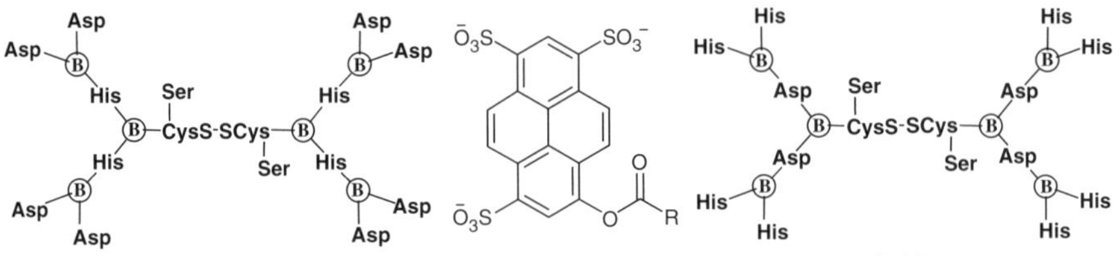

Scheme 5.9   Second-generation dendrimer **5.14** (with symmetrical 3,5-diamino-benzoate as the branching unit **B**) shows specific hydrolase activity (at pH 6.0) against cationic ester **5.12** (Scheme 5.8) while **5.14**, with surface histidines, is specific for anionic fluorogenic pyrene-sulfonate ester substrates **5.15**.[340]

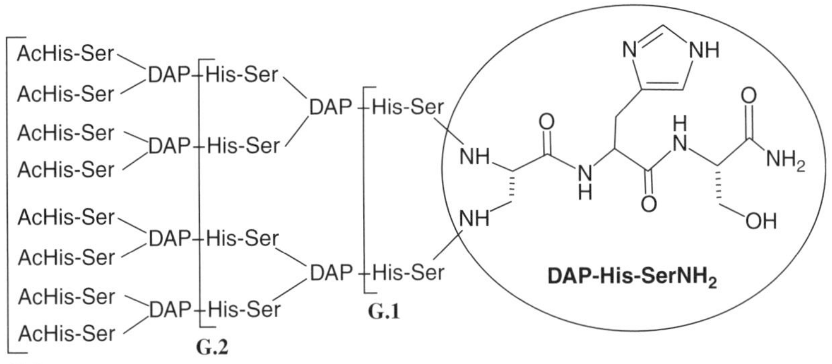

Scheme 5.10   The third-generation dendrimer based on the (His-Ser)$_2$DAP dendron, with the extents of the first- and second-generation structures indicated. Even this symmetrical structure defines up to 8 different amino-acid positions, the 4th generation up to 10, and so on.[342]

replaces the single amino-acids between branch points with dipeptides.[341,342] For example, **5.17(G3)** shown in Scheme 5.10, based on a repeating His-Ser dipeptide, is an efficient catalyst for the hydrolysis of pyrenesulfonate esters **5.15**. In particular, it is markedly more efficient than the corresponding first- and second-generation dendrimers.[342]

This *positive dendrimer effect*[343] indicates alternative possible ways to improve catalysis: either (i) to take advantage of the positive dendrimer effect, and go up a generation; or (ii) to try to optimize the amino-acid profile of the most promising member of the present generation. In this instance both approaches were successful. (i) The fourth-generation dendrimer (**G.4**) based on **5.17** (Scheme 5.10), with 16 external acetyl-His-Ser groups, and thus a total of 31 histidine imidazoles, is a significantly more effective Michaelis–Menten catalyst than **G.3**. The key kinetic parameters (Table 5.2), including both k_{cat} and k_M, and thus especially k_{cat}/K_M and its derivatives, are without exception

Table 5.2 Peptide dendrimers as model esterases. Catalytic parameters are for **5.16**-based dendrimers hydrolyzing nonanoate ester **5.15** (Scheme 5.9), unless indicated otherwise.

Dendrimer	k_{cat} (min^{-1})	k_{cat}/k_{uncat}	K_M (μM)	k_{cat}/K_M	$(k_{cat}/K_M)/k_{uncat}$	$(k_{cat}/K_M)/k_2$	$(k_{cat}/K_M)/k_2$ (per His)
5.17, G.1	0.099	4500	1600	62	2.8×0^6	130	44
5.17, G.2	0.096	4400	67	1430	6.6×10^7	3000	420
5.17, G.3	0.15	6700	13	11 500	5.2×10^8	23 000	1600
5.17, G.4	0.39	18 000	5.8	67 200	3.1×10^9	**140 000**	4500
5.17, G.3	0.15	6700	13	11 500	5.2×10^8	23 000	1600
5.17, G.3.ST[a]	0.035	16 000	40	875	4.0×10^8	20 000	1330
5.17, G.3.ST[a,b]	1.25	**90 000**	160	7800	5.6×10^8	15 000	1000
5.18	0.023	180	27	**860/920**		1800/1930	**860/920**

Notes. [a]Data for dendrimer based on the (His-Thr)$_2$DAP dendron (see the text, and Scheme 5.10).
[b]Data for the hydrolysis of the n-butyrate ester **5.15** (Scheme 5.9).

more favourable. More efficient also (ii) are various "mutants" of **G.3**. Five of 32 designed third-generation dendrimers based on the **5.17** structure were better catalysts in some way, notably **G.3.ST** (Table 5.2), based on the (His-Thr)$_2$DAP dendron, where each serine of **G.3** has been replaced by threonine.

This work had arrived at the point where combinatorial methods were clearly indicated. Combinatorial methods can in principle introduce a random element beyond design or intuition, but a complete scan of even the 8 individual positions of **5.17** using all 20 proteinogenic amino-acids is still not a practical proposition. A first split and mix combinatorial library based on the **5.17** structure was constructed using 4 carefully selected groups of 4 different (thus 16 in all) amino-acids at each position, to generate a library of $4^8 = 65\,536$ different dendrimers. Direct on-bead screening for ester hydrolysis using the fluorogenic butyrate ester **5.15** gave "a few" fluorescent beads, identifying catalytic hits: though none turned out to be as catalytically active as the most efficient dendrimers based on **5.17** (Table 5.2).[341]

The most significant result to emerge from the combinatorial approach comes from another 65 536-member library, based on the same basic structure as **5.17** but with catalytic functionality only at the centre. In this case the outer shells were built up from amino-acids with aromatic side-chains (Tyr, Phe, Trp) designed to support binding of a hydrophobic substrate, but no catalytic groups; while the set of amino-acids at the central positions included nucleo-philic (His, Cys) and cationic (Arg) residues expected to be involved in binding and catalysis.[344] The best third-generation catalyst, **5.18** (Scheme 5.11), cannot compete directly with the best dendrimer catalysts like **5.17**, with their 15 his-tidines: but in terms of catalytic activity its single, central histidine residue comes close to their per-histidine average (Table 5.2). Consistent with the design specifications, substrate binding is assisted by the pair of arginine resi-dues in the first-generation branch, and builds up generation by generation as

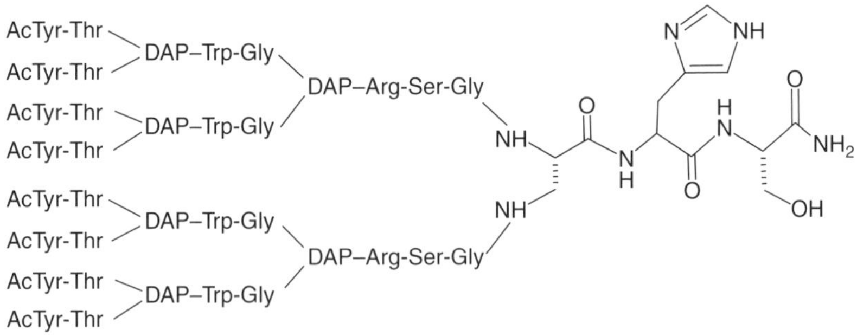

5.18

Scheme 5.11   Third-generation dendrimer with core active-site esterase activity against esters **5.15**. The data in Table 5.2 refer to the reaction with the n-butyrate.[344]

the outer shells are added, to generate a relatively compact conformation similar to a protein molten globule.

Though its efficiency is limited **5.18** qualifies as one of the better enzyme models, in that the outer amino-acid structure enhances the efficiency of a single group of active-site catalytic residues not by providing additional catalytic functionality but rather through such effects as directed binding of a specific substrate, the preorganization of catalytic residues, and no doubt polarization effects, all combining to stabilize the transition state.[344]

5.1.3 Molecular Imprinting

Molecularly imprinted polymers (MIPs) are synthetic polymers grown from or around a template, which leaves its molecular imprint on the resulting product. When the template is a transition-state analogue (TSA – see Section 3.5) the imprint – once the template is removed – is at least initially a cavity more or less complementary to the transition-state structure modeled by the TSA, and so in principle a potential catalyst for the reaction concerned. MIPs typically have much higher molecular weights than synzymes or dendrimers, and are often produced as solids, which have to be crushed or ground and sieved to achieve the high surface area necessary for effective access to substrate and other reactant molecules; though work on the development of soluble versions is in progress.[32]

The templating principle is illustrated by the example in Scheme 5.12. The glycoside template (not a TSA) is converted to a functional monomer by attaching one or more polymerizable groups. Radical polymerization is carried out under carefully controlled conditions, in the presence of large amounts of crosslinking agent, and inert solvent to act as a porogen. This produces polymers with a permanently open, porous structure, allowing access to an extensive "inner surface" area: and in particular to the encapsulated template molecules. The key step is the chemical removal of the template, which requires that the covalent bonds to the polymer be both accessible and easily cleaved (a requirement not easily achieved with high efficiency). If successful the result is a

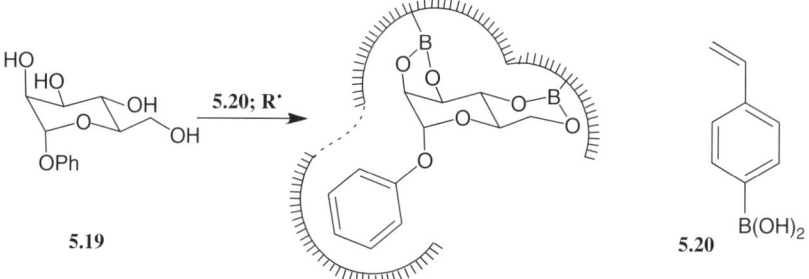

Scheme 5.12 Molecular imprinting. Phenyl α-D-mannopyranoside **5.19** acts as a template for the polymerization of 4-vinylbenzeneboronic acid **5.20** by forming covalent diester linkages to the monomer.[45]

polymer with multiple cavities made to match the template, which in this case (Scheme 5.12) could find use for example as an affinity column. The cavities as formed typically contain solvent, often resulting in swelling of the polymer: a process that can be reversed in the presence of the template.

Using TSAs as templates produces imprinted polymers that recognize and bind their template well, but are relatively inefficient as catalysts for the target reaction. (A reliable indication that catalysis is inefficient is that examples are limited almost exclusively to the reactions of familiar activated substrates.) The hydrolysis of aryl carbonates, for example, goes by way of a transition state close in energy, and thus in structure, to the tetrahedral intermediate (Scheme 5.13), making structurally corresponding phosphate diesters convenient TSAs (exactly as discussed for phosphonate TSAs in Section 3.5, see Scheme 3.6). But overall rate accelerations (measured as the ratio k_{impr}/k_{sol}[45]) rarely exceed a few thousand, and the theoretical limitations discussed in Section 3.5 are expected to apply to these simple systems also. As discussed frequently in these pages, one of the important properties of the active sites of natural enzymes is flexibility, and efforts continue to ameliorate the rigidity of the crosslinked polymer structures of molecularly imprinted polymers.

The most successful recent work has improved the catalytic efficiency of MIPs considerably by recruiting as cofactors metal cations, which can contribute their own ligand-ordering and exchange capabilities to catalysis. The nominal starting point was the mechanism of action of the zinc-containing peptidase carboxypeptidase A, the model reaction once again carbonate ester hydrolysis.[345] The target activated substrates were carbonates **5.23** and **5.24** (Scheme 5.14), and as TSA the phosphate diester **5.22**, mimicking the tetrahedral addition intermediate (**5.25**) formed by the addition of hydroxide anion to the ester carbonyl group. Phosphate ester anions in enzyme-active sites are often bound by hydrogen bonding to arginine guanidine residues; so the functional monomer **5.21** was equipped with two amidine groups, similar in geometry and basicity to guanidines, attached to a central ethylenediamine, in addition to two of the usual 4-vinylbenzenes needed for the polymerization process.

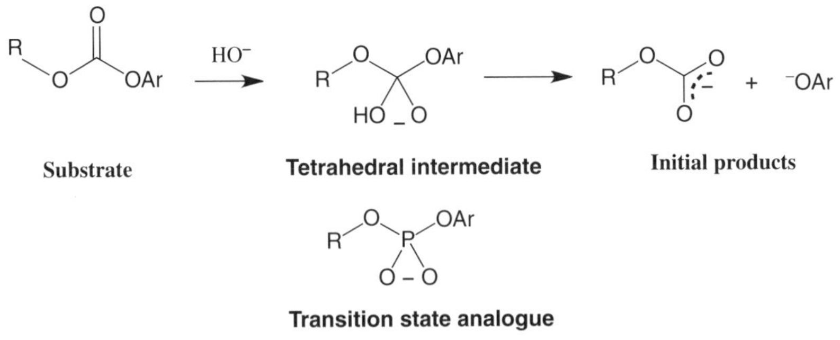

| Substrate | Tetrahedral intermediate | Initial products |

Transition state analogue

Scheme 5.13   Phosphate diesters make convenient TSAs for the alkaline hydrolysis of carbonate esters.

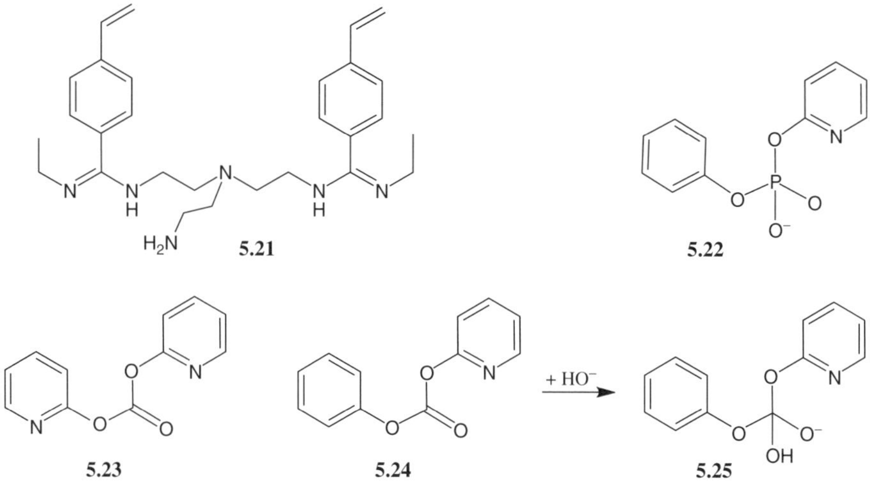

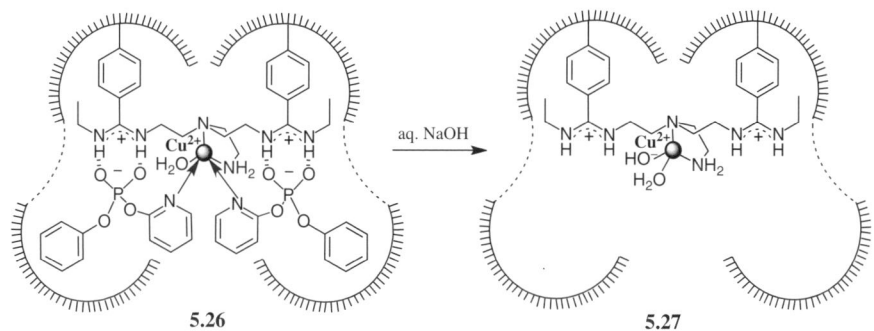

Scheme 5.14   The monomer **5.21** used in the preparation of the imprinted polymer (**5.27** in Scheme 5.15),[345] designed to bind the tetrahedral intermediate involved in the hydrolysis of the carbonate ester **5.24** by using the phosphate diester **5.22** as a TSA template.

Scheme 5.15   The template **5.22** is removed from the product of the polymerization of **5.21** (Scheme 5.14) in the presence of the template and Cu²⁺ ions, to give the imprinted catalysts **5.27**. This has two adjacent cavities tailored to bind one or two tetrahedral adducts **5.25**.

Monomer **5.21** was polymerized in the presence of the template **5.22** and Zn^{2+} or Cu^{2+}, using ethylene dimethacrylate as crosslinker. The product (*e.g.* **5.26**, Scheme 5.15) was ground to small particles and the template removed by washing with aqueous base to give **5.27**, with 75% of free cavities designed to accommodate one or two tetrahedral intermediates **5.25**, and the transition states leading to them.[345]

The imprinted polymer **5.27** shows strong catalytic activity (greater than that of the corresponding Zn^{2+}–based system),[345] with a rate enhancement

Table 5.3 Michaelis–Menten parameters for the hydrolysis of **5.23** by **5.27** and a control polymer.[345]

Polymer catalyst	k_{cat} (min^{-1})	k_{cat}/k_{uncat}	K_M (mM)	k_{cat}/K_M $(min^{-1}M^{-1})$	$(k_{cat}/K_M)/$ k_{uncat} (M^{-1})
Control[a]	1.15	4520	4.16	276	1.08×10^6
5.27	105	4.13×10^5	0.36	2.92×10^5	1.15×10^9

Note: [a]The control polymer was prepared exactly as for **5.27**, but with the template **5.22** absent.

k_{impr}/k_{soln}, corresponding to k_{cat}/k_{uncat}, close to 100 000 for the hydrolysis of substrate **5.24**: and twice that for the symmetrical ester **5.25**. Hydrolysis follows Michaelis–Menten kinetics, with $k_{cat}/k_{uncat} = 4.13 \times 10^5$ for the hydrolysis of **5.23**.[345] The Michaelis–Menten parameters are compared in Table 5.3 with those obtained for catalysis by a control polymer, prepared in the same way as **5.27**, and thus possessing the same catalytic functionality, but in the absence of the template phosphate diester **5.22**. This allows the best estimate of the effect of imprinting, which contributes a factor of about 1000 in this system. This makes **5.27** a respectable enzyme model, with reaction following Michaelis–Menten kinetics, a pH optimum in the region of 7-8 and competitive inhibition by the TSA.[345] The overall catalytic proficiency $(k_{cat}/k_{uncat})/K_M$ of $1.15 \times 10^9 \, M^{-1}$ is the highest so far obtained for an imprinted polymer catalyst, and compares favourably with catalytic antibodies designed to hydrolyze carbonate esters.[345]

5.2 Catalytic Antibodies

The highest profile, and perhaps the most successful use of transition-state analogues (TSAs) in the preparation of enzyme model catalysts has been in the development of catalytic antibodies. The principle of catalysis by antibodies raised against TSAs was introduced in Section 3.5, and their application to the catalysis of simple pericyclic processes discussed in Section 4.6. However, enzymes that catalyze pericyclic processes are rare special cases, and most work on antibody catalysis has concentrated on ionic reactions between the functional groups involved in familiar metabolic processes.

Catalytic antibodies (also known as *abzymes*) are proteins (Scheme 5.16), raised in biological systems against synthetic antigens (called haptens) designed as TSAs for a reaction of interest. Hapten design and synthesis involve only small molecule chemistry, except that the hapten needs to be conjugated to a suitable carrier protein to elicit a significant immunological response.[346] When challenged with a TSA-antigen the immune system produces vast numbers of different antibodies that recognize and bind the TSA, and these must be screened to identify candidate catalysts. So practical limitations are the accuracy of the hapten as a model for the transition state, and the effectiveness of screening the large numbers of antibodies produced for catalysis. Standard immunological methods screen very efficiently for *binding*, while direct

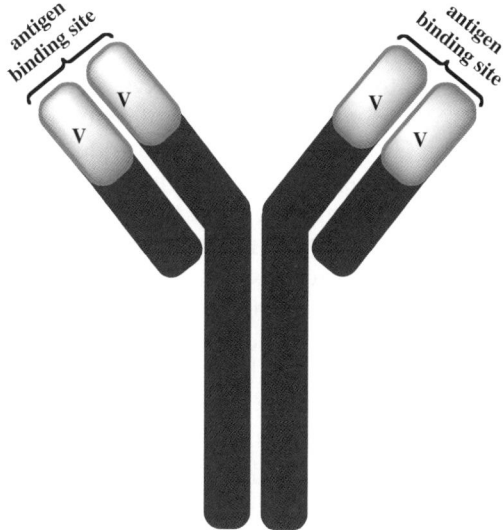

Scheme 5.16   Antibodies are immunoglobulins, with a common basic Y-shaped structure made up of four separate protein (actually glycoprotein) chains, held together by the usual noncovalent interactions reinforced by disulfide bonds. The primary amino-acid sequences of much of the two heavy and two light chains are constant for a given type of immunoglobulin, but they vary greatly at the N-terminal hypervariable regions **V**, which make up the antigen binding sites. The attractive feature of this protein fold is that it can accommodate multiple binding functions: antibodies raised against many different molecules retain this basic structure. Retaining the overall structure (only modifying a small part of it to build in a new function) avoids the complications with folding and stability involved in creating a completely new protein.

screening for catalysis is typically laborious and often specific to particular reactions.[347] This means that the selection for binding is from a very large pool ($>10^8$), whereas the screening for catalysis involves a much smaller sample of usually less than 1000 assay reactions. Effectively, this standard format for generating a catalytic antibody is a screening experiment, albeit one with a very efficient preselection. Integrating the selection for binding and the catalytic assay has been explored, by catELISA[347] or by generating a reactive product that acts as a suicide inhibitor,[348] to create a direct selection system for catalysis (see Section 5.4).

Though catalytic antibodies are currently perhaps the most successful enzyme mimics, most following the expected relationship $k_{cat}/k_{uncat} = K_S/K_{TSA}$,[311] they are not significantly more efficient than other well-developed models or mimics (Section 5.2.3). They catalyze the hydrolysis of activated esters when raised against phosphonate haptens, which model the structure of the tetrahedral transition states involved, as discussed in Section 3.5,[349] and there are reports involving the hydrolysis of benzyl[347] and other alkyl esters:[350]

but of not ordinary amides at a useful rate.[351] This work has been well reviewed,[311,349] and is conveniently summarized here in terms of the results shown in Table 5.4, for a few of the most efficient of the many hundreds of catalytic antibodies raised for reactions at carboxylic acid centres. The haptens (column 1) are all TSAs with a stable tetrahedral centre in the position corresponding to the target C=O centre of the substrate (column 2), with the addition of a linking chain used to attach the hapten to the necessary carrier protein. The substrates are arranged in order of decreasing intrinsic reactivity. Note that the catalytic proficiency does not increase significantly over the series, so that intrinsically very slow reactions – like amide hydrolysis – remain very slow even in the presence of the catalytic antibody.

In most cases these antibodies are thought to catalyze the attack of water or hydroxide ion on the C=O group of the substrate; though there is good evidence that the hydrolysis of the anilide substrate (Table 5.4, row 3, column 2) catalyzed by antibody 43C9 involves nucleophilic catalysis by a histidine imidazole side-chain of the antibody protein.[351] These antibody-catalyzed reactions are typically stereospecific, selecting one enantiomer or diastereoisomer of the substrate even when the hapten is racemic.[356] But the catalytic proficiencies (final column of Table 5.4) are unexceptional in comparison with those typical observed for most enzyme-catalyzed reactions (Figure 2.4), with transition states for antibody-catalyzed reactions typically bound some 1000 times more strongly than the substrates.[311]

5.2.1 Other Approaches

One response to the relatively low catalytic efficiencies of catalytic antibodies – and other types of enzyme models – raised against TSAs has been the development of new methods (including reactive immunization,[311] not discussed here) to take advantage of the possibilities offered by the immune response. A useful supplementary tactic, known as "bait and switch,"[311] is to equip a recognizable TSA with a charged group, in a position corresponding to one known to develop in the transition state of the reaction of interest, but of opposite polarity. The expectation is that antibodies raised against such a hapten will have charged groups complementary – and thus of opposite charge – to the group on the hapten. Thus, a positively charged group on the hapten should elicit antibodies with a negatively charged side-chain group in the corresponding position on the antibody. The anionic group concerned, which in a protein will almost certainly be carboxylate, is thus in position to act as a nucleophile or general base in the reaction of interest.

A simple example is the use of the hapten **5.28** (Scheme 5.17) to elicit antibodies designed to catalyze the cleavage of the RNA model **5.29**.[357] The enzyme reaction is known to involve the general base-catalyzed removal of the proton of the 2'-OH group (**5.30**), so an antibody that binds substrate **5.29** in appropriate close proximity to the induced group (**GB**) would be expected to catalyze the cyclization-cleavage process (**5.30** → **5.31**). The better of two antibodies

Table 5.4 Catalytic antibodies most effective in the hydrolysis of carboxylic acid derivatives.

Hapten (antibody, reference)	Substrate	k_{cat} (min^{-1})	K_M (mM)	k_{cat}/K_M	k_{cat}/k_{uncat}	$(k_{cat}/k_{uncat})/K_M$
48G7 [352]	p-Nitrophenyl ester	5.5	0.39	14 000	16 000	4.1×10^7
D.2.3 [353]	Benzyl ester	3.6	0.28	13 000	130 000	4.6×10^8
43C9 [354]	Anilide	0.08	0.57	140	250 000	4.4×10^8
BL25 [355]	Amide	0.003	0.15	20	~40 000	2.7×10^8
1a, 13D11 [356]	R-Amide [a]	1.0×10^{-5}	0.43	0.230	132	3.1×10^5

Notes: R_L indicates a linker chain of 4–6 carbon atoms connecting the hapten to the carrier protein, typically via an amide group. R in the substrate is usually the same or a similar chain with a terminal carboxylate. [a]The S-amide shows no reaction.

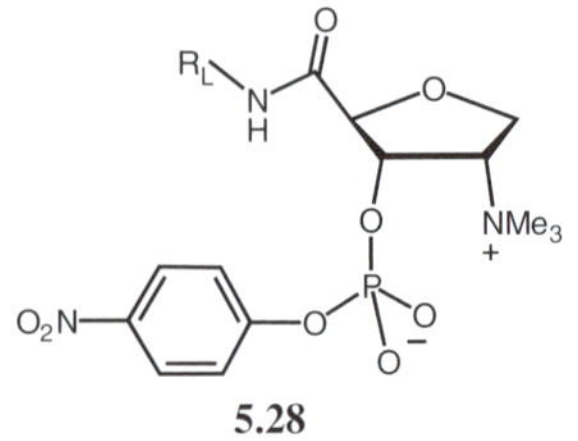

Scheme 5.17   The cleavage of the RNA model **5.29** is catalyzed by general bases (**GB**), including an anionic group elicited by the hapten **5.28** in antibody MATT.F1.[357]

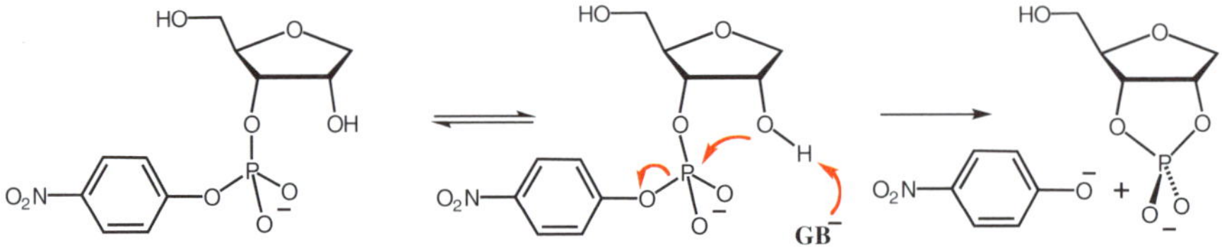

obtained catalyzed this reaction, with Michaelis–Menten kinetics, with k_{cat} $= 0.44\,min^{-1}$, $K_M = 0.10\,mM$, and a k_{cat}/k_{uncat} of 1650: and hapten **5.28** inhibited the catalytic activity of the antibody stoichiometrically. With a catalytic proficiency $(k_{cat}/K_M)/k_{uncat}$ of $1.6 \times 10^7\,M^{-1}$ this antibody is, nevertheless, one of the most efficient enzyme models for phosphodiester hydrolysis not using a metal. And for this particular activated, non-natural substrate, it is only some 1000 times less efficient a catalyst than RNAse A.

5.2.2   Proton Transfer from Carbon

Non-natural reactions define a major area of organic chemistry where catalytic antibodies might reasonably be expected, in the absence of competition from enzymes, to be a significant potential source of interesting catalysts. We have seen results based on the TSA approach for catalysis of simple pericyclic reactions in Section 4.6, and discuss here the application of the TSA-plus-charged group (bait and switch) strategy to a reaction well suited to act as a probe of mechanism and catalysis.

   The reaction is the Kemp elimination (Scheme 5.4), of mechanistic interest because it involves a simple one-step proton transfer from carbon that is extraordinarily sensitive to the polarity of the medium; and convenient in the context of antibody catalysis because the product phenolate differs significantly in geometry from the transition state (**TS**) leading to it. We know from work with intramolecular model systems (Section 1.4.1 and Section 4.3.3) that proton transfer catalysis can be efficient, but only given positioning of the donor and acceptor groups so precise that so far only a handful of carefully designed such systems meet the requirements.[358] High-resolution X-ray structures of enzymes can only suggest whether such geometries are or are not possible in bound transition states. So it is important to explore alternative ways of setting up the correct geometries.

Perhaps the most sophisticated way of bringing functional groups together outside enzyme-active sites is in the binding sites of antibodies, using in particular the bait and switch method, of eliciting complementary charged groups in the antibody binding site by the use of properly designed haptens, equipped with charged groups. Hilvert and coworkers[359] found that antibodies raised against the benzimidazolium hapten **5.32** catalyzed the Kemp elimination (Scheme 5.18, **5.3** → **5.33**) rather efficiently, with carboxylate groups acting as general bases, as intended.

The EM (effective molarity) of the carboxylate group of the most efficient antibody 34E4 in this reaction was estimated to be a remarkable 4×10^4 M, assuming catalysis to be due entirely to the positioning of the general base. However, such carboxylate-catalyzed Kemp eliminations are known to be highly sensitive to the polarity of the medium; the reaction catalyzed by 34E4 takes place in what is largely a hydrophobic binding site, and catalysis by antibodies is not usually remarkable for its efficiency. Testing a group of proteins known to possess the appropriate combination, of hydrophobic binding sites with potential general bases in close proximity, revealed that the same reaction of **5.3** (Scheme 5.18) was also catalyzed by serum albumens, also with remarkable efficiency but evidently only "accidental specificity".[360] It was concluded that EMs calculated (in the usual way, as k_{cat}/k_2) for catalysis by particular groups in antibody- – and especially enzyme- – active sites will normally contain contributions from a whole range of catalytic effects, of which group positioning is only one.

Catalysis by enzymes often involves contributions from multiple functional groups, while catalysis by antibodies often involves no more than stabilization

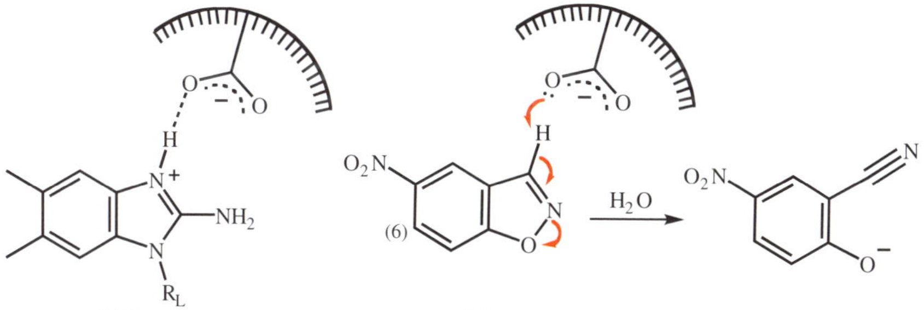

5.32   **5.3**   **5.33**

Scheme 5.18 Hapten **5.32** carries a positive charge, so antibodies elicited against it might be expected to have a side-chain carboxylate anion close enough for a favorable electrostatic interaction; which would contribute to binding most effectively if it also involved a hydrogen bond to the NH$^+$ group, as shown. The carboxylate would then be correctly placed to act as a general base in the Kemp elimination of a suitable substrate (**5.3**) bound in the same position.

of the transition state for the "uncatalyzed" process. The mechanism of the reaction catalyzed by antibody 34E4, elicited against hapten **5.32**, has been convincingly established as involving general base catalysis by the carboxylate group of Glu^{H50}, with the substrate held in position by a combination of H-bonding, π-stacking and van der Waals interactions. Mutagenesis experiments suggest that Glu^{H50} is as effective as carboxylate general bases in reactions of comparable enzymes, but the overall efficiency of the antibody is lower by many orders of magnitude. Proton transfer from carbon necessarily generates an anion, which in an enzyme-active site will be stabilized by protonation by, or at least H bonding to a general acid. The anion **5.33** (Scheme 5.18) is relatively stable (*i.e.* not strongly basic), but might benefit from H-bonding stabilization of the developing phenolate. So antibodies were raised against the modified hapten **5.34**, (Scheme 5.19), which has the hydrophobic linker group R_L attached to the benzene ring, exposing the polar imidazolium ring to a possible network of H-bonding groups in the protein.[361]

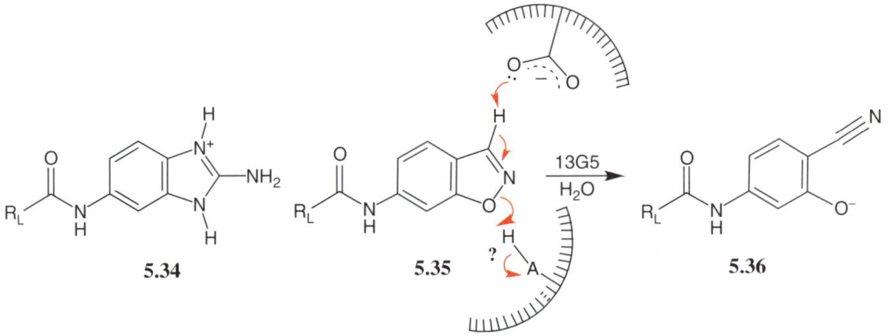

Scheme 5.19   Antibody 13G5 raised against hapten **5.34** catalyzes the ring-opening elimination of the unactivated benzisoxazole **5.35** efficiently (Table 5.5), and k_{cat}/K_M follows a bell-shaped pH–rate profile; as expected for a mechanism involving general acid and general base catalysis by separate groups (see the text).

Table 5.5   Antibody catalysis of the Kemp elimination reactions of more and less activated benzisoxazole substrates **5.3**, its 6-nitro isomer and **5.35**.[361]

Antibody	substrate	k_{cat} (min^{-1})	K_M (mM)	k_{cat}/K_M $(M^{-1} min^{-1})$	k_{cat}/k_{uncat}	$\frac{(k_{cat}/K_M)}{k_{uncat}(M^{-1})}$
47C4	**5.3**			1740 (*at pH max.*)		
13G5	6-nitro-iso-mer of **5.3**			372 (*at pH max.*)		
13G5	**5.35**	2.04 (lim)	0.84	2430	1.7×10^5	2.0×10^8

   Two antibodies were obtained, one catalyzing the Kemp elimination of the 5-nitrobenzimidazole **5.3**, and the other the reactions not only of the activated 6-nitro-derivative but also of the relatively unactivated benzisoxazole **5.35**. The Michaelis–Menten parameters are comparable with some of the most efficient catalytic antibodies, k_{cat}/K_M follows bell-shaped pH–rate profiles – as expected for reactions catalyzed by both a general base and a general acid (Section 1.3) – and both antibodies contain several candidate groups. However, the pH–rate profile for k_{cat} for the reaction of **5.35** is not bell-shaped but sigmoid, reaching a limiting value at high pH like those of the reactions catalyzed by 34E4 or the serum albumins, so the assignment of mechanism remains tentative.[361]

5.2.3   Conclusions

Antibody catalysis has been characterized for a much wider range of organic reactions than the handful of selected examples discussed in these pages.[311,362] But the conclusions that can be drawn from these systems are general. Antibodies can reasonably be regarded as the most successful enzyme mimics. They can match natural enzymes in selectivity and stereospecificity, but rate accelerations k_{cat}/k_{uncat} greater than about 10^6, corresponding to catalytic proficiencies greater than 10^{8-9}, are rare exceptions: as graphically illustrated by the plot shown in Figure 5.1. This is consistent with the level of the "glass

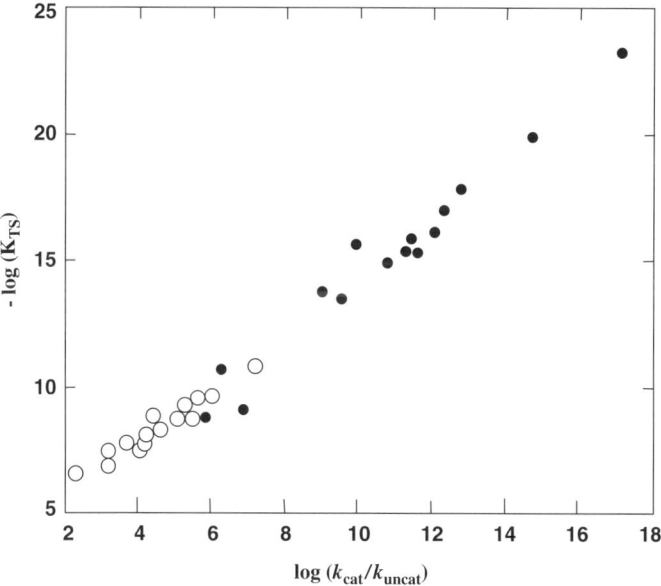

Figure 5.1   Catalytic proficiencies $(-\log(K_{TS}))$, and rate accelerations k_{cat}/k_{uncat} compared for a range of reactions catalyzed by enzymes (filled circles) and antibodies (open circles). The good correlation between rate acceleration and proficiency (and slope close to unity) is observed because K_M values are similar for both series of catalysts. (Plot based on Hilvert.[362])

ceiling" estimated for catalytic rate enhancements accessible from the transition-state analogue approach in Section 3.5.

The substantial body of systematic research on catalytic antibodies has undoubtedly informed further developments in the design of protein (and non-protein) catalysts, not least by establishing experimentally the limit in catalytic proficiency of the order of 10^{10} M^{-1} for these superior TSA-based systems. The reasons for this limitation, when compared with the proficiencies of natural enzymes, cover the complete range of factors that make enzymes so efficient: in addition to the following problems specific to antibody catalysis:[311,360]

(a) TSAs are at best poor mimics of true transition states: many antibodies raised against TSAs do not catalyze the target reaction significantly.
(b) Many reactions involve more than a single TS which will need stabilization.
(c) Most reactions catalyzed by catalytic antibodies do not involve an active-site functional group: covalent catalysis by antibody side-chain groups in particular is not supported.
(d) Only a tiny fraction of the immune response is typically sampled.
(e) Hapten binding rather than catalysis drives selection in most cases.

5.3   Nucleic Acids as Catalysts

Given the abundance of protein catalysts in Nature, it is perhaps not surprising that evidence for catalysis by nucleic acids was found only relatively recently, in the 1980s: in self-splicing of preribosomal RNA (rRNA) of the ciliate *Tetrahymena*[363] and in the RNA component of a ribonucleoprotein enzyme, ribonuclease (RNase) P.[364] These findings opened for the first time the possibility that another key biopolymer, rather than proteins alone, could provide catalysis. The immediate implications of these results stimulated the hypothesis of an RNA world, where RNA served as both the genetic material and the functional forerunner of protein enzymes.[365] This hypothetical world would have relied exclusively on RNA molecules, both to store genetic information and to carry out the entire repertoire of reactions necessary for survival or primitive organisms, with ribozymes directing a primitive metabolism before the evolutionary advent of proteins. The beauty of this concept is that Nature's functions would be served by just one type of molecule made up of four ribonucleotide building blocks, that DNA and proteins (with its 20 amino-acid alphabet) would not be necessary and that evolution would be facilitated by unifying genotype and phenotype.

5.3.1   Mechanisms of Nucleic-acid Catalysis[366–368]

The RNA world hypothesis makes assumptions about what RNA can do as a catalyst, considered here – in the spirit of this book – as a new category of enzyme model (or, better, model enzyme): namely that it could support the range of necessary reactions with rates high enough to support a coordinated network of chemical reactions. How reasonable is this? At first glance RNA does not look promising catalyst material: the backbone of RNA is more flexible than that of a protein (for lack of a conformationally constrained peptide bond), the repertoire of building blocks is limited to four (rather than 20, Section 1.1.2), their pK_a values are normally far from neutrality (making general acid base catalysis difficult, Section 1.3) and there are no obvious nucleophiles that could control reaction progress by forming covalent inter-mediates (as in many proteases; Section 1.4): thus many pieces of the catalytic jigsaw outlined in the previous chapters appear to be missing.

A quantitative analysis of ribozyme proficiency shows that these possible obstacles have been convincingly overcome by natural evolution. The known natural cellular and viral ribozymes catalyze only phosphodiester transfer chemistry efficiently. Typical kinetic data, showing submicromolar affinities and k_{cat} values of the order of $1 \, min^{-1}$, are summarized in Table 5.6. The second-order rate accelerations achieved range up to a remarkable 10^{11}, placing ribo-zymes on a par with less efficient natural enzymes (Figure 5.1), but well above most synthetic enzyme models for catalysis of what is a demanding chemical reaction. The first criterion for fulfilling an enzyme-like role – rate acceleration – is therefore met. In addition, natural ribozymes offer high substrate specificity: substrates with noncomplementary sequences are not accepted and minor substitutions – *e.g.* substituting sulfur for phosphoryl oxygens – led to rate decreases. A rare "promiscuous" amide-cleavage reaction involving attack on the carbonyl carbon of an aminoacyl ribonucleotide is catalyzed only very weakly (up to 15-fold).[369]

The first ribozymes to be identified catalyzed intramolecular reactions, such as the phosphate-transfer processes involved in self-splicing.[363] But it has proved possible to dispense with the covalent link and to incorporate proper catalyst–substrate recognition, rendering reactions truly *inter*molecular. Most ribozyme-catalyzed reactions suffer from strong product inhibition, as the great majority of catalyst–substrate interactions do not change during the course of the reaction: but specifically weakening these interactions by changing the nucleotide sequence can bring about multiple-turnover catalysis.

The efficiency of ribozyme catalysis of such phosphate-transfer reactions then raises the question how this respectable enhancement might occur. Initi-ally a relatively simple two-metal ion mechanism involving the large number of bound divalent metals was proposed. Ribozymes are replete with divalent metal ions (approximately one metal per 3–4 nucleotides), so the ribozyme is at least highly proficient at recruiting these useful cofactors to enhance its catalytic repertoire. The metals could catalyze phosphate-transfer reactions in the usual ways (Section 4.2.1.2), by lowering the pK_a of the nucleophilic OH and off-setting charge development on the leaving group and phosphoryl oxygens. These mechanisms (Scheme 5.20A) are reminiscent of the effects that protein enzymes use (see for example Scheme 4.41 in Section 4.2.1.2).

While all ribozymes are replete with metal cations, more recent results sug-gest that not all ribozymes need them for catalysis in the ways discussed in

Table 5.6 Summary of kinetic, structural and chemical data for selected examples of natural nucleic–acid enzymes and their *in-vitro* evolved counterparts. The cited values refer to experiments performed under a variety of conditions and are for orientation only.

Nucleic acid enzyme	Type of reaction	Size (nucleotides)	X-ray str.	Postulated catalytic mechanisms	k_{cat} [min^{-1}]	K_M [μM]	k_{cat}/K_M [min^{-1} M^{-1}]	k_{cat}/k_{uncat}	$(k_{cat}/K_M)/k_2$	Ref.
(1) Natural ribozymes										
Hammerhead ribozyme	phosphate transfer (self-scission of the RNA backbone)	38	411,412	Substrate orientation, metal-ion catalysis	0.6–9	0.05–3	~5×10^7	–	~10^6-fold	413
HDV		90	370	General acid base catalysis, positioning (EM: 10^3 M for the general acid), metal-ion catalysis	0.91	0.1–10	~10^7	–	10^6–10^7-fold	372,373,414
Group I intron		210	415,416	metal-ion catalysis, substrate destabilization, approximation (EM: ^{b} 2200 M for the nucleophile)	0.1	0.001	~9×10^7	–	~10^{11}-fold	417,418
Tetrahymena ribozyme	*Promiscuous* acyl transfer	247	374	metal-ion catalysis	–	–	–	5–15-fold	–	369
(2) In-vitro evolved ribozymes										
Oligonucleotideligases	phosphate transfer	220	419	metal-ion catalysis, possibly other factors	0.1–25	Up to 300	Up to 7×10^7	~10^7	~10^6-fold	385,386, 420–422
Continuously evolved oligonucleotide Ligases		175	–	metal-ion catalysis, possibly other factors	1–4	0.1–0.4	~10^7	–	–	387,423
Leadzyme (tRNA cleavage)		32	395	metal-ion catalysis (Pb^{2+} only)	0.1	^{f}	^{f}	1000-fold	–	393,394
γ-thio-ATP Kinase	thiophosphate transfer	93	–	metal-ion catalysis, possibly other factors	0.03–0.37	41–456	Up to 6×10^3	10^3-fold	10^9-fold	401
Oligonucleotide Diels-Alder reaction	Diels–Alder	49	323	approximation(EM: ^{b} 6.6 M)	–	370 (diene) 8000 (dienophile)	Up to 6×10^3	–	10^3-fold	322
Ribozyme	Porphyrin Metallation	35	–	–	2	16	Up to 2×10^3	460-fold	–	400
Flexizyme	acetylation of tRNAs	45	390	–	^{e}	^{e}	^{e}	^{e}	^{e}	389,424,425
Ribozyme 11D2	aldol reaction	~180	–	metal-ion catalysis (Zn^{2+} only)	1000	–	10^{-4}–10^{-5}	–	100–4000	402
(3) In-vitro evolved DNAzymes										
RNA cleavage	phosphate transfer	30	–	metal-ion catalysis (Pb^{2+}),^{h} possibly other factors	–	1	–	–	10^5-fold	426
DNA ligase	phosphate transfer	208	–	metal-ion catalysis, possibly other factors	0.1	0.0001^d	–	–	10^5-fold^g	397

^{a}Minimal catalytic unit. ^{b}EM: effective molarity (Section 1.4.1). ^{c}Transfer of the γ-thiophosphate of ATP-γ-S to the 5'-hydroxyl or to internal 2'-hydroxyls. ^{d}k_{obs}. ^{e}No kinetic data are available as flexizyme is used in an immobilized form on a resin. ^{f}Self-cleavage. ^{g}Based on observed rates k_{obs}. ^{h}In other DNAzymes metals such as Pb^{2+}, Pb^{2+}, Ca^{2+}, Cu^{2+}, Mg^{2+}, Zn^{2+} are catalytic.

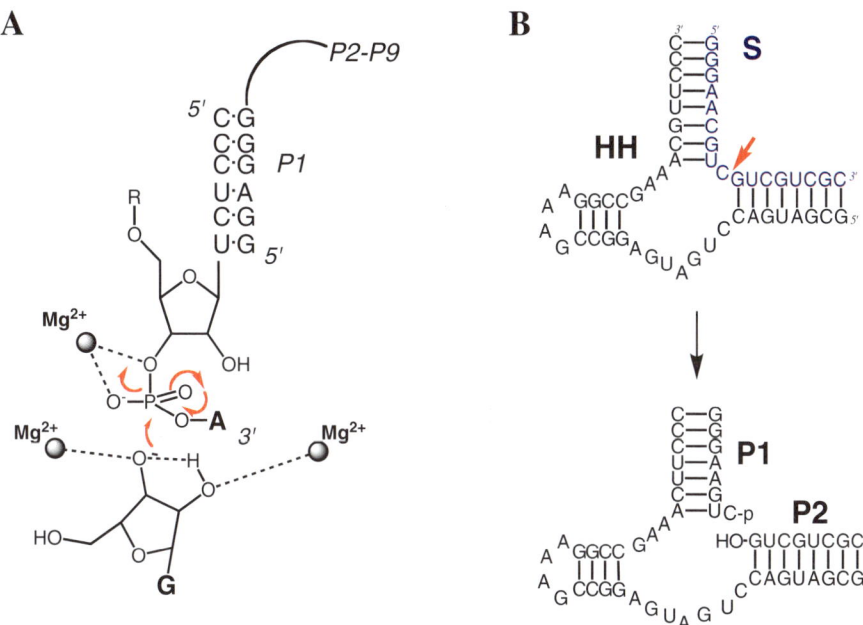

Scheme 5.20   Examples of ribozyme catalysts involving metal-ion catalysis. (**A**) Active-site arrangements of the *Tetrahymena* ribozyme, in which a guanine cofactor acts as the nucleophile. Nucleophilic attack, the departure of the leaving group and the charge change at the phosphoryl oxygens are facilitated by interactions with up to three metal cations. (**B**) In the presence of Mg^{2+} the 38 nucleotide Hammerhead ribozyme **HH**, derived from autocatalytically cleaving plant viroids, binds and cleaves the 17-nucleotide substrate **S** at the location indicated by the arrow, to give products **P1** and **P2**.

Section 4.2.1.2. In the HDV ribozyme, for example, the usual requirement for divalent metal ions disappears at high concentrations of monovalent cations: which cannot themselves support catalysis by forming metal hydroxides or by acting as Lewis acids. A crystal structure of this ribozyme showed a cytidine (C75, Scheme 5.21) in a metal ion-free cleft as the closest residue to the reaction centre.[370] Its functional role was confirmed by mutation and chemical rescue experiments and opened up the possibility of general acid–base catalysis by this residue. Subsequent functional studies (including solvent isotope effect and proton-inventory studies) further suggest that a chemical step involving proton transfer is indeed rate limiting in the HDV ribozyme.[371,372] The reaction catalyzed by this ribozyme displays a bell-shaped pH–rate profile, consistent with the involvement of two functional groups (Section 1.3.2) with apparent pK_a values of 6.5 and 9.[373] C75 is the prime candidate because of its proximity (and evidence for a large pK_a shift for this residue), with a metal-coordinated water providing the second functional group. The crystal structure suggests that this metal ion is ejected from the active site as the reaction product is generated.[370]

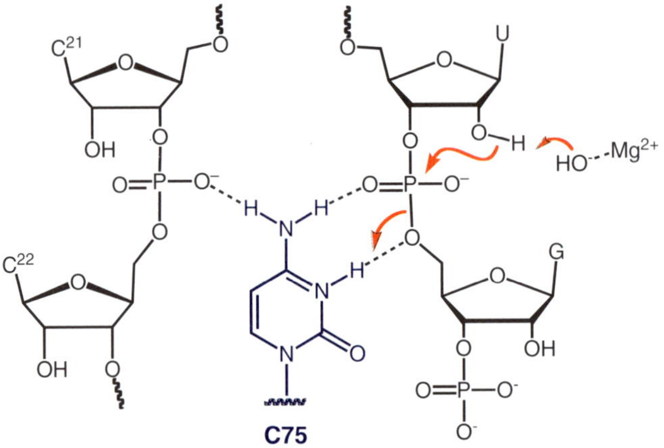

Scheme 5.21   The hepatitis δ virus (HDV) ribozyme forms a 2′,3′-cyclic intermediate in an identical reaction to the protein enzyme ribonuclease A (Scheme 4.56). This ribozyme employs general acid–base catalysis in addition to metal-ion catalysis. The available evidence is insufficient to resolve whether C75 acts as a general acid catalyst as shown, or exchanges roles with the metal-coordinated water and acts instead as a general base.

Currently, it is impossible to ascertain which of the two residues acts as the general acid and which as the general base; so kinetically equivalent (Section 1.3.3) alternatives, swapping the roles of metal ion and C75 exist, and a pH-dependent conformational change or more complicated scenarios cannot be excluded. Whatever further studies reveal, ribozymes do appear to be able to employ general acid catalysis as part of their repertoire.

The large number of ribozyme structures (see Table 5.6) that have become available as a result of the tremendous advances in the crystallization of RNA species in the last decade allow comparisons with the overall morphology of protein enzymes. The structures reveal that ribozyme active sites form in ways by no means fundamentally different from the process in proteins: close packing of helices creates a catalytic core in which the substrate can be tightly bound, by shape complementarity and hydrogen bonding. The relatively large 247-nucleotide Tetrahymena ribozyme appears to be largely preorganized for catalysis and has been compared to a globular protein enzyme.[374] Even relatively small ribozymes, such as a catalytic 49-mer[323] can form a hydrophobic core, introducing the possibility of local environmental effects.

Formation of a folded ribozyme core can also present problems, due to the conformational complexity of nucleic acids; and in some cases it has proved difficult to reconcile kinetic data with the picture presented by the X-ray structure. In particular, single-molecule studies have suggested that conformational changes (namely the formation of tertiary structure over timescales between milliseconds and minutes) can trap RNA in transient non-native or partially folded conformations, with one RNA molecule following multiple folding pathways.[375–377] Ribozymes are evidently not such primitive potential catalysts as their position in early evolution would make us believe: they deploy a full set of enzyme features that can be applied with surprising sophistication, at least in phosphate-transfer reactions, although the maximal rates achieved remain limited.[378,379] Most of all this makes us wonder how they emerged at such an early stage, anticipating many of the basic principles that proteins have later perfected. The explanation must lie in the way they developed: in which case the hint we should take from Nature is to turn to iterative systems of catalyst generation.

5.3.2 Selection as an Alternative to Design Strategies

Consistent with the hypothesis of an intermediate RNA world in evolutionary history is the idea that the process of evolution might itself be more straightforward for nucleic acids than for proteins: if the nucleic acids themselves have binding or catalytic activities, then genotype (*i.e.* the "code" of a library member) and phenotype (*i.e.* defined by the product) are identical. Indeed, nucleic acid selection experiments in the laboratory are much simpler in practical terms than the corresponding experiments with proteins, where genotype and phenotype have to be purposely linked. The importance of this conceptual simplicity will become evident when the rather complicated technologies necessary to combine genotype and phenotype in the selection of protein enzymes are discussed in the following section. The result is that, despite their late introduction, selections from nucleic-acid libraries have advanced much more rapidly than those for protein selections.

Szostak and Gold have translated this insight into an evolutionary scheme, called SELEX (systematic evolution of ligands by exponential enrichment),[380,381] that is now widely used.[382–384] The reaction product can be selected (*e.g.* by specific product-binding interactions) and capture of the product in this way provides the identity of the catalyst: since genotype and phenotype are identical. Scheme 5.22 outlines a typical SELEX experiment. Starting with a pool of random RNA sequences, molecules possessing a desired activity are isolated through successive cycles of activity selection. The most active "winners" are selected by removing them from the pool – in the case of ribozyme binders ("aptamers" – the nucleic acid equivalent of antibodies) by binding to an immobilized target molecule. The K_D (affinity) values for aptamers can be as low as picomolar to nanomolar. When the objective of the selection is to identify catalysts, the reaction product has to be detected. This is done most simply for synthetic self-modification reactions in which the ribozyme forms a new bond to a substrate labeled with a tag (*e.g.* biotin), that allows the product to be detected: see Scheme 5.22. Formation of the new bond distinguishes product physically from starting material and can be used to separate active catalysts from the pool: The attached tag can be an oligonucleotide, which is distinguishable in several ways: on the basis of its size (which affects its electrophoretic mobility); by its ability to hybridize with a complementary nucleic acid, which can lead to interaction with an affinity column,

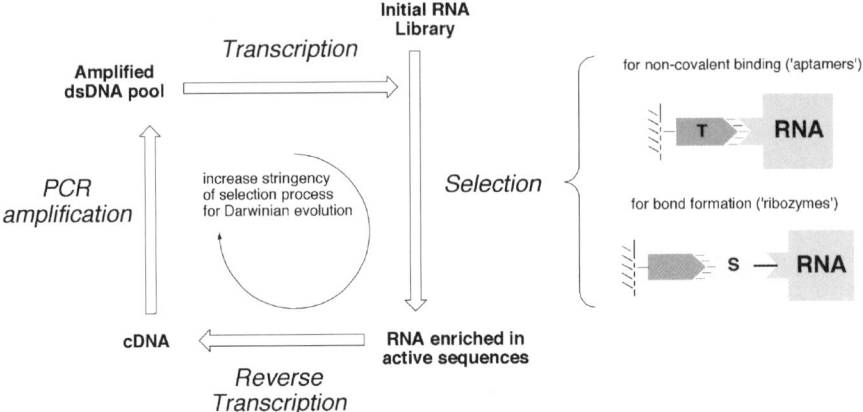

Scheme 5.22   Overview of a typical SELEX procedure (systematic evolution of ligands by exponential enrichment), to generate functional RNA molecules.[382,383] A library of RNA molecules (up to 10^15) is subjected to selection. When selecting for RNA binding the selection (of "aptamers") involves the removal of functional RNA molecules by interaction with the immobilized target molecule **T**. When selecting for catalysis synthesis of a new bond leads to covalent attachment of a substrate molecule **S** to the ribozyme. The substrate **S** is labeled with a tag (*e.g.* biotin) that allows detection of product formation and removal of active catalyst (*e.g.* by interaction of the biotin tag with streptavidin; alternative tagging strategies are discussed in the text). The enriched RNA pool is then reverse transcribed to give cDNA, that can be PCR amplified and the sequences of the selected clones obtained. For further rounds of selection the cDNA pool is randomized by error-prone PCR at the amplification stage, giving a double-stranded DNA library (dsDNA). This enriched pool is then transcribed to RNA to start a new round of selection.

or by its ability to be extended by a polymerase, if a specific primer is correctly recognized by an attached DNA sequence.

The selected RNA binder or catalyst is then translated into the more stable DNA version using the enzyme reverse transcriptase. Further rounds of evolution – normally 10 to 20 in total – can be initiated by amplifying the winner sequences (using the polymerase chain reaction, PCR). Even if very few molecules have the desired function, several rounds of amplification increase the number of winners (by a factor of ~ 10^6 per round) so that they eventually come to dominate the pool.

The amplification reaction is also the point where diversification of the pool is introduced: The PCR reaction can be carried out under conditions where random errors are introduced throughout the DNA sequence, leading to a diverse DNA library, which is converted once again to its RNA counterpart by transcription in readiness for a further round of selection. Library sizes in such experiments can be as large as 10^15, helping to make up for the lack of design in the way the libraries are generated.

How many of the possible combinations does such a number cover? The number of possible sequences for a nucleic acid of length n is 4^n. This means, for example, that a combinatorial library containing one copy of all the possible 25-mers would already contain about 10^{15} molecules (about a nanomole, roughly the amount of RNA that can be dissolved in 1 ml of water). Larger diversified ribozymes experimentally undersample the possible theoretical diversity – complete coverage of potential library diversity is impossible to achieve even for these relatively simple biomolecules. For proteins this problem grows even more rapidly out of hand.

5.3.3   Access to New Catalysts Using SELEX[382]

The SELEX methodology allows the iterative generation of catalysts that encompass a repertoire beyond that of the natural ribozymes,[321] and perform a variety of chemical transformations (Scheme 5.23). Initial targets were reactions crucial for evolution, starting with the self-replicating ribozymes that catalyze the ligation of RNA fragments.[385,386] Successful selection from RNA pools established that RNA can catalyze its own replication by generating a complementary RNA strand, although the fidelity and length of the copy required improvements for the process to be useful in a hypothetical RNA world. These ribozymes ligate two RNA molecules that are aligned on a common template, by catalyzing the attack of a 3′-hydroxyl on an adjacent 5′-triphosphate (a process illustrated for comparable DNAzymes in Scheme 5.24, and for a protein ligase in Scheme 5.30), with reaction rates $\sim 10^6$ times faster than the uncatalyzed reaction rate.[385] Such single-turnover catalysts can be engineered to act as true enzymes, catalyzing the multiple-turnover transformation of substrates into products.[386]

More recently, this procedure has been made continuous,[387] and even automated in microfluidic systems on a chip,[388] allowing an almost continuously

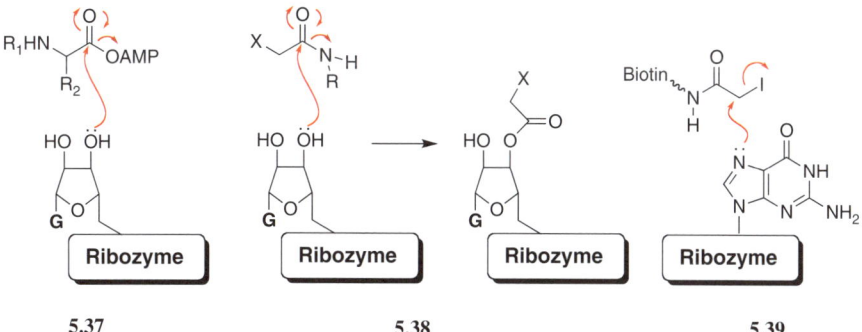

5.37   5.38   5.39

Scheme 5.23   Overview of the range of reactions catalyzed by ribozymes generated *via* SELEX with potential relevance in a primordial RNA world. Self-aminoacylation of the 3′-OH of a ribose (**5.37**); Hydrolysis of an amide bond (**5.38**, R = H or deoxyribonucleoside; X: deoxyribonucleoside or peptide).[369,391] and (**5.39**) the formation of a carbon-nitrogen bond in RNA alkylation.[392]

self-evolving system for iterative improvement. After multiple cycles of selection and amplification, a number of ligases have indeed emerged from the population, having outperformed nonligase competitors that were not replicated. Continuous evolution carried out for 80 successive cycles led to a 10^5-fold improvement in catalytic efficiency compared with the starting molecules.[387]

Further crucial reactions of evolutionary importance have been targeted. *In vitro* evolution by SELEX identified RNA catalysts for two steps of the protein synthesis pathway: formation of activated amino-acid adenylates and transfer of an amino-acid to a tRNA-like acceptor. The latter system ("flex-izyme")[389,390] can generate tRNA that is charged with non-natural amino-acids for incorporation into proteins at defined positions (Section 3.6.1).[56]

Evolution can also proceed under conditions that are not identical with those in cellular environments, enlarging the versatility of these catalysts. For example, selecting for catalysis in the presence of large amounts of lead gave rise to *leadzyme*, a ribozyme in which the usual magnesium cofactor has been exclusively replaced by Pb^{2+}.[393–395]

5.3.4 Changing the Catalyst Backbone: DNAzymes

Only protein enzymes and ribozymes are known to be responsible for natural biological catalysis, but deoxyribozymes with kinetic parameters that rival ribozymes can also be created in the laboratory. Most of the DNA in biological systems is double helical, restricting its catalytic potential: this could explain why there are no known examples of DNA catalysis in Nature. However, the SELEX scheme (Scheme 5.22) can be readily adapted – and indeed shortened – for DNA molecules. The selected "deoxyribozymes" (or DNAzymes) are single-stranded, and have the potential to form higher-order structures, just like the single-stranded RNA catalysts. The best examples of evolved DNA-zymes in Table 5.6 show maximum k_{cat} values similar to those obtained for RNA catalysts, but exhibit slightly lower second-order rate accelerations due to lower substrate affinities (albeit for the more difficult DNA hydrolysis).

A remarkably complex example is a DNA ligase that catalyzes the formation of a 3′,5′-phosphodiester bond in a two-step reaction (Scheme 5.24): the phosphate is first activated by the attachment of AMP via a pyrophosphate linkage with consumption of one molecule of ATP,[396] followed by purification before the second reaction is started in which a nucleophilic phosphate replaces the activated leaving group.[397] Both DNAzymes were generated by SELEX, suggesting that at least a very simple "biosynthetic" pathway can be built up by this methodology.

Clearly DNAzymes are good enzyme mimics without known equivalents in Nature, and serve as examples of the potential for catalysis even of DNA.

Despite the missing 2′-OH group the factors responsible for catalysis must be similar in DNAzymes and ribozymes. Thus Paul *et al.* have created a DNA version of an active RNA ligase ribozyme, preserving the analogous base sequence. This DNA version was initially inactive as an RNA ligase, but could readily be adapted by SELEX.[398] The resulting evolved DNAzyme had kinetic parameters similar to those of the ribozyme for catalysis of the same reaction. Converting the improved DNAzyme back to its analogous ribozyme resulted once again in the loss of activity: suggesting that the catalytic solutions found are specific and distinct.

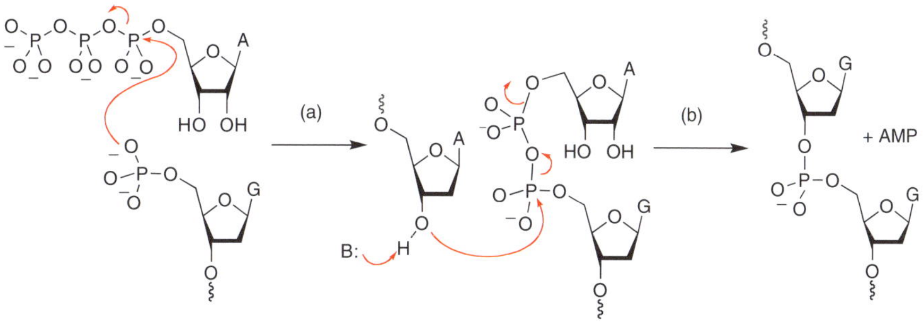

Scheme 5.24 A two-step mechanism for synthesis of a 3′,5′-phosphodiester linkage catalyzed by a DNAzyme. The reaction sequence involves (a) adenylation of DNA by ATP and (b) subsequent ligation of the 5′-activated DNA strand with a phosphate acceptor nucleotide with expulsion of the AMP.[396,397]

5.3.5 Nucleic-Acid Catalysis of Other Reactions

Catalysis by RNA is not limited to reactions likely to have played a role in the RNA world. SELEX has also been used to evolve enzymes catalyzing distinctly non-natural processes. Several groups have initially used a TSA approach, similar to that familiar from the generation of catalytic antibodies (Chapter Section 5.3).[399,400] The targets were reactions that involve a steric change as the transition state is approached (*e.g.* a porphyrin metallation reaction, **5.40**, Scheme 5.25). RNA aptamers were raised first against geometric TSAs and the selected binders screened for catalysis. The approach was successful as a proof of concept, but the rates achieved were more than eight orders of magnitude below those obtained with ribozymes evolved without mechanistic input, so that the immediate conclusion has to be that the power of the SELEX methodology exceeds the value of the mechanistic input.

Completely random selections for catalysis, involving SELEX in combination with a variety of tethering strategies, have been much more successful in enlarging the repertoire of nucleic-acid catalysis. The most efficient catalyst (with a second-order rate enhancement of 10^9) is for a kinase reaction that uses γ-thio-ATP, from which the thiophosphate group is transferred.[401] But other ribozymes significantly broaden the scope of RNA catalysis beyond their home turf of phosphate-transfer reactions. These examples include carbon–carbon bond forming reactions such as the aldol (**5.41**, Scheme 5.25),[402] Michael (**5.42**, Scheme 5.25)[403] and Diels–Alder reactions (**5.43**, Scheme 5.25).[322,404,405]

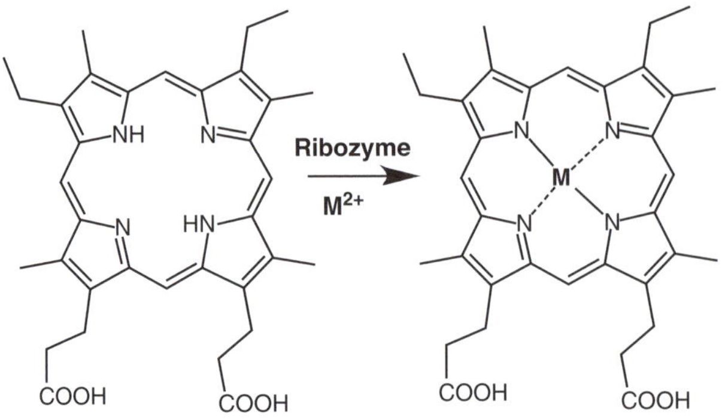

5.40

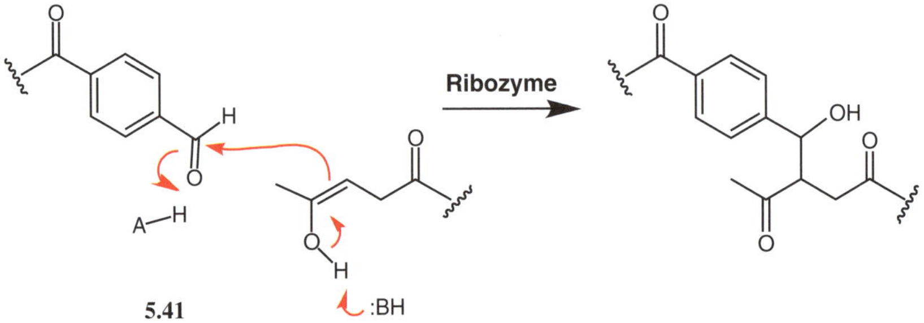

5.41

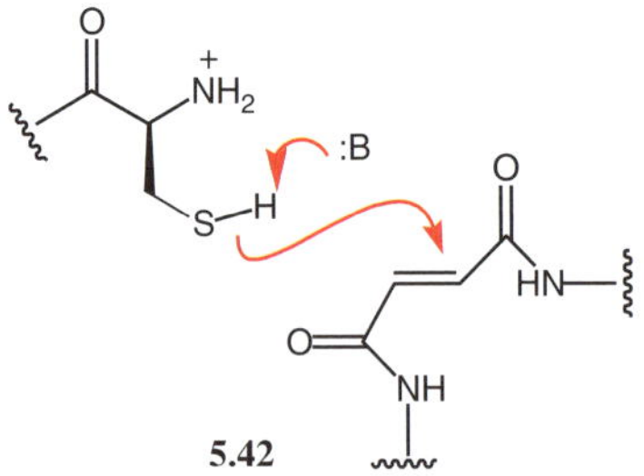

5.42

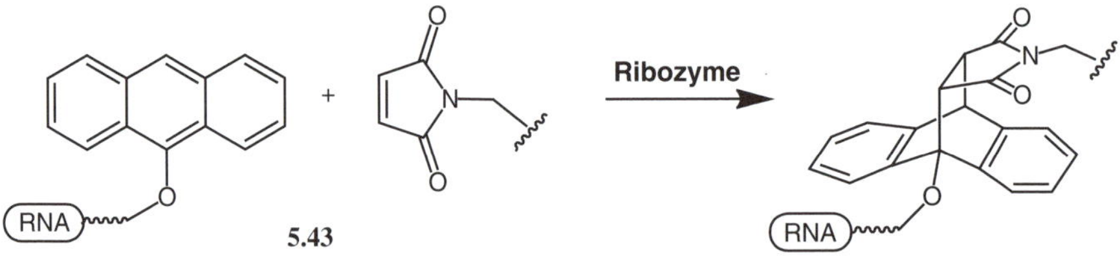

5.43

Scheme 5.25 The scope of RNA catalysis that can be explored by SELEX covers porphyrin metallation (**5.40**) and several examples of carbon–carbon bond forming reactions such as the aldol (**5.41**),[402] Michael (**5.42**)[403] and Diels–Alder reactions (**5.43**).[322,404,405]

Reactions that might be relevant to primitive biosynthetic pathways include an N-glycosylation, in the formation of a nucleotide from the base and the sugar),[406,407] the synthesis of an amide bond[408,409] and the formation of acyl-coenzyme A.[410]

The rate accelerations shown in Table 5.6 are similar to those for reasonable model systems (with the better rate accelerations around 10^3), but these results are most remarkable on account of their *lack* of mechanistic design.

Studies of the evolved catalysts have already provided,[367,368] and will provide in the future, many instructive lessons for the model builder, not least by enhancing our understanding of simple strategies for early evolution catalysts. But the high proficiencies of some of the catalysts in Table 5.6 also serve to show that catalyst discovery by iterative methods can be remarkably powerful, if the application of the effective SELEX system and the large library sizes (up to $\sim 10^{15}$) involved can achieve synergies sufficient to overcome the absence of mechanistic input. Unfortunately, accessible libraries for other types of catalysts are smaller – by many orders of magnitude for chemical compounds (see Sections 5.1 and 5.2) and by several orders of magnitude for proteins (see Section 5.4) – and their handling is significantly more complicated.

5.4 Improving Protein Enzymes

5.4.1 Challenges in Exploring Protein Catalysts

Proteins are unquestionably Nature's preferred catalysts: the best amongst them are up to 10 orders of magnitude more efficient than ribozymes (Table 2.1), even for the reactions ribozymes do best. But at least catalysts made from combinations of four nucleotide bases already give good catalysts whose activities could be created *de novo,* enhanced or altered by powerful cycles of directed evolution using the SELEX methodology. It should in principle be possible to emulate such success with proteins: the larger number of 20 amino-acid building blocks gives access to a larger and potentially more versatile repertoire of functional groups, the 3D structures of many enzymes suggest that their conformations play a productive role in catalysis (contrasting with the RNA-folding problem) and there are a host of efficient natural protein catalysts for us to look at, providing a potential basis for understanding their efficiency and informing us about catalytic strategies.

Despite this convincing advertising progress on protein evolution has been slow. A number of fundamental obstacles stand in the way of extrapolating from the enormous speed with which the relatively young discipline of nucleic-acid enzymes has learned to manipulate nucleic-acid catalysts. We may know more about protein than about nucleic-acid catalysts, but creating functional proteins *de novo* is still difficult – and in most cases remains elusive.

A number of challenges that have to be overcome before protein evolution becomes a straightforward, reliable procedure, offering a safe bet for generating an efficient model enzyme:

(i) The first and principal problem relates to the *"vastness of sequence space"*: The potential diversity of proteins, as a function of the larger number of building blocks, is much larger than for nucleic-acid enzymes. This larger potential repertoire may be advantageous for catalysis: there must be an optimal solution for any problem. But navigating through this vast parameter space is enormously more difficult. For a modest 100 amino-acid protein the hypothetical library of all possible sequences constructed from the 20 naturally occurring amino-acids would contain 20^{100} (*i.e.* more than 10^{130}) sequences. This is a number beyond imagination – compared with 10^{130} the (estimated) number of particles in the universe ($> 10^{80}$) is insignificant! Even in ribozyme selections (for nucleic acids of up to 25 nucleotides) the actual size of the library that could be screened (up to $\sim 10^{15}$) was smaller than the hypothetical diversity. For practical reasons protein libraries are smaller (see below) and in screening procedures, where each member of a library is characterized, the time taken for these measurements limits the accessible numbers. The discrepancy between hypothetical and selectable or screenable diversity is the dominant feature of combinatorial experiments with proteins. This means that shortcuts through "sequence space" are necessary: some of these may be provided by a fundamental understanding of catalysis.

(ii) A second general problem is the intrinsic delicacy of protein structure. Proteins are generally less robust than nucleic acids. The marginal stability of proteins[427,428] can lead to their denaturation and precipitation, rendering them inactive, unselectable – and useless! There are next to no successful cases (bar one example that proves the rule[429]) in protein evolution where the starting point was a random sequence library, as used successfully in numerous nucleic-acid selections. This makes it essential to identify the best structural starting points for evolution: the nature of the protein fold, its robustness and its evolvability will all play a role and must be balanced against the desired diversification of the library that will necessarily introduce instability.

This raises the question of how library design can take account of the delicacy of protein structure, a problem that could be ignored for ribozymes. Library synthesis for nucleic acids is straightforward by error-prone PCR or chemical synthesis of oligonucleotides – witness the examples of successful SELEX experiments not based on a design input. Given that protein 3D structure has to be preserved to some extent, the way the library is made will matter: a compromise will have to be struck between preserving the original fold (and hence protein stability) or venturing further into more diverse structures (potentially destabilizing, but covering a wider fraction of sequence space). As a rule the homologous recombination of genes offers a greater chance of producing folded proteins, but the heterologous recombination of genes encoding functionally diverse proteins promises readier access to new functions. Finally, there are the practicalities involved in generating a DNA library. The large variety of methods available for the generation of DNA libraries, from the familiar error-prone-PCR technique to the recombination of fragments of more or less similar proteins, is described in a comprehensive review.[430]

5.4.2 What Fraction of Diversity Space is Practically Accessible?

This is a fast-developing area of science that is unmistakeably technology driven.[431–433] The key to the success of nucleic-acid evolution was that selections could be carried out from relatively large libraries (up to $\sim 10^{15}$), even though the theoretical diversity was usually larger: more is generally better in this business. To find the proverbial needle in the haystack we need two pieces of information for each library member: how good it is as a catalyst, and its sequence. In protein evolution, in contrast to the position with nucleic acids, the genotype (see the introduction to this chapter) and the phenotype (*i.e.* the readout of function, for example a reaction product) are not identical. Thus, a primary practical consideration for a protein-evolution experiment is how the sequence of a potential catalyst can identified with its function. Technical solutions have been found that reliably and robustly combine the nucleic acid identifier and the protein encoded by it.[432] However, the necessity to work with protein *and* nucleic acid in a selection experiment means that the physical sizes of protein libraries will be orders of magnitude smaller.

Nature links genotype and phenotype by compartmentalizing genes in cells. Many early experiments in directed evolution have employed "colony screening", where an assay is conducted with a monoclonal population of cells. The spatial address of the clone (*e.g.* of colonies on an agar plate or a cell culture grown in a multiwell plate starting with a monoclonal cell sample) allows identification of the successful catalysts. This reliable method – still the workhorse of most directed evolution experiments – has relatively low throughput, of typically up to 10^4 members. The throughput can be increased – at least in an industrial setting – by up to two additional orders of magnitude with automated liquid-handling systems, but these robots are expensive.

To achieve larger libraries and higher throughput a genotype–phenotype link must be established that couples protein molecules to their encoding nucleic acids, for it to be possible for them to be amplified and identified. A range of technologies has been developed over the past two decades, and the key approaches are illustrated in Scheme 5.26.

(i) By far the best-established technology is *phage display*,[434] wherein the selected protein is fused to a phage coat protein (**5.44** in Scheme 5.26), providing the link to the viral DNA.

(ii) *Ribosome display* (**5.45** in Scheme 5.26),[435] links genotype and phenotype via the ribosome, by conserving the ternary complex formed by mRNA, the ribosome and the nascent polypeptide chain during translation. In the absence of a stop codon, the ribosome stalls at the

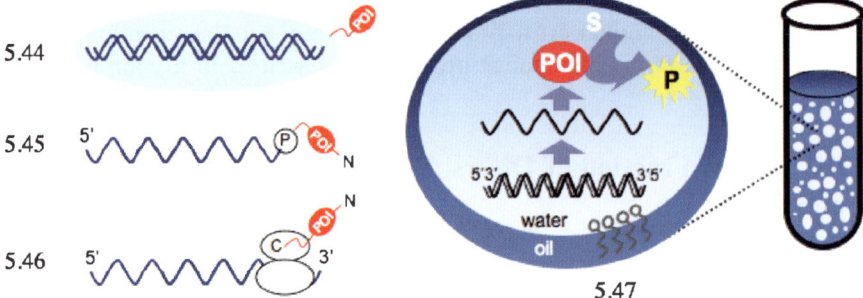

5.44

5.45

5.46

5.47

Scheme 5.26   Genotype–phenotype linkages employed in directed evolution experiments: In *phage display* (**5.44**) the protein of interest (POI) is fused to a phage coat protein to provide the link to the viral DNA. *Ribosome display* (**5.45**) links genotype and phenotype via the ribosome, and *mRNA display* (**5.46**) has a covalent connection in mRNA-puromycin(P)-protein fusions (see also Section 5.4.4). In *in-vitro* compartmentalization (IVC, **5.47**) single gene copies are compartmentalized in emulsion droplets (of micrometer diameter). Protein is produced using a commercially available *in vitro* expression system. The protein thus produced turns over substrate (**S**) to product (**P**), that is retained by the emulsion compartment and can be detected, *e.g.* by fluorescence measurements. In contrast to the three other selection methods, IVC allows screening (*i.e.* characterizing the entire library) rather than selecting winners and qualifiers only (Section 5.1):

   end of the mRNA template. Low temperatures and high magnesium ion concentrations further stabilize this ternary complex that can be used directly in affinity selections.

(iii) *mRNA display* (**5.46** in Scheme 5.26),[429,436,437] has a more robust covalent connection in mRNA-puromycin-protein fusions. Transcription of DNA is followed by the covalent attachment of a DNA-coupled puromycin, a ribosomal inhibitor, to the 3'-end of the mRNA in the ribosome *via* a peptide bond. A remarkable example of evolution of a ligase using this method is discussed in Section 5.4.4.

   In addition, there are a number of other display technologies (*e.g.* CIS-,[438] *MHaeIII*-,[439,440] or SNAP-tag-[441] display) each with different advantages, but so far – as newer inventions – with less of a track record.

   The main area of application of these technologies has been the evolution of binding molecules in the pharmaceutical industry, to create proteins that can interfere with biological processes. Such selections for binding rely on the process of "panning" (familiar from selections of nucleic-acid catalysts, Scheme 5.22): the binding target is immobilized on the solid phase and the protein binder (with the nucleic acid attached) pulled out, and its attached nucleic acid amplified and sequenced. Can this approach be adapted to select catalysts? Catalysis is defined in terms of binding to the elusive transition state rather than to a distinct molecule with a lifetime, and thus presents a different challenge. The methods of phage display[442,443] or ribosome display[444] can be extended to select for catalysis using the suicide inhibitor approach (Section 5.2). Just as in SELEX a positive "hit" requires just one catalytic turnover, so the evolved enzymes are likely to be stoichiometric reagents rather than true catalysts able to process more than one substrate molecule. Engineering multiple turnovers of protein enzymes is difficult, because, unlike nucleic-acid catalysts, they often form covalent intermediates, whose breakdown also has to be catalyzed.

The method of *in vitro* compartmentalization (IVC, **5.45** in Scheme 5.26) may remedy this shortcoming. IVC mimics natural cell compartments by creating artificial "cells" that perform only the one reaction to be selected: Cell-like emulsion droplets – readily prepared using oil, detergents and an inexpensive emulsifier – are used as the sole connection between genotype and phenotype. In a typical procedure, members of the nucleic-acid library are partitioned into separate microscopic compartments, so that on average one copy exists per droplet. *In vitro* expression generates multiple copies of the corresponding protein in the compartment, in which the formation of product (*e.g.* a fluorophore) can be detected. The separation of individual droplets provides for the selection of all enzymatic features simultaneously, namely substrate recognition, product formation, rate acceleration and turnover.

The throughput of each of these methods is determined by how much space the selected DNA-protein species occupies: a typical reaction volume of <1 ml can accommodate no more than 10^{10} emulsion droplets, but more ribosomes (up to 10^{12}), more mRNA-puromycin fusions (up to 10^{14}), and even more RNA molecules (up to 10^{15}, see Section 5.3).

An attraction of the cell-free methods is that they are performed partly or entirely *in vitro*, where selections can be carried out with non-natural amino-acids, at extremes of pH or temperature and, if desired, under nonphysiological conditions. *In vitro* experiments avoid potentially inefficient cloning and transformation (in a typical electroporation experiment we can get the DNA coding for the protein to be evolved into no more than 10^8 electrocompetent *E. coli* cells).[432] This means that library sizes in *in vitro* experiments can be much larger, allowing the exploration of a larger fraction of sequence space. The inability to select under conditions different from the cellular environment, problems with selection of proteins that are toxic to the host, the low dynamic range and cells circumventing selection pressure, *etc.*, often limit the stringency of selection processes *in vivo* (*i.e.* how directly one can select for the desired property).

The last decade has seen enormous progress in addressing each of these issues. To illustrate the nature of the systematic approaches that have been developed, the following paragraphs discuss – only very briefly – some themes of the research involved in this vast and rapidly growing area. We focus on how attempts to classify the proteome can be based on mechanistic principles (Section 5.4.1), and on the underlying technical issues. Finally, we describe two remarkable examples of creating functional proteins from scratch that give a taste of the emerging possibilities.

5.4.3 Mapping Enzyme Function in the Proteome: Protein Superfamilies as a Basis for Understanding Functional Links

Compared to the enormous combinatorial diversity of protein sequences the number of functional active sites is small. Convergent evolution of proteins shows how efficient functional patterns are repeated, *e.g.* in the serine proteases (Section 4.1.1, Figure 4.1). Mapping the larger number of proteins that are now known at the level of sequence, structure and function, gives rise to super-families (Section 1.8): that we can expect to lead – eventually – to a systematic understanding of protein function. The hope is that the patterns recognized can provide the basis for a more systematic understanding of proteins at all levels of sequence, structure and function. These considerations may allow us to decipher the molecular logic of evolution, *e.g.* to detect evolutionary shortcuts that Nature must have used to switch from one function to another, defying the "vastness of sequence space".

The mapping of the protein universe can occur at several levels. Schemes such as SCOP[445–447](http://scop.mrc-lmb.cam.ac.uk/scop) or CATH[448–450] assign proteins to structural superfamilies based on available sequence and 3D struc-tural data. The SUPERFAMILY database provides a web portal access to this information[451] (http://supfam.mrc-lmb.cam.ac.uk/SUPERFAMILY/). Other schemes supplement structural assignments with information on typical catalytic residues in enzymes, *e.g.* in the Catalytic Site Atlas[452] (http://www.ebi.ac.uk/thornton-srv/databases/CSA/). There are also databases that archive available information on mechanistic functional data on enzymes, *e.g.* EzCatDB[453] (http://mbs.cbrc.jp/EzCatDB/), MaCie[454] (http://www.ebi.ac.uk/thornton-srv/databases/MACiE/) or BRENDA[455] (http://www.brenda.uni-koeln.de).

A great deal of structural and mechanistic data on enzymes is thus readily available on the web. But the detailed explanation of functional links still requires careful integration of functional and structural data and mechanistic interpreta-tion. Thus, common reaction steps or the ability to stabilize specific intermediates or transition states have been used to define *catalytic* superfamilies – and to trace the build up of a mechanism in their members. The common set of relevant cat-alytic residues may also provide the basis for the stabilization of alternative transition states or for the rerouting of a common reactive intermediate in a dif-ferent direction, to obtain other products. Starting with a set of congruent che-mical functionalities as well as a shared elementary step will make finding a new activity more likely in a template that is already partly set up for multiple tasks.

5.4.3.1 The Enolase Superfamily

The enolase superfamily (discussed previously in Section 4.4.1.4.3) is one of the most extensively described catalytic superfamilies.[456,457] All superfamily mem-bers are TIM-barrels and bear a metal ion binding site with almost universally conserved metal-binding residues. This metal ion carries out essential functions across the chemically diverse net reactions (*i.e.* isomerization, elimination, lactone hydrolysis, dehydration, epimerization) depicted in Scheme 5.27.

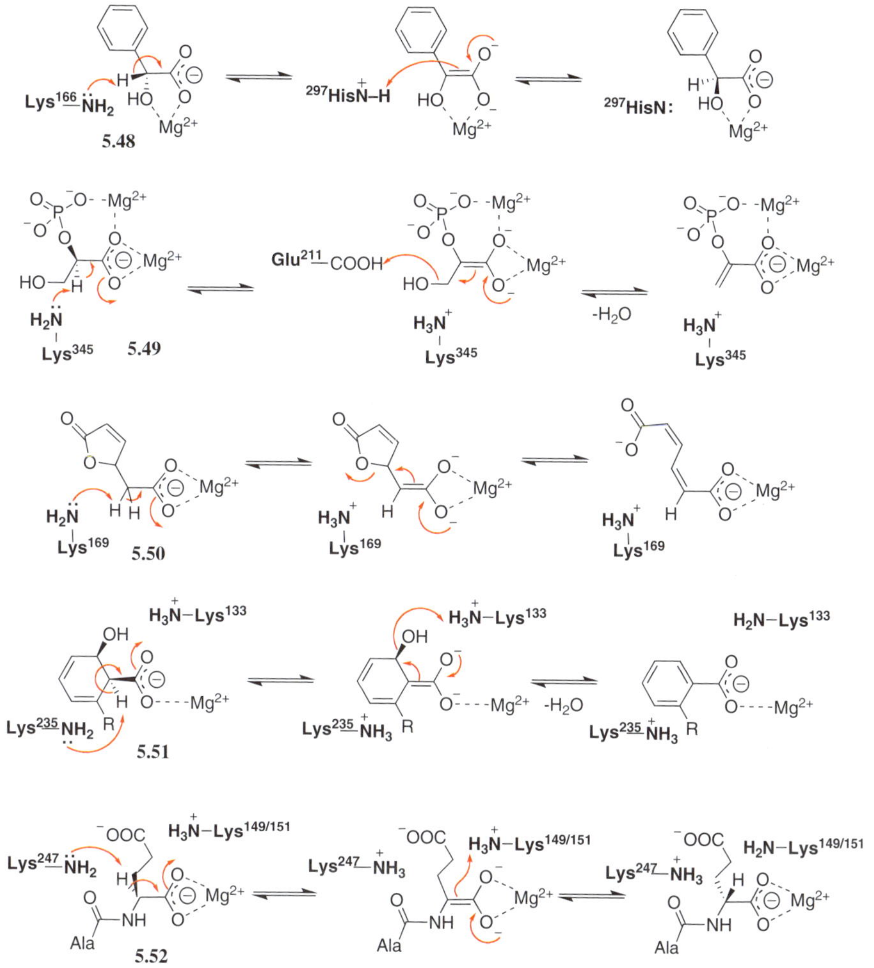

Scheme 5.27   Representative members of the enolase superfamily and their schematic reaction mechanisms compared. Only residues directly involved in the reactions are depicted. (**5.48**) Mandelate racemase (MR) from *P. putida*[458] (MR subgroup members are characterized by a His-Asp dyad, His acting as a general base); (**5.49**) Enolase from yeast[459] (enolase subgroup members are characterized by a conserved Lys acting as a general base); (**5.50**) Muconate lactonizing enzyme (MLE) from *P. putida*[460,461] (MLE subgroup members are characterized by two Lys one acting as a general base, one as a general acid). (**5.51**) *o*-succinylbenzoate synthase (OSBS) from *E. coli*[462] (member of the MLE subgroup); (**5.52**) L-Ala-D/L-Glu epimerase (AEE) from *E. coli* [463] (member of the MLE subgroup). In each case a magnesium ion binds the carboxylate ground state and, more strongly, the dianionic intermediates (middle); the second unifying feature is the catalytic base (lysine/histidine) acting as a general base catalyst that abstracts a hydrogen from the carbon atom in the α-position to the carboxylate.

Combined with the essential magnesium ion is a catalytic base, whose position is varied efficiently to abstract a proton (Section 4.4.1) in the α-position to a carboxylate group bound to the metal. For each reaction in Scheme 5.27, the charge on the carboxylate increases as the transition state is approached and an enediolate intermediate is formed (see Scheme 4.110). This means that the metal ion has to be tuned to bind the dianionic intermediate more strongly than the monoanionic ground state. Partitioning of the intermediate towards different products is achieved by variable residues that are presented from a conserved scaffold in different positions.

Such interrelated scaffolds bear the potential for functional interconversion: superfamily members occasionally exhibit promiscuous activities corresponding to another member of the same family. A member of the enolase superfamily, originally identified as an *N*-acylamino-acid racemase, exhibits OSBS activity (*o*-succinylbenzoate synthase, Scheme 5.27).[464] The crystal structure of OSBS shows how the interactions of the substrate with the enzyme allow room for manoeuvre: just a few direct hydrogen bonds tie the substrate down and most interactions are hydrophobic or mediated through water molecules, leaving potential wiggle room. This lack of restriction might explain the catalytic promiscuity displayed by OSBS.[464] The alternative substrate, an N-acetyl amino-acid, can bind in the same active site, analogous to the native substrate, positioning it ideally for the proton abstraction and donation in the promiscuous racemization reaction.

5.4.3.2 The Alkaline Phosphatase Superfamily

The alkaline phosphatase superfamily (Table 3.2) is also defined by its protein fold,[465] but exhibits an extensive and unusually strong network of crossover promiscuous activities that together suggest that its members are evolutionarily related.[84,466] It exemplifies ways by which reactive groups in the active sites of related superfamily members can deal with a number of substrates by virtue of the high reactivity of the active-site residues.

The eponymous founder member of this superfamily, alkaline phosphatase (AP) is a nonspecific phosphate monoester hydrolase, introduced in Section 4.2.1.2. Other related members include nucleotide pyrophosphatases/phosphodiesterases (NPP),[467] cofactor-independent phosphoglycerate mutases (iPGM), phosphonoacetate hydrolases, phosphonate monoester hydrolases (PMH)[468] and a large number of sulfatases, such as arylsulfatases (AS).[469,470] The superfamily members share very limited sequence homology, so alignments are based on broad structural similarities (discussed below). Other characteristics do not immediately suggest a close relationship: AP has three metal ions (Scheme 4.41), of which at least two are involved in the catalytic mechanism, while other members such as the arylsulfatases use only one metal ion. Specifically the active-site residues vary: besides the metal or metals and the nucleophilic group, the residues lining the pocket diverge considerably. However, the *character* of the active-site residues is conserved: for example, the second metal ion in AP is substituted in arylsulfatase A by a cationic group in a very similar position.[471] This functional rather than literal conservation of residues suggests the common theme is the similar catalytic tasks carried out by this superfamily: substrates (Scheme 5.28) are predominantly negatively charged, with phosphorus- and sulfur-containing groups and hydrolytic group-transfer reactions dominant. So, perhaps an equally definitive common characteristic of the AP superfamily is the similarity of its functional catalytic features.

AP has additional promiscuous activities, namely sulfate monoesterase,[84] phosphate diesterase and phosphonate monoesterase[85,86] activities in addition to its original phosphate monoesterase activity. More recently, an additional phosphite-dependent hydrogenase activity has been found,[87] giving AP no less than five activities. Remarkably, many of the rate accelerations involved are substantial, with second-order rate enhancements $(k_{cat}/K_M)/k_2$ ranging up to 10^{11} (Table 5.7). The additional reactions mirror the functions of other members of this superfamily: thus the arylsulfatase from *Pseudomonas aeruginosa* (PAS), accelerates the hydrolysis of phosphate monoesters (**5.53**, (Scheme 5.28)) by 10^{13},[80] and the hydrolysis of phosphate diesters (**5.54**) by factors as large as 10^{15} to 10^{18} (all in $(k_{cat}/K_M)/k_2$), depending on the structure of the diester substrate.[79] A further member of the AP superfamily, a phosphonate monoester hydrolase (PMH) shows activity not only towards the originally

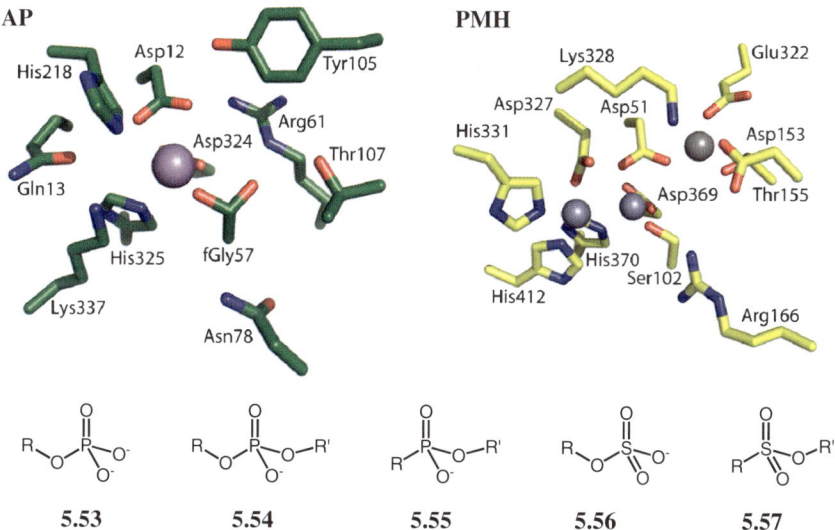

Scheme 5.28 The structures of two members of the alkaline phosphates superfamily (**AP**: see Scheme 4.41, and **PMH**) showing the congruence of functional active site features. The hydrolytic breakdown of a large number of substrates including phosphate monoesters (**5.53**), phosphate diesters (**5.54**), phosphonate monoesters (**5.55**), sulfate monoesters (**5.56**) and sulfonates (**5.57**) is catalyzed by enzymes in the alkaline phosphatase superfamily.

Table 5.7 Catalytically promiscuous enzymes of the alkaline phosphatase superfamily with their respective first- and second-order rate accelerations.

Enzyme	Activity	k_{uncat} (s^{-1})	k_{cat} (s^{-1})	k_{cat}/K_M (M^{-1} s^{-1})	k_{cat}/k_{uncat}	K_{tx} (M)	$(k_{cat}/K_M)/k_2$	Ref
Alkaline phosphatase (AP)	**Phosphomonoesterase**	$\mathbf{2.8 \times 10^{-9}}$	**36**	$\mathbf{3.3 \times 10^{7}}$	$\mathbf{1 \times 10^{10}}$	$\mathbf{8 \times 10^{-17}}$	$\mathbf{7 \times 10^{17}}$	84–86
	Phosphodiesterase	1.1×10^{-11}	n.d.	5×10^{-2}	n.d.	2×10^{-10}	3×10^{11}	
	Phosphonatemonoesterase	4×10^{-11}	n.d.	3×10^{-2}	n.d.	1×10^{-9}	4×10^{10}	
	Sulfatase	5.1×10^{-10}	n.d.	1×10^{-2}	n.d.	5×10^{-8}	1×10^{9}	84
Phosphonate monoester hydrolase (PMH)/ phosphodiesterase	**Phosphodiesterase**	$\mathbf{2.6 \times 10^{-13}}$	**5.8**	$\mathbf{9.2 \times 10^{3}}$	$\mathbf{2 \times 10^{13}}$	$\mathbf{3 \times 10^{-17}}$	$\mathbf{2 \times 10^{18}}$	468–472
	Phosphonate monoesterase	1.7×10^{-11}	2.7	1.5×10^{4}	2×10^{11}	1×10^{-15}	5×10^{16}	
	Sulfonatemonoesterase	5.5×10^{-9}	0.012	4.9×10^{1}	2×10^{6}	1×10^{-10}	5×10^{11}	
	Phosphomonoesterase	4.3×10^{-9}	7.7×10^{-3}	2.2×10^{1}	2×10^{5}	2×10^{-10}	3×10^{11}	
	Sulfatase	1.1×10^{-9}	0.04	5.6×10^{-1}	4×10^{7}	2×10^{-9}	3×10^{10}	
Arylsulfatase (AS)	**Sulfatase**	$\mathbf{5.1 \times 10^{-10}}$	**14**	$\mathbf{4.9 \times 10^{7}}$	$\mathbf{3 \times 10^{10}}$	$\mathbf{1 \times 10^{-17}}$	$\mathbf{5 \times 10^{18}}$	79
	Phosphodiesterase	1.1×10^{-11}	0.55	2.5×10^{5}	5×10^{10}	4×10^{-17}	1×10^{18}	80
	Phosphomonoesterase	2.8×10^{-9}	0.023	7.9×10^{2}	8×10^{6}	4×10^{-12}	2×10^{13}	
Nucleotide Pyrophosphatase (NPP)/ Phosphodiesterase	**Phosphodiesterase**	$\mathbf{1.1 \times 10^{-11}}$	**n.d.**	$\mathbf{2.3 \times 10^{3}}$	**n.d.**	$\mathbf{5 \times 10^{-15}}$	$\mathbf{1 \times 10^{16}}$	467
	Phosphomonoesterase	2.8×10^{-9}	n.d.	1.1	n.d.	3×10^{-9}	2×10^{9}	
	Sulfatase	5.1×10^{-10}	n.d.	2×10^{-5}	n.d.	3×10^{-5}	2×10^{6}	476

assigned substrate phosphonate monoester (**5.55**, (Scheme 5.28)), but also towards phosphate mono- (**5.53**) and diester (**5.54**) as well as sulfates (**5.56**) and sulfonates (**5.57**, Scheme 5.28)). The respective rate accelerations (($k_{cat}/K_M)/k_2$), ranging from 10^{10} to as high as 10^{18}, are summarized in Table 5.7.[472]

As well as the common net hydrolysis reaction of the AP superfamily, this collection encompasses a range of diverse substrates, with charges ranging from zero to –2. The reactions concerned involve attack at two different reaction centers (P and S), and diverse intrinsic reactivities (half-lives from 200 days to 10^5 years under near neutral conditions); and a spectrum of bond-making and - breaking events (leading to associative or dissociative transition states).[467,473,474]

The rate accelerations observed for PMH with promiscuous substrates surpass the accelerations achieved by moderately efficient natural enzymes.[26] Members of this superfamily are remarkably versatile chemical machines that manage to catalyze difficult chemical reactions without seeming to forfeit catalytic efficiency for the very broad observed specificity. This could render PMH a useful scavenging enzyme, offering a "sweeper" function to hydrolyze a wide range of compounds.

These results also have evolutionary implications: For PAS the k_{cat}/K_M values fall within a range of 10^5 and k_{cat} values within 10^3 for all three activities. When a promiscuous activity acquires a new function, it must confer a selective advantage. Although it is difficult to predict exactly what level of activity would endow an organism with a selective advantage, the transition to a new activity will be facilitated if the promiscuous activity is high enough and at least similar to the original activity. This suggests that these promiscuous enzymes may be a useful starting point for evolution of new functions.

The comparison of the catalytically promiscuous members of this superfamily suggests a list of possibly general criteria:

(i) *Native and promiscuous reactions share at least some key features.* The reactions are all hydrolytic, involve a trigonal-bipyramidal geometry at the reaction centre and benefit from the possibility of nucleophilic and general acid/base catalysis. The detailed nature of the transition state appears to be relatively unimportant: both AP and PAS achieve a higher rate acceleration for phosphate diester hydrolysis (with an associative transition state) than for their promiscuous sulfate or phosphate monoesterase reactions (with dissociative transition states resembling the transition states of their cognate, native reactions).[79,80,86] This means that complete coincidence of the catalytic factors is not important – a subset of similarity can be sufficient.

(ii) *A reactive active site.* The observed catalysis must be due in no small part to the intrinsic reactivity of the metal ion bound in the active site. The catalytic motif of two metal ions has been shown in model systems (Section 4.2.1.2) to accelerate hydrolytic reactions by lowering the pK_a of the nucleophile and thus increasing the concentration of reactive, deprotonated nucleophile available for the reaction. Notably several highly promiscuous enzymes (such as serum paroxonase, carbonic

anhydrase[475] and phosphotriesterase[89]) as well as many enzyme models
(Section 4.2.1.2.1) feature metal ions. The availability of a nucleophile
activated by a metal ion will invariably accelerate hydrolytic reactions,
as long as the substrate is brought into "appropriate proximity", thus
reducing the problem of catalysis to some extent to a problem of
binding (*i.e.* orientation and positioning of the substrate with respect to
the reactive nucleophile).

(iii) *A sufficiently spacious active site.* AP, PAS and PMH all possess relatively accessible binding sites that can accommodate substrates with
different steric demands, perhaps by allowing different binding modes
in a large binding pocket.

(iv) *A recyclable nucleophile.* In the case of PMH and PAS the rates of
native and promiscuous reactions cover a relatively narrow range. This
suggests that the reaction mechanisms of these enzymes may be set up
for particularly efficient promiscuity. The enzyme takes advantage of
nucleophilic catalysis by the unusual *geminal diol* nucleophile (Scheme
5.29) instead of the much more common side-chain nucleophiles of
serine, threonine, tyrosine or cysteine.

It is clearly thermodynamically advantageous to use this nucleophile for sul-
fate-transfer hydrolysis, because an enzyme sulfate ester intermediate would be
difficult to hydrolyze *via* cleavage of the S–O bond. Furthermore, the breakdown
of the intermediate involves the same C–O bond, rather than S–O or P–O, for
both native and promiscuous reactions. Chemically, this second step (Scheme
5.29) then differs only in the leaving group, and the requirements for general

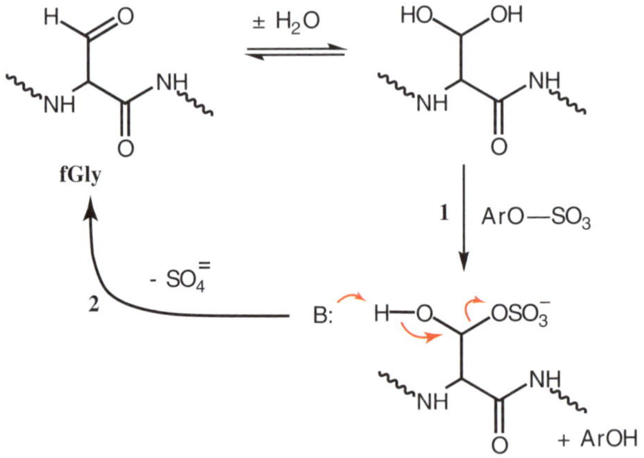

Scheme 5.29   The effective nucleophile in the active site of PAS and PMH (see the
text) is an OH group of the (metal-coordinated) gem-diol hydrate of
formyl glycine (**fGly**: formed by post-translational modification of Ser
or Cys residues). Acylation of one OH (step 1) converts OH into a good
leaving group, readily eliminated (step 2) to regenerate **fGly**.

acid–base catalysis of hemiacetal cleavage are identical. Thus, the formylglycine nucleophile unifies the second step of the catalytic cycle for all its reactions.

Further work is needed to rank and quantify the contributions of these various criteria and to ascertain their generality, but for the moment they provide a working hypothesis for explaining the unusual promiscuity of the AP superfamily. More generally, the criteria discussed provide a chemical dimension to the evolution of function, based on the economical conservation of catalytic features.

5.4.4 Challenging Chance by Design and Directed Evolution[477,478]

While changing the specificity of an existing enzyme by directed evolution has often been done by starting from an existing – sometimes promiscuous – activity, the *de novo* creation of protein enzymes is a much more ambitious goal, rarely achieved. Two recent examples show that such daring projects are just becoming realistic, benefiting from technological advances on the one hand, and from design input on the other.

Seelig and Szostak[437] evolved a completely new enzyme *in vitro* from a random library, without mechanistic or detailed structural rationale. They relied primarily on selection from a very large pool of randomized proteins by mRNA display (one of the genotype–phenotype linkages mentioned above, Scheme 5.26). The target reaction was a DNA ligation reaction, reminiscent of ribozyme selections in Section 5.3: the bond formed between nucleic acid substrate and the encoding nucleic acid distinguishes a winner in the selection process, so that one turnover is enough to label a catalyst as a hit – the only difference being that in this case protein is attached to the coding RNA. The power of this selection experiment lies in the *in vitro* setup that allows selection from a very large library, with $>10^{12}$ members. The chances of finding a catalyst were further increased by: (i) adding an RNA splint to the reaction solution, to pair with substrate and coding nucleic acids, so that the reaction centres were brought into close proximity (Scheme 5.30): (ii) using a good (pyrophosphate) leaving group to lower the catalytic barrier, and (iii) using a starting sequence corresponding to a simple, structured protein motif (a zinc finger) rather than a completely random sequence. It is likely that these decisions were necessary to emerge successfully from the "vastness of sequence space". In the event, 17 rounds of randomization and "panning" yielded an active protein RNA ligase, with a k_{obs} of $36\,min^{-1}$ and a estimated rate acceleration of $>10^6$.

Navigating the vastness of sequence space successfully should be further facilitated by some rational input to the nature of the scaffold on which functional groups are mounted, and the desired interactions between substrate and protein. Protein design alone – creating a model enzyme from scratch – is currently too difficult, but an element of design can improve the chances of success in hybrid approaches. A recent example combines mechanistic analysis

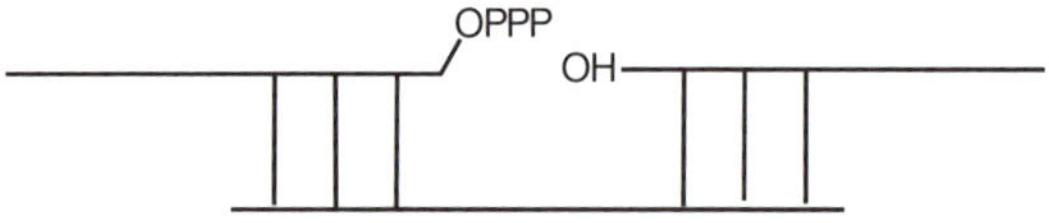

5.58

Splint

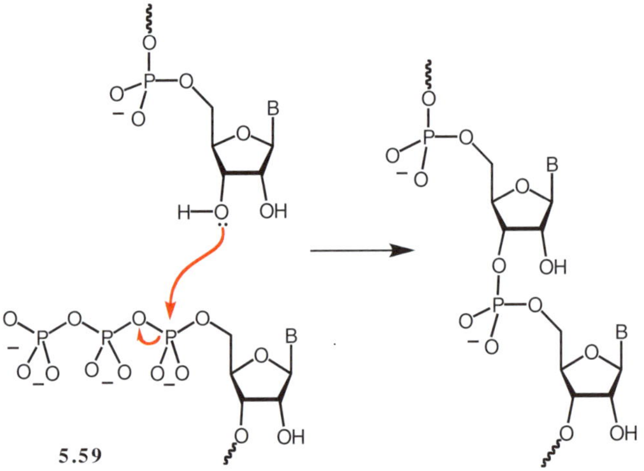

5.59

Scheme 5.30   The RNA ligase reaction catalyzed by a protein selected by mRNA
display (Scheme 5.26). A RNA "splint" serves as a reaction template
(**5.58**), facilitating the recognition of the two substrates and bringing
together the activated 5′- and 3′-hydroxyl groups of two RNA chains.
The reaction catalyzed by the evolved ligase shown in (**5.59**) involves
attack by a 3′-ribose hydroxyl nucleophile on the α-phosphate group of
the (activated) 5′-triphosphate electrophile.

with protein design and directed evolution. Röthlisberger et al.[479] have calculated the transition state for the well-known Kemp elimination reaction
(Scheme 5.4: discussed previously in Section 5.1.1 and Section 5.2.2).

Calculations and structural analysis were used to find a folded protein
expected to have a chance of catalyzing this reaction, while minimizing the
problem of protein unfolding often encountered in *de novo* design approaches.
Quantum-mechanical methods predicted where the catalytic base and other
catalytic groups should ideally be placed (delocalizing charge by π-stacking or
offsetting charge development elsewhere in the transition state **TS** (Scheme 5.4)).
The idealized geometry calculated for the transition state was then implanted *in
silico* into existing, structurally known protein folds, using a hashing algorithm
called RosettaMatch to predict sequences that would accommodate this geometry. The benefit of large numbers shows up here, too: the virtual screening
obviates the need for a screening *experiment* at this stage of the project.

The protein structures that came up were predominantly TIM barrels, a very
common catalytic scaffold in Nature, that directs catalytic side-chains towards
the catalytic pocket from central β-strands. TIM barrels are used for a variety

of chemically distinct reactions in Nature, so perhaps it is no wonder that this familiar fold should be selected. The corresponding mutants of existing enzymes (with their own, different specificities) were then expressed, and of these 59 selected proteins eight were active catalysts. with k_{cat} and k_{cat}/k_{uncat} values up to $0.005\,min^{-1}$ and 10^5, respectively. This design phase was followed up by directed evolution, in which the designed catalysts were randomized using error-prone PCR. Seven rounds of titre-plate screening, of 10^3 mutants per round, improved the designed catalysts by up to 200-fold. The best catalyst to emerge had a rate acceleration k_{cat}/k_{uncat} of 10^6. This is undoubtedly a success for a novel way of integrating mechanistic data and virtual docking: and it will be even more interesting to see whether the predicted stabilizing interactions are indeed influencing the TS in the desired way.

There is reason to be optimistic, and to expect further progress in this area, as several groups are now employing hybrid approaches. For example Hermann *et al.*[480] have used virtual screening of transition states derived from a large number of potential substrates against protein structures. Sequencing and structural genomic initiatives have provided data on large numbers of potential enzymes, though their actual function remains undetermined in the majority of cases. Hermann *et al.* were thus able to match a TS against one of the many orphan structures from structural genomics exercises, and experimentally verify that prediction, so experiment and calculation are gratifyingly complementary.

Combining design, to increase the chances of finding active catalysts by virtual screening, with efficient selection technologies that make possible the actual screening of large numbers of candidates, is a promising prospect. The success of such an approach can be evaluated against the body of evidence provided by previous models of all sorts. The Kemp elimination has long been a target for model builders, so a comparison with other systems is uniquely possible in this case. Data for a variety of systems, catalytic antibodies, off-the-shelf proteins, a host–guest system and synzymes (modified PEIs, Section 5.1.1), as well as the designed proteins of Röthlisberger *et al.* are available (and summarized in Table 5.8). The rate acceleration k_{cat}/k_{uncat} for the latter designed and evolved catalyst is similar to that for the synzyme (Section 5.1.1) and its k_{cat} or k_{cat}/K_M values are comparable to catalytic antibodies and off-the-shelf proteins. This brief analysis suggests that synthetic and protein catalyst converge in their rate-enhancing effects.

Much elegant ingenuity has gone into the design of enzyme models and mimics, and into devising schemes to evolve catalysts. Remarkably, the best *improvements* seen in different systems – the Kemp elimination is a fitting example – do not diverge a great deal. Kemp's insightful original study of this elimination reaction has set the basis for detailed understanding of how the rate enhancements can be achieved in different ways. One raison d'être for an enzyme mimic in this situation is whether the rationales behind its conception have been translated into the desired catalytic effect. The conclusions may be consensual – *e.g.* that synzymes achieve their catalysis through medium effects – or sometimes controversial – *e.g.* whether the catalytic antibody can or cannot position a catalytic base efficiently.[360,481,482] The crucial test for any model must be whether the protein's function responds convincingly to the imperatives set

Table 5.8   Kinetic data and rate accelerations for selected enzyme models and model enzymes catalyzing the Kemp elimination (Scheme 5.18).

Catalyst	k_{cat} $[min^{-1}]$	K_M $[\mu M]$	k_{cat}/K_M $[min^{-1}M^{-1}]$	k_{cat}/k_{uncat}	$(k_{cat}/K_M/k_2)$
Antibodies					
34E4 [359]	36	120	3.3×10^5	9700	3.8×10^7
35F10 [359]	21	630	3.4×10^4		9.1×10^6
13G5 [361]				10^5	
Off-the-shelf proteins					
BSA [360,481] pH 8	0.7	2000	0.35	$\sim 10^3 - 10^4$	–
pH 9	14.6	700	21	$\sim 10^3 - 10^4$	–
Synzymes					
7D/1.1 [481]	370^a	9600	–		–
7D/2.1 [481]	120^a	7700	1.6×10^{4a}		
	1.4^a		181^b	2.9×10^{5b}	7.2×10^{4b}
7D.2.2 [330]	40^a	4200			
	$>5^b$			$>10^{6b}$	3.5×10^{5b}
Designed TIM-barrel [479]					
KE59	17	1800	9780	2.5×10^5	–
KE07 wt	1	1400	730	1.6×10^4	–
KE07 mutant R7 2/5B^c	70	860	8.3×10^4	1×10^6	–
Designed host–guest system[483]	0.04	10 000		1000	4

Notes: ^{a}Overall;
^{b}per site;
^{c}adapted by rounds of directed evolution

by the designer, and we look forward to further controversy and an emerging consensus. In each case an understanding of rules for each effect provides the basis for the next, crucial challenge. Putting different catalytic effects together is often more difficult than emulating a single effect in isolation (as discussed in the introduction to Chapter 3).

Creating diversity and emulating natural evolution are strategies that can synergize with design, to make progress more rapid. However, the best improvements delivered by each approach – and even, so far, their combinations – are still limited to relatively small changes compared to the total catalytic package that Nature achieves routinely: and gives us some idea of the limitations in our current understanding. It is the systematic analysis and development of natural and unnatural, synthetic and biological systems that will teach us better how to take advantage of the multiple leads Nature offers.

Recommended Further Reading

I. M. Klotz, J. Suh, Evolution of Synthetic Polymers with Enzyme-like Catalytic Activities. In *Artificial Enzymes*, R. Breslow, ed.; Wiley-VCH: Weinheim, 2005.

B. Helms, J. M. J. Frechet, The dendrimer effect in homogeneous catalysis. *Adv. Synth. & Catal.* **2006**, *348*, 1125–1148.

J. Kofoed, J. L. Reymond, Dendrimers as artificial enzymes. *Curr. Opin. Chem. Biol.* **2005**, *9*, 656–664.

D. Hilvert, Critical analysis of antibody catalysis. *Ann. Rev. Biochem.* **2000**, 69, 751–93.

Binding Energy and Catalysis: Implications for TS Analogs and Catalytic Antibodies. M. M. Mader, P. A. Bartlett, *Chem. Rev.* **1997**, *97*, 1281–1301.

Y. Xu, N. Yamamoto, K. D. Janda, Catalytic antibodies: hapten design strategies and screening methods. *Bioorg. Med. Chem.* **2004**, *12*, 5247–5268.

J. A. Doudna, T. R. Cech, The chemical repertoire of natural ribozymes. *Nature* **2002**, 418, (6894), 222–8.

S. Hammes-Schiffer, S. J. Benkovic, Relating protein motion to catalysis. *Ann. Rev. Biochem.* **2006**, 75, 519–41.

C. Jackel, P. Kast, D. Hilvert, Protein design by directed evolution. *Ann. Rev. Biophys.* **2008**, 37, 153–73.

M. E. Glasner, J. A. Gerlt, P. C. Babbitt, Mechanisms of protein evolution and their application to protein engineering. *Adv. Enzymol. Relat. Areas Mol. Biol.* **2007**, 75, 193–239, xii–xiii.

References

1. *Enzymes in Industry*; 3 edn.; Aehle, W., ed.; Wiley-VCH: Weinheim, 2007.
2. *The Lock and Key Principle*; Behr, J.-P., ed.; Wiley: Chichester, 1994.
3. H. Maskill, *The Physical Basis of Organic Chemistry*; Oxford University Press: Oxford, New York, 1985.
4. W. P. Jencks, *J. Am. Chem. Soc.*, 1972, **94**, 4731.
5. W. P. Jencks, *Chem. Rev.*, 1972, **72**, 705–718.
6. A. J. Kirby and P. W. Lancaster, *J. Chem. Soc. Perkin Trans. 2*, 1972, **1206**.
7. A. J. Kirby, R. S. McDonald and C. R. Smith, *J.C.S. Perkin Trans. 2*, 1974, 1495–1504.
8. A. J. Kirby, *Adv. Phys. Org. Chem.*, 1980, **17**, 183–278.
9. W. P. Jencks, *Adv. Enzymol.*, 1975, **43**, 219–410.
10. F. M. Menger, *Accts. Chem. Res.*, 1985, **18**, 28–33.
11. L. Mandolini, *Adv. Phys. Org. Chem.*, 1986, **22**, 1.
12. A. J. Kirby and G. J. Lloyd, *J. Chem. Soc. Perkin Trans. 2*, 1976, 1753.
13. L. B. Spector, *Covalent Catalysis by Enzymes*, Springer-Verlag, New York, 1982.
14. X. Y. Zhang and K. N. Houk, *Accts. Chem. Res.*, 2005, **38**, 379–385.
15. K.-E. Jaeger and T. Eggert, *Curr. Opin. Biotechnol.*, 2004, **15**, 305–313.
16. K. N. Houk, A. G. Leach, S. P. Kim and X. Y. Zhang, *Angew. Chem., Int. Ed. Engl.*, 2003, **42**, 4872–4897.
17. W. Blokzijl and J. Engberts, *Angew. Chem., Int. Ed. Engl.*, 1993, **32** 1545–1579.
18. S. Otto and J. B. F. Engberts, *N. Org. Biomol. Chem.*, 2003, **1**, 2809–2820.
19. R. Mannhold and R. F. Rekker, *Perspec. Drug Discov. Des.*, 2000, **18** 1–18.
20. J. DeChancie and K. N. Houk, *J. Am. Chem. Soc.*, 2007, **129**, 5419–5429.
21. W. P. Jencks, *Ann. Rev. Biochem.*, 1997, **66**, 1–18.

From Enzyme Models to Model Enzymes
By Anthony J. Kirby and Florian Hollfelder
© Anthony J. Kirby and Florian Hollfelder 2009
Published by the Royal Society of Chemistry, www.rsc.org

22. A. C. Eliot and J. F. Kirsch, *Ann. Rev Biochem*, 2004, **73**, 383–415.

23. H. Decker, T. Schweikardt and F. Tuczek, *Angew. Chem., Int. Ed. Engl.*, 2006, **45**, 4546–4550.

24. T. Imoto, L. N. Johnson, A. C. T. North, D. C. Phillips, J. A. Rupley, In: *The Enzymes*, P. D. Boyer (Ed.), 1972, Vol. VII, p. 665.

25. B. Capon, M. C. Smith, E. Anderson, R. H. Dahm and G. H. Sankey, *J. Chem. Soc. B*, 1969, 1038–1047.

26. R. Wolfenden, *Chem. Rev.*, 2006, **106**, 3379–3396.

27. G. K. Schroeder, C. Lad, P. Wyman, N. H. Williams and R. Wolfenden, *Proc. Nat. Acad. Sci. USA*, 2006, **103**, 4052–4055.

28. A. R. Fersht, *Structure and Mechanism in Protein Science*, Freeman, New York, 1999.

29. A. Williams, *Free-Energy Relationships in Organic and Bio-Organic Chemistry*, Royal Society of Chemistry: Cambridge, 2003.

30. M. F. Perutz, *Proc. Roy. Soc. B*, 1967, **167**, 349.

31. P. A. Frey and A. D. Hegeman, *Enzymatic Reaction Mechanisms*, Chapter 2, Oxford, New York, 2007.

32. H. Li, A. D. Robertson and J. H. Jensen, *Proteins-Struct. Funct. Bioinform.*, 2005, **61**, 704–721.

33. D. S. Kemp, D. D. Cox and K. G. Paul, *J. Am. Chem. Soc.*, 1975, **97**, 7312–7318.

34. A. Warshel, *Computer Modelling of Chemical Reactions in Enzymes and in Solutions*, Wiley, Chichester, 1991.

35. M. J. Dewar and D. M. Storch, *Proc. Natl. Acad. Sci. USA*, 1985, **82**, 2225–9.

36. W. R. Cannon and S. J. Benkovic, *J. Biol. Chem.*, 1998, **273**, 26257–60.

37. A. Warshel, *J. Biol. Chem.*, 1998, **273**, 27035–8.

38. A. Warshel, P. K. Sharma, M. Kato, Y. Xiang, H. Liu and M. H. Olsson, *Chem. Rev.*, 2006, **106**, 3210–35.

39. F. Cramer, *Einschlussverbindungen*, Springer-Verlag, Berlin, 1954.

40. R. Breslow and S. D. Dong, *Chem. Rev.*, 1998, **98**, 1997–2011.

41. O. S. Tee, M. Bozzi, J. J. Hoeven and T. A. Gadosy, *J. Am. Chem. Soc.*, 1993, **115**, 8990–8998.

42. R. L. VanEtten, J. F. Sebastian, G. A. Clowes and M. L. Bender, *J. Am. Chem. Soc.*, 1967, **89**, 3242–3253.

43. P. J. Bartlett and C. K. Marlowe, *Biochemistry*, 1983, **22**, 4618–4624.

44. M. M. Mader and P. A. Bartlett, *Chem. Rev.*, 1997, **97**, 1281–1301.

45. G. Wulff, *Chem. Rev.*, 2002, **102**, 1–28.

46. P. T. Corbett, J. Leclaire, L. Vial, K. R. West, J. L. Wietor, J. K. M. Sanders and S. Otto, *Chem. Rev.*, 2006, **106**, 3652–3711.

47. I. M. Bell, M. L. Fisher, Z. P. Wu and D. Hilvert, *Biochemistry*, 1993, **32**, 3754–62.

48. D. Haring, B. Hubert, E. Schuler and P. Schreier, *Arch. Biochem. Biophys.*, 1998, **354**, 263–9.

49. G. Lian, L. Ding, M. Chen, Z. Liu, D. Zhao and J. J. Ni, *J. Biol. Chem.*, 2001, **276**, 28037–41.

50. G. J. Bernardes, J. M. Chalker, J. C. Errey and B. G. Davis, *J. Am. Chem. Soc.*, 2008, **130**, 5052–3.

51. P. M. Rendle, A. Seger, J. Rodrigues, N. J. Oldham, R. R. Bott, J. B. Jones, M. M. Cowan and B. G. Davis, *J. Am. Chem. Soc.*, 2004, **126**, 4750–1.

52. B. G. Davis, K. Khumtaveeporn, R. R. Bott and J. B. Jones, *Bioorg. Med. Chem.*, 1999, **7**, 2303–11.

53. B. G. Davis, X. Shang, G. DeSantis, R. R. Bott and J. B. Jones, *Bioorg. Med. Chem.*, 1999, **7**, 2293–301.

54. T. L. Hendrickson, V. de Crecy-Lagard and P. Schimmel, *Ann. Rev. Biochem.*, 2004, **73**, 147–76.

55. L. Wang and P. G. Schultz, *Angew. Chem., Int. Ed. Engl.*, 2004, **44**, 34–66.

56. L. Wang, J. Xie and P. G. Schultz, *Ann. Rev. Biophys. Biomol. Struct.*, 2006, **35**, 225–49.

57. C. Khosla and P. Harbury, *Nature*, 2001, **409**, 247–252.

58. L. Panella, J. Broos, J. Jin, M. W. Fraaije, D. B. Janssen, M. Jeronimus-Stratingh, B. L. Feringa, A. J. Minnaard and J. G. de Vries, *J. Chem. Soc. Chem. Commun.*, 2005, 5656–8.

59. D. R. Corey and P. G. Schultz, *Science*, 1987, **238**, 1401–3.

60. D. Pei, D. R. Corey and P. G. Schultz, *Proc. Nat. Acad. Sci. USA*, 1990, **87**, 9858–62.

61. G. M. Culver and H. F. Noller, *Methods Enzymol.*, 2000, **318**, 461–75.

62. M. E. Wilson and G. M. Whitesides, *J. Am. Chem. Soc.*, 1978, **100**, 306.

63. M. Skander, N. Humbert, J. Collot, J. Gradinaru, G. Klein, A. Loosli, J. Sauser, A. Zocchi, F. Gilardoni and T. R. Ward, *J. Am. Chem. Soc.*, 2004, **126**, 14411–8.

64. J. Collot, J. Gradinaru, N. Humbert, M. Skander, A. Zocchi and T. R. Ward, *J. Am. Chem. Soc.*, 2003, **125**, 9030–1.

65. M. D. Toscano, K. J. Woycechowsky and D. Hilvert, *Angew. Chem. Int. Ed. Engl.*, 2007, **46**, 3212–36.

66. M. J. Wishart, J. M. Denu, J. A. Williams and J. E. Dixon, *J. Biol. Chem.*, 1995, **270**, 26782–5.

67. A. Brik, L. J. D'Souza, E. Keinan, F. Grynszpan and P. E. Dawson, *ChemBiochem.*, 2002, **3**, 845–51.

68. K. Johnsson, R. K. Allemann, H. Widmer and S. A. Benner, *Nature*, 1993, **365**, 530–2.

69. N. V. Raghavan and D. L. Leussing, *J. Am. Chem. Soc.*, 1977, **99** 2188–95.

70. D. L. Leussing and N. V. Raghavan, *J. Am. Chem. Soc.*, 1980, **102**, 5635.

71. T. K. Harris, R. M. Czerwinski, W. H. Johnson Jr, P. M. Legler, C. Abeygunawardana, M. A. Massiah, J. T. Stivers, C. P. Whitman and A. S. Mildvan, *Biochemistry*, 1999, **38**, 12343–57.

72. U. T. Bornscheuer, C. Bessler, R. Srinivas and S. H. Krishna, *Trends Biotechnol.*, 2002, **20**, 433–7.

73. U. T. Bornscheuer and R. J. Kazlauskas, *Hydrolases in Organic Synthesis*, Wiley-VCH, Weinheim, 1999.

74. O. Khersonsky, C. Roodveldt and D. S. Tawfik, *Curr. Opin. Chem. Biol.*, 2006, **10**, 498–508.

75. R. K. Andrews, A. Dexter, R. L. Blakeley and B. J. Zerner, *Am. Chem. Soc.*, 1986, **108**, 7124–7125.

76. P. J. O'Brien and D. Herschlag, *Chem. Biol.*, 1999, **6**, R91–R105.

77. U. T. Bornscheuer and R. J. Kazlauskas, *Angew. Chem., Int. Ed. Engl.*, 2004, **43**, 6032–40.

78. S. Jonas and F. Hollfelder, In *The Handbook of Protein Engineering*, ed. U. T. Bornscheuer and S. Lutz, Wiley-VCH, Chichester, 2008, **1**, pp. 47–72.

79. A. C. Babtie, S. Bandyopadhyay, L. Olguin, F. Hollfelder, *Angew. Chem., Int. Ed. Engl.* 2009, **48**, 3692–3694.

80. L. F. Olguin, S. E. Askew, A. C. O'Donoghue and F. Hollfelder, *J. Am. Chem. Soc.*, 2008, **130**, 16547–16555.

81. R. A. Jensen, *Ann. Rev. Microbiol.*, 1976, **30**, 409–25.

82. S. Ohno, *Evolution by Gene Duplication*, Springer, New York, 1970.

83. R. H. Abeles, P. A. Frey and W. P. Jencks, *Biochemistry*, Jones and Bartlett, Boston, MA, 1992.

84. P. J. O'Brien and D. Herschlag, *J. Am. Chem. Soc.*, 1998, **120**, 12369–12370.

85. J. G. Zalatan and D. Herschlag, *J. Am. Chem. Soc.*, 2006, **128**, 1293–303.

86. P. J. O'Brien and D. Herschlag, *Biochemistry*, 2001, **40**, 5691–9.

87. K. Yang and W. W. Metcalf, *Proc. Nat. Acad. Sci. USA*, 2004, **101**, 7919–24.

88. P. J. O'Brien, *Chem. Rev.*, 2006, **106**, 720–752.

89. H. Shim, S. B. Hong and F. M. Raushel, *J. Biol. Chem.*, 1998, **273**, 17445–50.

90. C. Roodveldt and D. S. Tawfik, *Biochemistry*, 2005, **44**, 12728–12736.

91. V. Carnevale, S. Raugei, C. Micheletti and P. Carloni, *J. Am. Chem. Soc.*, 2006, **128**, 9766–9772.

92. D. M. Quinn and S. R. Feaster, In *Comprehensive Biological Catalysis*, ed. M. L. Sinnott, Academic Press, London, 1998, **Vol. 1**, p. 455–482.

93. A. M. Vandersteen and K. D. Janda, *J. Am. Chem. Soc.*, 1996, **118** 8787–8790.

94. C. J. Belke, S. C. K. Su and J. A. Shafer, *J. Am. Chem. Soc.*, 1971, **93**, 4552.

95. K. N. G. Chiong, S. D. Lewis and J. A. Shafer, *J. Am. Chem. Soc.*, 1975, **97**, 418–423.

96. S. M. Felton and T. C. Bruice, *J. Am. Chem. Soc.*, 1969, **91**, 6721–6732.

97. G. A. Rogers and T. C. Bruice, *J. Am. Chem. Soc.*, 1974, **96**, 2473–2481.

98. P. Carter and J. A. Wells, *Nature*, 1988, **332**, 564–568.

99. W. L. Mock, *Perkin. 2*, 1995, 2069–2074.

100. T. P. O'Connell, R. M. Day, E. V. Torchilin, W. W. Bachovchin and J. P. G. Malthouse, *Biochem. J.*, 1997, **326**, 861–866.

101. G. M. Hübner, J. Gläser, C. and F. Vögtle, *Angew. Chem. Int. Ed.*, 1999, **38**, 383–385.

102. R. Breslow, G. Trainor and A. Ueno, *J. Am. Chem. Soc.*, 1983, **105**, 2739–2744.

103. F. M. Menger and M. Ladika, *J. Am. Chem. Soc.*, 1987, **109**, 3145–3146.

104. K. Rama Rao, T. N. Srinivasan, N. Bhanumathi and P. B. Sattur, *J. Chem. Soc. Chem. Commun.*, 1990, 10–11.

105. D. Q. Yuan, Y. Kitagawa, K. Aoyama, T. Douke, M. Fukudome and K. Fujita, *Angew. Chem., Int. Ed. Engl.*, 2007, **46**, 5024–5027.

106. F. Diederich, G. Schurmann and I. Chao, *J. Org. Chem.*, 1988, **53**, 2744–2757.

107. D. J. Cram, P. Y. Lam and S. P. Ho, *J. Am. Chem. Soc.*, 1986, **108**, 839–841.

108. R. Menard, C. Plouffe, P. Laflamme, T. Vernet, D. C. Tessier, D. Y. Thomas and A. C. Storer, *Biochemistry*, 1995, **34**, 464–471.

109. P. Ghosh, G. Federwisch, M. Kogej, C. A. Schalley, D. Haase, W. Saak, A. Lutzen and R. M. Gschwind, *Org. Biomol. Chem.*, 2005, **3**, 2691–2700.

110. J. W. Keillor and R. S. Brown, *J. Am. Chem. Soc.*, 1992, **114**, 7983–7989.

111. R. S. McDonald, P. Patterson and A. Stevenswhalley, *Can. J. Chem.*, 1983, **61**, 1846–1852.

112. Y. Chao, G. R. Weisman, G. D. Y. Sogah and D. J. Cram, *J. Am. Chem. Soc.*, 1979, **101**, 4948–4958.

113. M. L. Bender, Y.-L. Chow and F. Chloupek, *J. Am. Chem. Soc.*, 1958, **80**, 5380–5384.

114. M. F. Aldersley, A. J. Kirby, P. W. Lancaster, R. S. McDonald and C. R. Smith, *J. Chem. Soc., Perkin Trans. 2*, 1974, 1487.

115. N. S. Andreeva and L. D. Rumsh, *Protein Science*, 2001, **10**, 2439–2450.

116. B. M. Dunn, *Chem. Rev.*, 2002, **102**, 4431–4458.

117. G. Iliadis, B. Brzezinski and G. Zundel, *Biophys. J.*, 1996, **71**, 2840–2847.

118. B. Swoboda, M. Beltowska-Brzezinska, G. Schroeder, B. Brzezinski and G. Zundel, *J. Phys. Org. Chem.*, 2001, **14**, 103–108.

119. G. Parkin, *Chem. Rev.*, 2004, **104**, 699–767.

120. R. Breslow, *Rec. Trav. Chim.*, 1995, **113**, 493.

121. J. Xia, Y. Shi, Y. Zhang, Q. Miao and W. Tang, *Inorg. Chem.*, 2003, **42**, 70–77.

122. R. Alsfasser, R. Ruf and H. Vahrenkamp, *Chem. Ber.*, 1993, **126**, 703–710.

123. A. Looney, R. Han, K. McNeill and G. Parkin, *J. Am. Chem. Soc.*, 1993, **115**, 4690–4697.

124. M. Ruf and H. Vahrenkamp, *Chem. Ber.*, 1996, **129**, 1025–1028.

125. E. Kimura, T. Shiiota, T. Koike, M. Shiro and M. Kodama, *J. Am. Chem. Soc.*, 1990, **112**, 5805–5811.

126. C. Bazzicalupi, A. Bencini, E. Berni, A. Bianchi, P. Fornasari, C. Giorgi and B. Valtancoli, *Eur. J. Inorg. Chem.*, 2003, **1974**, 1983.

127. E. Kimura, T. Koike, Y. Kodama and M. Shiro, *J. Am. Chem. Soc.*, 1994, **116**, 4764–4771.

128. J. T. Groves and R. R. Chambers, *J. Am. Chem. Soc.*, 1984, **106**, 630–638.

129. G. B. Zhang and R. Breslow, *J. Am. Chem. Soc.*, 1997, **119**, 1676–1681.

130. Z. G. Wang, W. Fast, A. M. Valentine and S. Benkovic, *J. Curr. Opin. Chem. Biol.*, 1999, **3**, 614–622.

131. J. Weston, *Chem. Rev.*, 2005, **105**, 2151–2174.

132. M. T. B. Luiz, B. Szpoganicz, M. Rizzoto, M. G. Basallote and A. E. Martell, *Inorg. Chim. Acta*, 1999, **287**, 134–141.

133. E. F. Hammel Jr and S. Glasstone, *J. Am. Chem. Soc.*, 1954, **76**, 3741–5.
134. H. Sakiyama, Y. Igarashi, Y. Nakayama, M. J. Hossain, K. Unoura and Y. Nishida, *Inorg. Chim. Acta. 351, 256.*, 2003, **351**, 256–260.
135. H. Sakiyama, R. Mochizuki, A. Sugawara, M. Sakamoto, Y. Nishida and M. Yamasaki, *J. Chem. Soc., Dalton Trans.*, 1999, 997–1000.
136. N. V. Kaminskaia, B. Spingler and S. J. Lippard, *J. Am. Chem. Soc.*, 2000, **122**, 6411–6422.
137. N. V. Kaminskaia, B. Spingler and S. J. Lippard, *J. Am. Chem. Soc.*, 2001, **123**, 6555–6563.
138. A. Tamilselvi, M. Nethaji and G. Mugesh, *Chem. Eur. J.*, 2006, **12**, 7797–7806.
139. W. W. Cleland and A. C. Hengge, *Chem. Rev.*, 2006, **106**, 3252–3278.
140. P. J. O'Brien and D. Herschlag, *Biochemistry*, 2002, **41**, 3207–3225.
141. A. J. Kirby and A. G. Varvoglis, *J. Chem. Soc. Section B*, 1968, 135.
142. W. P. Jencks, *Accts. Chem. Res.*, 1980, **13**, 161–169.
143. A. J. Kirby and A. G. Varvoglis, *J. Am. Chem. Soc.*, 1967, **89**, 415.
144. D. Asthagiri, T. Q. Liu, L. Noodleman, R. L. Van Etten and D. Bashfordt, *J. Am. Chem. Soc.*, 2004, **126**, 12677–12684.
145. R. H. Hoff, A. C. Hengge, L. Wu, Y.-F. Keng and Z.-Y. Zhang, *Biochemistry*, 2000, **39**, 46–54.
146. A. C. Hengge, In *Adv Phys. Org. Chem.* 2005, 40, p. 49–108.
147. C. Alhambra, L. Wu, Z.-Y. Zhang and J. Gao, *J. Am. Chem. Soc.*, 1998, **120**, 3858–3866.
148. D. Asthagiri, V. Dillet, T. Q. Liu, L. Noodleman, R. L. Van Etten and D. Bashford, *J. Am. Chem. Soc.*, 2002, **124**, 10225–10235.
149. A. J. Kirby, N. Dutta-Roy, D. da Silva, J. M. Goodman, M. F. Lima, C. D. Roussev and F. Nome, *J. Am. Chem. Soc.*, 2005, **127**, 7033–7040.
150. R. H. Bromilow and A. J. Kirby, *J. Chem. Soc. B*, 1972, 149.
151. M. W. Hosseini, J.-M. Lehn, K. C. Jones, K. E. Plute, K. B. Mertes and M. P. Mertes, *J. Am. Chem. Soc.*, 1989, **111**, 6330–5.
152. M. W. Hosseini and J. M. Lehn, *J. Am. Chem. Soc.*, 1987, **109**, 7047–58.
153. W. H. Chapman and R. Breslow, *J. Am. Chem. Soc.*, 1995, **117**, 5462–5469.
154. T. Koike, M. Inoue, E. Kimura and M. Shiro, *J. Am. Chem. Soc.*, 1996, **118**, 3091–3099.
155. T. T. Simopoulos and W. P. Jencks, *Biochemistry*, 1994, **33**, 10375–10380.
156. N. H. Williams, *Biochim. Biophys. Acta*, 2004, **1697**, 279–287.
157. A. J. Kirby and M. Younas, *J. Chem. Soc. Section B*, 1970, 1165–1172.
158. G. R. J. Thatcher and R. Kluger, *Adv. Phys. Org. Chem.*, 1996, **25**, 99–265.
159. R. H. Bromilow, S. A. Khan and A. J. Kirby, *J. Chem. Soc.B*, 1972, 911.
160. D. M. Perreault and E. V. Anslyn, *Angew. Chem. Intl. Ed. Engl.*, 1997, **36**, 432–450.
161. H. Lönnberg, R. Stromberg and A. Williams, *Org. Biomol. Chem.*, 2004, **2**, 2165–2167.
162. A. M. Davis, A. D. Hall and A. Williams, *J. Am. Chem. Soc.*, 1988, **110**, 5105–8.

163. C. Beckmann, A. J. Kirby, S. Kuusela and D. C. Tickle, *J. Chem. Soc. Perkin Trans. 2*, 1998, 573–582.

164. M. Oivanen, S. Kuusela and H. Lönnberg, *Chem. Rev.*, 1998, **98**, 961–990.

165. N. H. Williams, In *Comprehensive Biological Catalysis*, ed. M. L. Sinnott, Academic Press, London, 1998, **Vol. 1**, pp. 543–561.

166. S. Grazulis, E. Manakova, M. Roessle, M. Bochtler, G. Tamulaitiene, R. Huber and V. Siksnys, *Proc. Nat. Acad. Sci. USA*, 2005, **102**, 15797–15802.

167. J. A. Stuckey and J. E. Dixon, *Nat. Struct. Biol.*, 1999, **6**, 278–284.

168. J. C. Wang, *Ann. Rev. Biochem.*, 1996, **65**, 635–692.

169. H. Interthal, J. J. Pouliot and J. J. Champoux, *Proc. Nat. Acad. Sci. USA*, 2001, **98**, 12009–12014.

170. Y.-C. Cheng, C.-C. Hsueh, S.-C. Lu and T.-H. Liao, *Biochem. J.*, 2006, **398**, 177–185.

171. A. J. Kirby, M. F. Lima, D. da Silva, C. D. Roussev and F. Nome, *J. Am. Chem. Soc.*, 2006, **128**, 16944–16952.

172. K. W. Y. Abell and A. J. Kirby, *J. Chem. Soc. Perkin Trans. 2*, 1983, 1171–1174.

173. R. T. Raines, *Chem. Rev.*, 1998, **98**, 1045–1065.

174. C. Mihai, A. V. Kravchuk, M.-D. Tsai and K. S. Bruzik, *J. Am. Chem. Soc.*, 2003, **125**, 3236–3242.

175. L. Zhao, H. Liao and M. D. Tsai, *J. Biol. Chem.*, 2004, **279**, 31995–32000.

176. E. Anslyn and R. Breslow, *J. Am. Chem. Soc*, 1989, **111**, 5972–3.

177. J. C. Verheijen, B. A. L. M. Deiman, E. Yeheskiely, G. A. v. d. Marel and J. H. v. Boom, *Angew. Chem. Int. Ed. Engl.*, 2000, **39**, 369–372.

178. M. Komiyama, T. Inokawa and K. Yoshinar, *J. Chem. Soc. Chem. Commun.*, 1995, 77–78.

179. T. Niittymaki and H. Lönnberg, *Org. Biomol. Chem.*, 2006, **4**, 15–25.

180. S. Franklin, *USA Patent 7,091,026 of 15.08.06* 2006.

181. T. A. Steitz and J. A. Steitz, *Proc. Nat. Acad. Sci. USA*, 1993, **90**, 6498–6502.

182. M. Nowotny and W. Yang, *EMBO Journal*, 2006, **25**, 1924–1933.

183. T. A. Steitz, *Nature*, 1998, **391**, 231–232.

184. C. Romier, R. Dominguez, A. Lahm, O. Dahl and D. Suck, *Proteins: Struct., Funct. and Genet.*, 1998, **32**, 414–424.

185. S. F. Martin and P. J. Hergenrother, *Biochemistry*, 1999, **38**, 4403–4408.

186. F. Mancin and P. Tecilla, *New J. Chem.*, 2007, **31**, 800–817.

187. P. Molenveld, J. F. J. Engbersen and D. N. Reinhoudt, *Chem. Soc. Rev.*, 2000, **29**, 75–86.

188. M. Y. Yang, J. P. Richard and J. R. Morrow, *J. Chem. Soc. Chem. Commun.*, 2003, 2832–2833.

189. K. Worm, F. Y. Chu, K. Matsumoto, M. D. Best, V. Lynch and E. V. Anslyn, *Chem. Eur. J.*, 2003, **9**, 741–747.

190. R. Cacciapaglia, A. Casnati, L. Mandolini, A. Peracchi, D. N. Reinhoudt, R. Salvio, A. Sartori and R. Ungaro, *J. Am. Chem. Soc.*, 2007, **129**, 12512–12520.

191. A. O'Donoghue, S. Y. Pyun, M. Y. Yang, J. R. Morrow and J. P. Richard, *J. Am. Chem. Soc.*, 2006, **128**, 1615–1621.

192. G. Q. Feng, D. Natale, R. Prabaharan, J. C. Mareque-Rivas and N. H. Williams, *Angew. Chem. Int. Ed. Engl.*, 2006, **45**, 7056–7059.

193. M. Komiyama, N. Takeda, Y. Takahashi, H. Uchida, T. Shiiba, T. Kodama and M. Yashiro, *J. Chem. Soc. Perkin Trans. 2*, 1995, 269–274.

194. Y. Yamamoto and M. Komiyama, In *Artificial Enzymes*, ed. R. Breslow, Wiley-VCH, Weinheim, 2005, p. 159–175.

195. J. S. Tsang, A. A. Neverov and R. S. Brown, *J. Am. Chem. Soc.*, 2003, **125**, 1559–1566.

196. K. Ichikawa, M. Tarnai, M. K. Uddin, K. Nakata and S. Sato, *J. Inorg. Biochem.*, 2002, **91**, 437–450.

197. E. Boseggia, M. Gatos, L. Lucatello, F. Mancin, S. Moro, M. Palumbo, C. Sissi, P. Tecilla, U. Tonellato and G. Zagotto, *J. Am. Chem. Soc.*, 2004, **126**, 4543–4549.

198. F. D. Urnov, J. C. Miller, Y. L. Lee, C. M. Beausejour, J. M. Rock, S. Augustus, A. C. Jamieson, M. H. Porteus, P. D. Gregory and M. C. Holmes, *Nature*, 2005, **435**, 646–651.

199. G.-A. Craze, A. J. Kirby and R. Osborne, *J. Chem. Soc. Perkin Trans. 2*, 1978, 357–368.

200. N. S. Banait and W. P. Jencks, *J. Am. Chem. Soc.*, 1991, **113**, 7951–7958.

201. A. J. Kirby, *Stereoelectronic Effects*, Oxford University Press, Oxford, 1996.

202. A. J. Kirby, *The Anomeric Effect and Related Stereoelectronic Effects at Oxygen*, Springer-Verlag: Berlin and Heidelberg, 1983.

203. A. J. Briggs, C. M. Evans, R. Glenn and A. J. Kirby, *J. Chem. Soc. Perkin Trans. 2*, 1983, 1637.

204. S. Chandrasekhar, A. J. Kirby and R. J. Martin, *J. Chem. Soc. Perkin Trans. 2*, 1983, 1619.

205. R. Wolfenden, X. D. Lu and G. Young, *J. Am. Chem. Soc.*, 1998, **120**, 6814–6815.

206. A. T. N. Belarmino, S. Froehner, D. Zanette, J. P. S. Farah, C. A. Bunton and L. S. Romsted, *J. Org. Chem.*, 2003, **68**, 706–717.

207. T. H. Fife and L. H. Brod, *J. Am. Chem. Soc.*, 1970, **92**, 1681–1684.

208. E. Anderson and T. H. Fife, *J. Am. Chem. Soc.*, 1969, **91**, 7163–7166.

209. K. E. S. Dean, A. J. Kirby and I. V. Komarov, *J. Chem. Soc. Perkin Trans. 2*, 2002, **1**, 337–341.

210. G. Davies, M. L. Sinnott and S. G. Withers, *In Comprehensive Biological Catalysis*, Academic Press, London, 1998, Vol. 1, p. 118–209.

211. J. D. McCarter and S. G. Withers, *Curr. Opin. Struct. Biol*, 1994, **4** 885–892.

212. C. C. F. Blake, D. F. Koenig, G. A. Mair, A. C. T. North, D. C. Phillips and V. R. Sarma, *Nature*, 1965, **206**, 757–761.

213. D. J. Vocadlo, G. J. Davies, R. Laine and S. G. Withers, *Nature*, 2001, 835–838.

214. B. L. Mark and M. N. G. James, *Can. J. Chem.*, 2002, **80**, 1064–1074.

215. S. J. Williams, B. L. Mark, D. J. Vocadlo, M. N. G. James and S. G. Withers, *J. Biol. Chem.*, 2002, **277**, 40055–40065.

216. S. G. Withers, *Carbohydrate Polymers*, 2001, **44**, 325–337.
217. C. Goedl, R. Griessler, A. Schwarz and B. Nidetzky, *Biochem. J.*, 2006, **397**, 491–500.
218. M. L. Sinnott and W. P. Jencks, *J. Am. Chem. Soc.*, 1980, **102**, 2026–2032.
219. L. L. Lairson and S. G. Withers, *J. Chem. Soc. Chem. Commun.*, 2004, 2243–2248.
220. S. G. Withers, W. W. Wakarchuk and N. C. Strynadka, *J. Chem. Biol.*, 2002, **9**, 1270–1273.
221. R. P. Gibson, J. P. Turkenburg, S. J. Charnock, R. Lloyd and G. Davies, *J. Chem. Biol.*, 2002, **9**, 1337–1346.
222. G.-A. Craze and A. J. Kirby, *J. Chem. Soc. Perkin Trans. 2*, 1974, 61.
223. F. H. Allen and A. J. Kirby, *J. Am. Chem. Soc.*, 1991, **113**, 8829–31.
224. X. M. Cherian, S. A. Vanarman and A. W. Czarnik, *J. Am. Chem. Soc.*, 1988, **110**, 6566–6568.
225. A. J. Kirby and A. Parkinson, *J. Chem. Soc., Chem. Commun.*, 1994, 707–8.
226. E. Hartwell, D. R. W. Hodgson and A. J. Kirby, *J. Am. Chem. Soc.*, 2000, **122**, 9326–9327.
227. D. Piszkiewicz and T. C. Bruice, *J. Am. Chem. Soc.*, 1968, **90**, 2156–2163.
228. E. Anderson and T. H. Fife, *J. Am. Chem. Soc.*, 1973, 95.
229. K. E. S. Dean and A. J. Kirby, *J. Chem. Soc. Perkin Trans. 2*, 2002, 428–432.
230. T. H. Fife and T. J. Przystas, *J. Am. Chem. Soc.*, 1979, **101**, 1202–1210.
231. D. T. H. Chou, J. Zhu, X. Huang and A. J. Bennet, *J. Chem. Soc., Perkin Trans. 2*, 2001, 83–89.
232. C. Rousseau, N. Nielsen and M. Bols, *Tetrahedron Lett.*, 2004, **45**, 8709–8711.
233. F. Ortega-Caballero, J. Bjerre, L. S. Laustsen and M. J. Bols, *Org. Chem.*, 2005, **70**, 7217–7226.
234. F. Ortega-Caballero, C. Rousseau, B. Christensen, T. E. Petersen and M. Bols, *J. Am. Chem. Soc.*, 2005, **127**, 3238–3239.
235. W.-M. Ching and R. G. Kallen, *J. Am. Chem. Soc.*, 1978, **100**, 6119–24.
236. J. Bjerre, T. H. Fenger, L. G. Marinescu and M. Bols, *Eur. J. Org. Chem.*, 2007, 704–710.
237. *Hydrogen-Transfer Reactions*, J. T. Hynes, J. P. Klinman, H. Limbach, R. L. Schowen, ed., Wiley-VCH Verlag GmbH & Co.: Weinheim, Germany, 2007.
238. T. D. H. Bugg, *An Introduction to Enzyme and Coenzyme Chemistry*, Blackwell: Oxford, 1997.
239. J. P. Richard and T. L. Amyes, *Curr. Opin. Chem. Biol.*, 2001, **5**, 626–633.
240. M. Eigen, *Angew. Chem., Intl. Ed. Engl.*, 1964, **3**, 1–72.
241. C. F. Bernasconi, *Adv. Phys. Org. Chem.*, 1992, **27**, 119–238.
242. R. A. Bednar and W. P. Jencks, *J. Am. Chem. Soc.*, 1985, **107**, 7117–7126.
243. T. L. Amyes and J. P. Richard, In *Hydrogen-Transfer Reactions*, ed. J. T. Hynes, J. P. Klinman, H. Limbach, R. L. Schowen, Wiley-VCH Verlag GmbH & Co., Weinheim, Germany, 2007, **Vol. 3**, p. 949–973.
244. T. L. Amyes and J. P. Richard, *J. Am. Chem. Soc.*, 1996, **118**, 3129–3141.

245. A. F. Hegarty, J. P. Dowling, S. J. Eustace and M. McGarraghy, *J. Am. Chem. Soc.*, 1998, **120**, 2290–2296.
246. K. P. Shelly, S. Venimadhavan, K. Nagarajan and R. Stewart, *Can. J. Chem.*, 1989, **67**, 1274–82.
247. A. J. Kirby and F. O'Carroll, *J. Chem. Soc., Perkin Trans.* 2, 1994, 649–655.
248. E. Kimura, T. Gotoh, T. Koike and M. Shiro, *J. Am. Chem. Soc.*, 1999, **121**, 1267–1274.
249. C. D. Gutsche, R. S. Buriks, K. Kowotny and H. Grassner, *J. Am. Chem. Soc.*, 1962, **84**, 3775–3777.
250. T. D. Fenn, D. Ringe and G. A. Petsko, *Biochemistry*, 2004, **43**, 6464–6474.
251. T. L. Amyes and J. P. Richard, *Biochemistry*, 2007, **46**, 5841–5854.
252. J. R. Knowles, *Philos. Trans. Roy. Soc. London, Ser. B*, 1991, **332**, 115–21.
253. W. J. Albery and J. R. Knowles, *Biochemistry*, 1976, **15**, 5627–5631.
254. S. C. Blacklow, R. T. Raines, W. A. Lim, P. D. Zamore and J. R. Knowles, *Biochemistry*, 1988, **27**, 1158–1167.
255. A. J. Mulholland, P. D. Lyne and M. Karplus, *J. Am. Chem. Soc.*, 2000, **122**, 534–535.
256. J. A. Gerlt and P. C. Babbitt, *Ann. Rev. Biochem. 70*, 2001, **70**, 209–246.
257. J. P. Richard, G. Williams, A. C. O'Donoghue and T. L. Amyes, *J. Am. Chem. Soc.*, 2002, **124**, 2957–2968.
258. J. A. Gerlt, P. C. Babbitt and I. Rayment, *Arch. Biochem. Biophys.*, 2005, **433**, 59–70.
259. G. L. Kenyon, J. A. Gerlt, G. A. Petsko and J. W. Kozarich, *Accts. Chem. Res.*, 1995, **28**, 178–186.
260. D. T. Lin, V. M. Powers, L. J. Reynolds, C. P. Whitman, J. W. Kozarich and G. L. Kenyon, *J. Am. Chem. Soc.*, 1988, **110**, 323–324.
261. S. L. Bearne and R. Wolfenden, *Biochemistry*, 1997, **36**, 1646–1656.
262. Y. Chiang, A. J. Kresge, V. V. Popik and N. P. Schepp, *J. Am. Chem. Soc.*, 1997, **119**, 10203–10212.
263. J. B. Thoden, A. D. Hegeman, G. Wesenberg, M. C. Chapeau, P. A. Frey and H. M. Holden, *Biochemistry*, 1997, **36**, 6294–6304.
264. J. Kvassman, L. Larsson and G. Pettersson, *Eur. J. Biochem.*, 1981, **114**, 555–563.
265. N. Kanomata and T. Nakata, *J. Am. Chem. Soc.*, 2000, **122**, 4563–4568.
266. P. van Eikeren and D. L. Grier, *J. Am. Chem. Soc.*, 1976, **98**, 4655–4657.
267. D. J. Norris and R. Stewart, *Can. J. Chem.*, 1977, **55**, 1687–1695.
268. C. I. F. Watt, *Adv. Phys. Org. Chem.*, 1988, **24**, 57–112.
269. J. W. Verhoeven, W. Vangerresheim, F. M. Martens and S. M. Vanderkerk, *Tetrahedron*, 1986, **42**, 975–992.
270. D. C. Dittmer and B. B. Blidner, *J. Org. Chem.*, 1973, **38**, 2873–2882.
271. A. J. Kirby and D. R. Walwyn, *Tetrahedron Lett.*, 1987, **28**, 2421–2424.
272. A. J. Kirby and D. R. Walwyn, *Gazz. Chim. Ital.*, 1987, **117**, 67.
273. R. S. McDonald and C. E. Sibley, *Can. J. Chem.*, 1981, **59**, 1061–1067.
274. A. M. Davis, M. I. Page, S. C. Mason and I. Watt, *J. Chem. Soc. Chem. Commun.*, 1984, 1671–1672.

275. R. Cernik, G.-A. Craze, O. S. Mills and C. I. F. Watt, *J. Chem. Soc. Perkin Trans. 2*, 1982, 361–367.

276. P. A. Frey, *Ann. Rev. Biochem.*, 2001, **70**, 121–148.

277. S. Y. Reece, J. M. Hodgkiss, J. Stubbe and D. G. Nocera, *Philos. Trans. Roy. Soc. B*, 2006, **361**, 1351–1364.

278. P. A. Frey, A. D. Hegeman and G. H. Reed, *Chem. Rev.*, 2006, **106** 3302–3316.

279. P. A. Frey, A. D. Hegeman, *Enzymatic Reaction Mechanisms, Chapters 3 and 4*; Oxford University Press, New York, 2007.

280. P. A. Frey and O. T. Magnusson, *Chem. Rev.*, 2003, **103**, 2129–2148.

281. S. C. Wang and P. A. Frey, *Trends Biochem. Sci.*, 2007, **32**, 101–110.

282. A. Becker and W. Kabsch, *J. Biol. Chem.*, 2002, **277**, 40036–40042.

283. M. Uppsten, M. Farnegardh, V. Domkin and U. Uhlin, *J. Mol. Biol.*, 2006, **359**, 365–377.

284. S. Licht and J. Stubbe, *Compr. Nat. Prod. Chem.*, 1999, **5**, 163–203.

285. K.-B. Cho, F. G. A. Himo and P. E. M. Siegbahn, *J. Phys. Chem. B*, 2001, 6445–6452.

286. M. D. Sintchak, G. Arjara, B. A. Kellogg, J. Stubbe and C. L. Drennan, *Nature Struct. Biol.*, 2002, **9**, 293–300.

287. P. Nordlund and P. Reichard, *Ann. Rev. Biochem.*, 2006, **75**, 681–706.

288. J. Stubbe, *Curr. Opin. Struct. Biol.*, 2000, **10**, 731–736.

289. J. Stubbe and W. A. van der Donk, *Chem. Rev.*, 1998, **98**, 705–762.

290. J. M. Sirovatka and R. G. Finke, *J. Inorg. Biochem.*, 2000, **78**, 149–160.

291. J. E. Baldwin, D. Brown, P. H. Scudder and M. E. Wood, *Tetrahedron Lett.*, 1995, **36**, 2105–2108.

292. M. L. Steigerwald, W. A. Goddard and D. A. Evans, *J. Am. Chem. Soc.*, 1979, **101**, 1994–1997.

293. M. J. Robins and G. J. Ewing, *J. Am. Chem. Soc.*, 1999, **121**, 5823–5824.

294. R. Lenz and B. Giese, *J. Am. Chem. Soc.*, 1997, **119**, 2784–2794.

295. M. S. Akhlaq and C. Vonsonntag, *Zeit. Naturforsch. C*, 1987, **42**, 134–140.

296. D. L. Reid, G. V. Shustov, D. A. Armstrong, A. Rauk, M. N. Schuchmann, M. S. Akhlaq and C. von Sonntag, *Phys. Chem. Chem. Phys.*, 2002, **4**, 2965–2974.

297. E. Mulliez, S. Ollagnier, M. Fontecave, R. Eliasson and P. Reichard, *Proc. Nat. Acad. Sci. USA*, 1995, **92**, 8759–8762.

298. H. Oikawa and T. Tokiwano, *Nat. Prod. Rep.*, 2004, **21**, 321–352.

299. E. Wilson, *Chem. Eng. News*, 2005, **83**, 38–38.

300. J. S. Rhoads, In *Molecular Rearrangements*, ed. P. de Mayo, Interscience, New York, 1963, **Vol. 1**, p. 655–706.

301. J. Zaitseva, J. Lu, K. L. Olechoski and A. L. Lamb, *J. Biol. Chem.*, 2006, **281**, 33441–33449.

302. M. S. DeClue, K. K. Baldridge, P. Kast and D. Hilvert, *J. Am. Chem. Soc.*, 2006, **128**, 2043–2051.

303. X. D. Zhang, X. H. Zhang and T. C. Bruice, *Biochemistry*, 2005, **44**, 10443–10448.

304. S. D. Copley and J. R. Knowles, *J. Am. Chem. Soc.*, 1987, **109**, 5008–5013.

305. T. Ishida, D. G. Fedorov and K. Kitaura, *J. Phys. Chem. B*, 2006, **110**, 1457–1463.

306. S. K. Wright, M. S. DeClue, A. Mandal, L. Lee, O. Wiest, W. W. Cleland and D. Hilvert, *J. Am. Chem. Soc.*, 2005, **127**, 12957–12964.

307. J. J. Gajewski, *Accts. Chem. Res.*, 1997, **30**, 219–25.

308. B. Ganem, *Angew. Chem., Int. Ed. Engl.*, 1996, **35**, 936–945.

309. H. Wade and T. S. Scanlan, *Ann. Rev. Biophys. Biomol. Str.*, 1997, **26**, 461–493.

310. D. Jackson, M. N. Liang, P. A. Bartlett and P. G. Schulz, *Angew. Chem. Int. Ed. Engl.*, 1992, **31**, 182–83.

311. Y. Xu, N. Yamamoto and K. D. Janda, *Bioorg. Med. Chem.*, 2004, **12**, 5247–5268.

312. D. Hilvert, *Accts. Chem. Res.*, 1993, **26**, 552–558.

313. J. Lagona, P. Mukhopadhyay, S. Chakrabarti and L. Isaacs, *Angew. Chem., Int. Ed. Engl.*, 2005, **44**, 4844–4870.

314. W. L. Mock, T. A. Irra, J. P. Wepsiec and M. Adhya, *J. Org. Chem.*, 1989, **54**, 5302–5308.

315. R. Cacciapaglia, S. Di Stefano and L. Mandolini, *Accts. Chem. Res.*, 2004, **37**, 113–122.

316. S. P. Kim, A. G. Leach and K. N. Houk, *J. Org. Chem.*, 2002, **67**, 4250–4260.

317. H.-J. Schneider and N. K. Sangwan, *Chem. Commun.*, 1986, 1787–1789.

318. C. J. Walter and J. K. M. Sanders, *Angew. Chem., Int. Ed. Engl.*, 1995, **34**, 217–219.

319. C. J. Walter, H. L. Anderson and J. K. M. Sanders, *J. Chem. Soc., Chem. Commun.*, 1993, 458.

320. B. Brisig, J. K. M. Sanders and S. Otto, *Angew. Chem., Int. Ed. Engl.*, 2003, **42**, 1270–1273.

321. R. Fiammengo and A. Jaschke, *Curr. Opin. Biotechnol.*, 2005, **16**, 614–621.

322. B. Seelig, S. Keiper, F. Stuhlmann and A. Jaschke, *Angew. Chem., Int. Ed.*, 2000, **39**, 4576–4579.

323. A. Serganov, S. Keiper, L. Malinina, V. Tereshko, E. Skripkin, C. Höbartner, A. Polonskaia, A. T. Phan, R. Wombacher, R. Micura, Z. Dauter, A. Jäschke and D. Patel, *J. Nature Str. Mol. Biol.*, 2005, **12**, 218–224.

324. C. G. Overberger and J. C. Salamone, *Accts. Chem. Res.*, 1969, **2**, 217–224.

325. H. Dugas, *Bioorganic Chemistry*, Springer-Verlag, New York, 1996.

326. H. C. Kiefer, W. I. Congdon, I. Scarpa and I. Klotz, *Proc. Natl. Acad. Sci. USA*, 1972, **69**, 2155–2159.

327. J. Suh, I. S. Scarpa and I. M. Klotz, *J. Am. Chem. Soc.*, 1976, **98**, 7060–7064.

328. I. M. Klotz and J. Suh, In *Artificial Enzymes*, ed. R. Breslow, Wiley-VCH, Weinheim, 2005.

329. F. Hollfelder, A. J. Kirby and D. S. Tawfik, *J. Am. Chem. Soc.*, 1997, **119**, 9578–9579.

330. F. Hollfelder, A. J. Kirby and D. S. Tawfik, *J. Org. Chem.*, 2001, **66**, 5866–5874.

331. F. Avenier, J. B. Domingos, L. D. Van Vliet and F. Hollfelder, *J. Am. Chem. Soc.*, 2007, **129**, 7611–7619.

332. J. Kofoed and J. L. Reymond, *Curr. Opin. Chem. Biol.*, 2005, **9**, 656–664.

333. B. Helms and J. M. Frechet, *J. Adv. Synth. Catal.*, 2006, **348**, 1125–1148.

334. L. J. Twyman, A. S. King and I. K. Martin, *Chem. Soc. Rev.*, 2002, **31**, 69–82.

335. L. Liu and R. Breslow, *Bioorg. Med. Chem.*, 2004, **12**, 3277–3287.

336. L. Liu and R. Breslow, *J. Am. Chem. Soc.*, 2003, **125**, 12110–12111.

337. A. Esposito, E. Delort, D. Lagnoux, F. Djojo and J. L. Reymond, *Angew. Chem., Int. Ed. Engl.*, 2003, **42**, 1381–1383.

338. J. Kofoed, T. Darbre and J. L. Reymond, *Org. Biomol. Chem.*, 2006, **4**, 3268–3281.

339. K. S. Broo, H. Nilsson and J. L. B. Nilsson, *J. Am. Chem. Soc.*, 1998, **120**, 10287–10295.

340. C. Douat-Casassus, T. Darbre and J. L. Reymond, *J. Am. Chem. Soc.*, 2004, **126**, 7817–7826.

341. T. Darbre and J. L. Reymond, *Accts. Chem. Res.*, 2006, **39**, 925–934.

342. E. Delort, T. Darbre and J.-L. Reymond, *J. Am. Chem. Soc.*, 2004, **126**, 15642–15643.

343. G. Zaupa, P. Scrimin and L. J. Prins, *J. Am. Chem. Soc.*, 2008, **130**, 5699–709.

344. S. Javor, E. Delort, T. Darbre and J. L. Reymond, *J. Am. Chem. Soc.*, 2007, **129**, 13238–13246.

345. J. Q. Liu and G. Wulff, *J. Am. Chem. Soc.*, 2008, **130**, 8044–8054.

346. R. A. Lerner, S. J. Benkovic and P. G. Schultz, *Science*, 1991, **252**, 659–667.

347. D. S. Tawfik, B. S. Green, R. Chap, M. Sela and Z. Eshhar, *Proc. Nat. Acad. Sci. USA*, 1993, **90**, 373–377.

348. K. D. Janda, L. C. Lo, C. H. Lo, M. M. Sim, R. Wang, C. H. Wong and R. A. Lerner, *Science*, 1997, **275**, 945–8.

349. F. Tanaka, *Chem. Rev.*, 2002, **102**, 4885–4906.

350. S. J. Pollack, P. Hsiun and P. G. Schultz, *J. Am. Chem. Soc.*, 1989, **111**, 5962–5964.

351. M. M. Thayer, E. H. Olender, A. S. Arvai, C. K. Koike, I. L. Canestrelli, J. D. Stewart, S. J. Benkovic, E. D. Getzoff and V. A. Roberts, *J. Mol. Biol.*, 1999, **291**, 329–345.

352. G. J. Wedemayer, P. A. Patten, L. H. Wang, P. G. Schultz and R. C. Stevens, *Science*, 1997, **276**, 1665–1669.

353. J.-B. Charbonnier, B. Golinelli-Pimpaneau, B. Gigant, D. S. Tawik, R. Chap, D. G. Schindler, S.-H. Kim, B. S. Green, Z. Eshhar and M. Knossow, *Science*, 1997, **275**, 1140–1142.

354. K. D. Janda, R. A. Lerner and A. J. Tramontano, *Am. Chem. Soc.*, 1988, **110**, 4835–4837.

355. C. Gao, B. J. Lavey, C.-H. L. Lo, A. Datta, P. J. Wentworth and K. D. Janda, *J. Am. Chem. Soc.*, 1998, **120**, 2211–2217.

356. M. T. Martin, T. S. Angeles, R. Sugasawara, N. I. Aman, A. D. Napper, J. M. Darsley, R. I. Sanchez, P. Booth and R. C. Titmas, *J. Am. Chem. Soc.*, 1994, **116**, 6508–6512.

357. P. Wentworth, Y. Liu, A. D. Wentworth, P. Fan, M. J. Foley and K. D. Janda, *Proc. Natl. Acad. Sci. USA*, 1998, 5971–5975.

358. A. J. Kirby, *Accts. Chem. Res.*, 1997, **30**, 290–296.

359. S. N. Thorn, R. G. Daniels, M.-T. M. Auditor and D. N. Hilvert, *Nature*, 1995, **373**, 228–230.

360. F. Hollfelder, A. J. Kirby, D. S. Tawfik, K. Kikuchi and D. Hilvert, *J. Am. Chem. Soc.*, 2000, **124**, 1022–1029.

361. K. Kikuchi, R. B. Hannak, M. J. Guo, A. J. Kirby and D. Hilvert, *Bioorg. Med. Chem.*, 2006, **14**, 6189–6196.

362. D. Hilvert, *Ann. Rev. Biochem.*, 2000, **69**, 751–793.

363. K. Kruger, P. J. Grabowski, A. J. Zaug, J. Sands, D. E. Gottschling and T. R. Cech, *Cell*, 1982, **31**, 147–57.

364. C. Guerrier-Takada, K. Gardiner, T. Marsh, N. Pace and S. Altman, *Cell*, 1983, **35**, 849–57.

365. *The RNA World*, 3rd edn., R. F. Gesteland, T. R. Cech, J. F. E. Atkins, ed. Cold Spring Harbor Laboratory Press: Cold Spring Harbor, 2006.

366. J. A. Doudna and T. R. Cech, *Nature*, 2002, **418**, 222–8.

367. J. A. Doudna and J. R. Lorsch, *Nat. Struct. Mol. Biol.*, 2005, **12**, 395–402.

368. G. J. Narlikar and D. Herschlag, *Ann. Rev. Biochem.*, 1997, **66**, 19–59.

369. J. A. Piccirilli, T. S. McConnell, A. J. Zaug, H. F. Noller and T. R. Cech, *Science*, 1992, **256**, 1420–4.

370. A. Ke, K. Zhou, F. Ding, J. H. Cate and J. A. Doudna, *Nature*, 2004, **429**, 201–5.

371. P. C. Bevilacqua and R. Yajima, *Curr. Opin. Chem. Biol.*, 2006, **10**, 455–464.

372. I. H. Shih and M. D. Been, *Ann. Rev. Biochem.*, 2002, **71**, 887–917.

373. I. H. Shih and M. D. Been, *Proc. Nat. Acad. Sci. USA*, 2001, **98**, 1489–94.

374. B. L. Golden, A. R. Gooding, E. R. Podell and T. R. Cech, *Science*, 1998, **282**, 259–64.

375. J. Liphardt, B. Onoa, S. B. Smith, I. J. Tinoco and C. Bustamante, *Science*, 2001, **292**, 733–7.

376. R. Russell, X. Zhuang, H. P. Babcock, I. S. Millett, S. Doniach, S. Chu and D. Herschlag, *Proc. Nat. Acad. Sci. USA*, 2002, **99**, 155–60.

377. X. Zhuang, L. E. Bartley, H. P. Babcock, R. Russell, T. Ha, D. Herschlag and S. Chu, *Science*, 2000, **288**, 2048–51.

378. R. R. Breaker, G. M. Emilsson, D. Lazarev, S. Nakamura, I. J. Puskarz, A. Roth and N. Sudarsan, *RNA*, 2003, **9**, 949–57.

379. G. M. Emilsson, S. Nakamura, A. Roth and R. R. Breaker, *RNA*, 2003, **9**, 907–18.

380. A. D. Ellington and J. W. Szostak, *Nature*, 1990, **346**, 818–22.

381. C. Tuerk and L. Gold, *Science*, 1990, **249**, 505–10.

382. G. F. Joyce, *Ann. Rev. Biochem.*, 2004, **73**, 791–836.

383. J. R. Lorsch and J. W. Szostak, *Accts. Chem. Res.*, 1996, **29**, 103–10.

384. R. Stoltenburg, C. Reinemann and B. Strehlitz, *Biomol. Eng.*, 2007, **24**, 381–403.

385. D. P. Bartel and J. W. Szostak, *Science*, 1993, **261**, 1411–8.

386. E. H. Ekland, J. W. Szostak and D. P. Bartel, *Science*, 1995, **269**, 364–70.
387. S. B. Voytek and G. F. Joyce, *Proc. Nat. Acad. Sci. USA*, 2007, **104**, 15288–93.
388. B. M. Paegel and G. F. Joyce, *PLoS Biol*, 2008, **6**, e85.
389. M. Ohuchi, H. Murakami and H. Suga, *Curr. Opin. Chem. Biol.*, 2007, **11**, 537–42.
390. H. Xiao, H. Murakami, H. Suga and A. R. Ferre-D'Amare, *Nature*, 2008, **454**, 358–61.
391. X. Dai, A. De Mesmaeker and G. F. Joyce, *Science*, 1995, **267**, 237–40.
392. C. Wilson and J. W. Szostak, *Nature*, 1995, **374**, 777–82.
393. T. Ohmichi, Y. Okumoto and N. Sugimoto, *Nucleic Acids Res.*, 1998, **26**, 5655–61.
394. T. Pan, B. Dichtl and O. C. Uhlenbeck, *Biochemistry*, 1994, **33**, 9561–5.
395. J. E. Wedekind and D. B. McKay, *Biochemistry*, 2003, **42**, 9554–63.
396. Y. Li, Y. Liu and R. R. Breaker, *Biochemistry*, 2000, **39**, 3106–14.
397. A. Sreedhara, Y. Li and R. R. Breaker, *J. Am. Chem. Soc.*, 2004, **126**, 3454–60.
398. N. Paul, G. Springsteen and G. F. Joyce, *Chem. Biol.*, 2006, **13**, 329–38.
399. M. C. Conn, J. R. Prudent and P. G. Schlutz, *J. Am. Chem. Soc.*, 1996, **118**, 7012–7013.
400. J. R. Prudent, T. Uno and P. G. Schultz, *Science*, 1994, **264**, 1924–7.
401. J. R. Lorsch and J. W. Szostak, *Nature*, 1994, **371**, 31–6.
402. S. Fusz, A. Eisenfuhr, S. G. Srivatsan, A. Heckel and M. Famulok, *Chem. Biol.*, 2005, **12**, 941–50.
403. G. Sengle, A. Eisenfuhr, P. S. Arora, J. S. Nowick and M. Famulok, *Chem. Biol.*, 2001, **8**, 459–73.
404. T. M. Tarasow, S. L. Tarasow and B. E. Eaton, *Nature*, 1997, **389**, 54–7.
405. J. J. Agresti, B. T. Kelly, A. Jaschke and A. D. Griffiths, *Proc. Nat. Acad. Sci. USA*, 2005, **102**, 16170–16175.
406. P. J. Unrau and D. P. Bartel, *Nature*, 1998, **395**, 260–3.
407. P. J. Unrau and D. P. Bartel, *Proc. Nat. Acad. Sci. USA*, 2003, **100**, 15393–7.
408. P. A. Lohse and J. W. Szostak, *Nature*, 1996, **381**, 442–4.
409. T. W. Wiegand, R. C. Janssen and B. E. Eaton, *Chem. Biol.*, 1997, **4**, 675–83.
410. V. R. Jadhav and M. Yarus, *Biochemistry*, 2002, **41**, 723–9.
411. J. A. Nelson and O. C. Uhlenbeck, *RNA*, 2008, **14**, 605–15.
412. H. W. Pley, K. M. Flaherty and D. B. McKay, *Nature*, 1994, **372**, 68–74.
413. T. K. Stage-Zimmermann and O. C. Uhlenbeck, *RNA*, 1998, **4**, 875–89.
414. I. Shih and M. D. Been, *Biochemistry*, 2000, **39**, 9055–66.
415. P. L. Adams, M. R. Stahley, A. B. Kosek, J. Wang and S. A. Strobel, *Nature*, 2004, **430**, 45–50.
416. J. H. Cate, A. R. Gooding, E. Podell, K. Zhou, B. L. Golden, C. E. Kundrot, T. R. Cech and J. A. Doudna, *Science*, 1996, **273**, 1678–85.
417. D. Herschlag and T. R. Cech, *Biochemistry*, 1990, **29**, 10159–71.

418. G. J. Narlikar, V. Gopalakrishnan, T. S. McConnell, N. Usman and D. Herschlag, *Proc. Nat. Acad. Sci. USA*, 1995, **92**, 3668–72.
419. M. P. Robertson and W. G. Scott, *Science*, 2007, **315**, 1549–53.
420. N. H. Bergman, W. K. Johnston and D. P. Bartel, *Biochemistry*, 2000, **39**, 3115–23.
421. G. F. Joyce, In *The RNA World*, ed. R. F. Gesteland, T. R. Cech, J. F. Atkins, Cold Spring Harbor Laboratory Press: Cold Spring Harbor, 2006, Appendix 3.
422. M. P. Robertson and A. D. Ellington, *Nat. Biotechnol.*, 1999, **17**, 62–6.
423. A. D. Ellington, *PLoS Biol*, 2008, **6**, e132.
424. D. Kourouklis, H. Murakami and H. Suga, *Methods*, 2005, **36**, 239–44.
425. H. Murakami, A. Ohta, H. Ashigai and H. Suga, *Nat. Methods*, 2006, **3**, 357–9.
426. R. R. Breaker and G. F. Joyce, *Chem. Biol.*, 1994, **1**, 223–9.
427. D. D. Axe, *J. Mol. Biol.*, 2004, **341**, 1295–315.
428. D. M. Taverna and R. A. Goldstein, *J. Mol. Biol.*, 2002, **315**, 479–84.
429. A. D. Keefe and J. W. Szostak, *Nature*, 2001, **410**, 715–8.
430. C. Neylon, *Nucl. Acids Res.*, 2004, **32**, 1448–1459.
431. H. Lin and V. W. Cornish, *Angew. Chem., Int. Ed. Engl.*, 2002, **41**, 4402–25.
432. H. Leemhuis, V. Stein, A. D. Griffiths and F. Hollfelder, *Curr. Opin. Str. Biol.*, 2005, **15**, 472–8.
433. H. R. Hoogenboom, *Nat. Biotechnol.*, 2005, **23**, 1105–16.
434. *Phage Display in Biotechnology and Drug Discovery*, ed. S. S. Sidhu, CRC Press: Boca Raton, 2005.
435. C. Zahnd, P. Amstutz and A. Plückthun, *Nat. Methods*, 2007, **4**, 269–79.
436. D. S. Wilson, A. D. Keefe and J. W. Szostak, *Proc. Nat. Acad. Sci. USA*, 2001, **98**, 3750–5.
437. B. Seelig and J. W. Szostak, *Nature*, 2007, **448**, 828–31.
438. R. Odegrip, D. Coomber, B. Eldridge, R. Hederer, P. A. Kuhlman, C. Ullman, K. FitzGerald and D. McGregor, *Proc. Nat. Acad. Sci. USA*, 2004, **101**, 2806–10.
439. J. Bertschinger, D. Grabulovski and D. Neri, *Protein Eng. Des. Sel.*, 2007, **20**, 57–68.
440. J. Bertschinger and D. Neri, *Protein Eng Des Sel*, 2004, **17**, 699–707.
441. V. Stein, I. Sielaff, K. Johnsson and F. Hollfelder, *ChemBiochem.*, 2007, **8**, 2191–4.
442. D. Legendre, N. Laraki, T. Graslund, M. E. Bjornvad, M. Bouchet, P. A. Nygren, T. V. Borchert and J. J. Fastrez, *Mol. Biol.*, 2000, **296**, 87–102.
443. S. Cesaro-Tadic, D. Lagos, A. Honegger, J. H. Rickard, L. J. Partridge, G. M. Blackburn and A. Pluckthun, *Nat. Biotechnol.*, 2003, **21**, 679–85.
444. P. Amstutz, J. N. Pelletier, A. Guggisberg, L. Jermutus, S. Cesaro-Tadic, C. Zahnd and A. Plückthun, *J. Am. Chem. Soc.*, 2002, **124**, 9396–403.

445. A. Andreeva, D. Howorth, J. M. Chandonia, S. E. Brenner, T. J. Hubbard, C. Chothia, A. G. Murzin, Nucl. Acids Res. 2008, 36, D419-25 (http://scop.mrc-lmb.cam.ac.uk/scop).

446. T. J. Hubbard, A. G. Murzin, S. E. Brenner and C. Chothia, *Nucl. Acids Res.*, 1997, **25**, 236–9.

447. A. G. Murzin, S. E. Brenner, T. Hubbard and C. Chothia, *J. Mol. Biol.*, 1995, **247**, 536–40.

448. A. L. Cuff, I. Sillitoe, T. Lewis, O. C. Redfern, R. Garratt, J. Thornton and C. A. Orengo, *Nucl. Acids Res.*, 2009, **37**, D310–4.

449. J. E. Bray, A. E. Todd, F. M. Pearl, J. M. Thornton and C. A. Orengo, *Protein Eng.*, 2000, **13**, 153–65.

450. C. A. Orengo, A. D. Michie, S. Jones, D. T. Jones, M. B. Swindells and J. M. Thornton, *Structure*, 1997, **5**, 1093–108.

451. D. Wilson, M. Madera, C. Vogel, C. Chothia and J. Gough, *Nucl. Acids Res.*, 2007, **35**, D308–13.

452. C. T. Porter, G. J. Bartlett and J. M. Thornton, *Nucl. Acids Res.*, 2004, **32**, D129–33.

453. N. Nagano, *Nucl. Acids Res.* 2005, 33, D407–12 (http://mbs.cbrc.jp/EzCatDB/).

454. G. L. Holliday, D. E. Almonacid, G. J. Bartlett, N. M. O'Boyle, J. W. Torrance, Murray-P. Rust, J. B. Mitchell, J. M. Thornton, *Nucl. Acids Res.* 2007, 35, D515–20 (http://www.ebi.ac.uk/thornton-srv/databases/MACiE/).

455. I. Schomburg, A. Chang, C. Ebeling, M. Gremse, C. Heldt, G. Huhn, D. Schomburg, *Nucl. Acids Res.* 2004, 32, D431–3 (http://www.brenda.uni-koeln.de).

456. J. A. Gerlt, P. C. Babbitt and I. Rayment, *Arch. Biochem. Biophys.*, 2005, **433**, 59–70.

457. M. E. Glasner, J. A. Gerlt, P. C. Babbitt, *Adv. Enzymol. Relat. Areas Mol. Biol.* 2007, **75**, 193–239, xii–xiii.

458. D. J. Neidhart, P. L. Howell, G. A. Petsko, V. M. Powers, R. S. Li, G. L. Kenyon and J. A. Gerlt, *Biochemistry*, 1991, **30**, 9264–73.

459. T. M. Larsen, J. E. Wedekind, I. Rayment and G. H. Reed, *Biochemistry*, 1996, **35**, 4349–58.

460. M. S. Hasson, I. Schlichting, J. Moulai, K. Taylor, W. Barrett, G. L. Kenyon, P. C. Babbitt, J. A. Gerlt, G. A. Petsko and D. Ringe, *Proc. Nat. Acad. Sci. USA*, 1998, **95**, 10396–401.

461. S. Helin, P. C. Kahn, B. L. Guha, D. G. Mallows and A. J. Goldman, *Mol. Biol.*, 1995, **254**, 918–41.

462. T. B. Thompson, J. B. Garrett, E. A. Taylor, R. Meganathan, J. A. Gerlt and I. Rayment, *Biochemistry*, 2000, **39**, 10662–76.

463. A. M. Gulick, D. M. Schmidt, J. A. Gerlt and I. Rayment, *Biochemistry*, 2001, **40**, 15716–24.

464. D. R. Palmer, J. B. Garrett, V. Sharma, R. Meganathan, P. C. Babbitt and J. A. Gerlt, *Biochemistry*, 1999, **38**, 4252–8.

465. M. Y. Galperin, A. Bairoch and E. V. Koonin, *Protein Sci.*, 1998, **7**, 1829–35.

466. S. Jonas, F. Hollfelder, *Pure Appl. Chem.* 2009, **81**, 731–742.

467. J. G. Zalatan, T. D. Fenn, A. T. Brunger and D. Herschlag, *Biochemistry*, 2006, **45**, 9788–803.

468. S. Jonas, B. van Loo, M. Hyvonen and F. Hollfelder, *J. Mol. Biol.*, 2008, **384**, 120–36.

469. D. Ghosh, *Methods Enzymol.*, 2005, **400**, 273–93.

470. S. R. Hanson, M. D. Best and C. H. Wong, *Angew. Chem., Int. Ed. Engl.*, 2004, **43**, 5736–63.

471. G. Lukatela, N. Krauss, K. Theis, T. Selmer, V. Gieselmann, K. von Figura and W. Saenger, *Biochemistry*, 1998, **37**, 3654–64.

472. B. van Loo, S. Jonas, A. C. Babtie, O. Berteau, M. Hyvonen, F. Hollfelder, in preparation.

473. A. J. Kirby and A. G. Vargolis, *J. Am. Chem. Soc.*, 1967, **89**, 415–423.

474. F. Hollfelder and D. Herschlag, *Biochemistry*, 1995, **34**, 12255–64.

475. A. Aharoni, L. Gaidukov, O. Khersonsky, Q. G. S. Mc, C. Roodveldt and D. S. Tawfik, *Nat. Genet.*, 2005, **37**, 73–6.

476. J. K. Lassila, D. Herschlag, *Biochemistry* 2008, **47**, 12853–12859.

477. K. J. Woycechowsky, K. Vamvaca, D. Hilvert, *Adv. Enzymol. Relat. Areas Mol. Biol.* 2007, 75, 241–94, xiii.

478. C. Jackel, P. Kast and D. Hilvert, *Ann. Rev. Biophys.*, 2008, **37**, 153–73.

479. D. Röthlisberger, O. Khersonsky, A. M. Wollacott, L. Jiang, J. DeChancie, J. Betker, J. L. Gallaher, E. A. Althoff, A. Zanghellini, O. Dym, S. Albeck, K. N. Houk, D. S. Tawfik and D. Baker, *Nature*, 2008, **453**, 190–5.

480. J. C. Hermann, R. Marti-Arbona, A. A. Fedorov, E. Fedorov, S. C. Almo, B. K. Shoichet and F. M. Raushel, *Nature*, 2007, **448**, 775–9.

481. F. Hollfelder, A. J. Kirby and D. S. Tawfik, *Nature*, 1996, **383**, 60–2.

482. F. P. Seebeck and D. Hilvert, *J. Am. Chem. Soc.*, 2005, **127**, 1307–12.

483. A. J. Kennan and H. W. Whitlock, *J. Am. Chem. Soc.*, 1996, **118**, 3027–3028.

Subject Index

Note: page numbers in *italic* refer to schemes, figures or tables

abzymes *198*, 212
acetals 26, 126–9
  glycosyl transfer models 129–32, 138,
    140–2, 145
acetyl coenzyme A 155
acid proteinases *see* aspartic
    proteinases
acid–base catalysis 111, 116–18, 122,
    146, 150, 159, 204
  general *vs* specific 8–9
  intramolecular reactions 15–16
  and kinetic equivalence 13–14
  mechanisms 9–13
  and pH-rate profiles 7–8
  see also general acid, base catalysis
activated substrates 31
activation energy 13
activation enthalpy 13
active-sites 24, 37, 43–8, 50, 52–6, 59, 62–4,
    66–8, 69, 75–7, 79, 82, 90, 98, 100,
    114, 118–20, 133–7, 147–8, 156–7,
    159, 162, 166, 168, 171–2, 180, 182–3,
    202, 208–10, 217, 223–4, 238–9
  α-chymotrypsin *63*
  combining with new functionality 53
  pK_a values in 44
  and promiscuous enzymes 242–3
  and protein mapping 236
acyl transfer
  aspartic proteinases 79–83
  metallopeptides/amide hydrolases
    83–95

serine proteases 62–74
SH hydrolases 75–9
3′-adenosyl radical 179
adenosylcobalamin 168–71, 176
S-adenosylmethionine 168–70
aldol reaction 229, *230*
alkaline phosphatase superfamily *58*,
    103–6, 238–43
amide hydrolases *see*
    metallopeptidases
amino-acids
  and minimalist protein redesign 56–7
  non-natural 54
  side chains/functional groups 4–6,
    43–4, *45*
aminopeptidases 90–1
  models 92–3
aminothiols 76, 78
anthraquinones 125
antibiotics 91, 94–5
antibodies 50–2
  binding free energy 22–3
  iterative methods for enzyme
    mimics *196*, 212–14
  pericyclic reactions 185–91
  structure *213*
artificial enzyme defined 27
arylsulfatases 238, *240*, 241, 242
aspartic proteinases 79–80
  intramolecular models 80–3
ATP 26, 102–3, 106
avidin–biotin complex 23

background rate constants 36–7
bait and switch tactic 214
 and proton transfer from carbon
  216–19
benzimidazoles 200–1
binding 20–1, 39
 and constructing enzyme models 47–52
 equilibrium 33
 hydrophobic binding 22–3
 see also transition state
biotin 55
 biotin–avidin complex 23
bond dissociation energies 176
bound hydroxide 84–5, 103
Brønsted coefficient 111, 144, 148–9
Brønsted equation 9

carbohydrates 126
carbonic anhydrase 87
carboxylic acyl transfer *see* acyl
 transfer
carboxypeptidases 86, 90
2-carboxyphenyl β-D-glucoside 127
catalysis, principles of 6–8
 general acid–base 9–13
catalytic antibodies *198*, 212–20
 see also abzymes
catalytic efficiency 29–40
 assessing an enzyme model *39*
 efficiencies tabulated and compared
  40–1, 207, 212, 215, 218, 222,
  240, *246*
 rate measurements
  enzyme catalyzed reactions 32–5
  expressing rate acceleration 35–40
  uncatalyzed reactions 30–2, 34–5
catalytic proficiency
 defined 35, 37, *39*
 enzymes compared *40*, 41
 metal-complex catalysis 122–3
 and promiscuous reactions 59
catalytic promiscuity 57–9, 238–43
 alkaline phosphatase superfamily
  58, 238–43
catalytic triad 63–5
 modeling 68

cerium cation 124
chemical libraries *see* libraries
chorismate mutase 180, 181–2
 catalysis by antibodies 182–5
chymotrypsins 26, 48, 64
 active site *63*
 key steps in peptide cleavage *65*
citrate synthase 155
Claisen rearrangement 181–2, 183
coenzymes 25, 54
 adenosylcobalamin as radical
  source 168–71
 and citrate synthase 155
 NAD⁺/NADH system 159–65
cofactors 24–5, 54, 55–6, 210
combinatorial methods 208
 see also libraries
conjugate acids/bases 9
 see also pK_a values
convergent evolution 26, 64, 236
copper cation 90, 211
covalent design 47
cross-reactivity 57–9
 see also catalytic promiscuity
crown ether 70, 79, *92*, *103*, *105*
curcurbituril 187–8
cyclization reactions 14–17, 66–7
 see also pericyclic reactions
cyclodextrins 23
 acylation by nitrophenyl acetates
  48–9
 bis-imidazole models for RNAse 117
 and Diels–Alder reactions 188–9
 mimics for glycosidases 143–5
 mimics for serine proteases 69–73
 twinned metal-chelated 90
cyclophane derivative 74
cystein proteases 76
cysteinyl radical 174
cystine disulfide radical anion 179–80

dehydrogenases 161, *163*, *164*
dendrimers 202–9
5'-deoxyadenosyl radical 168–71
deoxyribonucleoside 172, *173*
deoxyribozymes 228–9

Diels–Alder reaction 180, 183
  catalyzed by antibodies 185–6
  catalyzed by RNA 191–3, 229, *230*
  supramolecular catalysis 187–91
diffusion, rate limiting 4, 6, 32, 66
  see also rate determining step
dihydroxyacetone phosphate 152–4
directed evolution 52, *196, 198*, 196–7
  coupled with design 243–6
dissociation constants
  enzyme kinetics 32–5
  ionization *see* pK_a values
divergent evolution 26, 64
diversity space 233–5
DNA cleavage
  DNAses 113, *114*, 118
  supramolecular models 123–6
DNA ligase reaction 228, *229*
DNAzymes *198, 222*, 228–9

effective molarity 17–18, *39*, 66, 80,
    108, 165, 217, 223
  for enolization 149–50
  hydride transfer 166
endopeptidases 90
energetics 18–20, 34–5
energy-profile diagrams 19, 34–5
enol ethers 8
enolase superfamily 155–7, 236
enolization 9, 13, 146–59, *237*
  see also proton transfer
enzyme mimic defined 27
enzyme model scorecard 39
enzyme models 42–3
  general strategies for constructing
      46–52
    and solvent catalysis 43–6
    see also iterative methods
  modifying existing
      enzymes/proteins
    adding functionality 54–6
    advantages and disadvantages 52–4
    catalytic promiscuity 57–9
    site-directed mutation 56–60
    see also protein catalysts
enzyme promiscuity *see* catalytic
    promiscuity

enzymes, introduction to 1–6
ethyl trifluoroacetate 10, *11*

ferrocenyl acrylic acid 73
Fischer's model 4
free-energy profile 34–5
fumaric acid diamide derivative 82–3
functional groups/amino-acid side
      chains 4–6
  and hydrophobic perturbation 43–4, *45*

general acid–base catalysis *see* acid–
      base catalysis
general acid catalysis 78, 79–82, 101–2,
      111–18, 120, 127, 130–1, 133–5,
      138–42, 144–5, 148–50,
    154, 157–9, 164–5, 174, 179–80, 204,
      218–19, 222, 224, 237
  intramolecular 101–2, 113–15, 127,
      138–42, 149–50
general base catalysis 66–8, 75, 81–3,
      101, 110–11, 115, 117, 120, 133–4,
      147–53, 155, 157, 159–63, 174–5,
      177, 179, 201, 216, 219, 224, 237
  intramolecular 67–8, 75, 81–2, 149–50,
      177
genotypes 196–7, 233, 235
α-D-glucopyranosyl fluoride 128
glucosides 26–7, 130, 144
  p-nitrophenyl derivative 143–4, *145*
  salicyl derivative 140
glyceraldehyde-3-phosphate 152–4
glycogen phophorylase 136–7
glycosidases 26–7, 129, 132–5
glycoside as template 209
glycosides 129–30, 131–2
  2-acetamido derivatives *134*, 135, *136*
glycosyl fluorides 136
glycosyl transfer 126–9
  enzyme mimics 142–5
  glycoside hydrolases 26–7, 129, 132–5
  glycoside transferases 132–3, 135–8
  models
    intramolecular 138–42
    simple 129–32
glycyl radicals 177, *178*
glycyl glycine 92

ground state 39, 45, 47–50
Groves' complex 89

haptens described 212
HDV ribozyme 223–4
hexaazadioxacyclotetracosane 92
HPNP 120–3, 125, 201–2
hydride transfer 146, 159–66
  dehydrogenases 161, *163*, *164*
  models 161, 164–6
  UDP-galactose-4-epimerase 160–1, *162*
hydrogen atom transfer 146, 167–8
  radical reactions 176–80
hydrogen bonding 22, 43–4, 49, 64, 68–70,
  74–5, 81–4, 87, 92, 101–4, 113, 116,
  122–4, 136, 139–40, 142, 144, 150,
  153–5, 187, 201, 210, 217, 224, 238
  and Kemp decarboxylation 44, *46*
hydrogen transfer *see* hydride transfer;
  hydrogen atom transfer; proton
  transfer
hydrolysis reaction mechanisms 8–13
hydronium ion 8–9
hydrophobic binding 21–3, 65
  solvent effects 43–6
hydrophobic pocket 64
2-hydroxypropyl-*p*-nitrophenyl
  phosphate 120–3, 125, 201–2

imidazole catalysis 67–8, 73
in vitro compartmentalization 235
inhibitors 50
intermolecular *vs* intramolecular
  catalysis 17, 46–7
intramolecularity 14–18, 66
  see also effective molarity
isomerases 152–5
iterative methods
  background and summary 195–7, *198*
  catalyic polymers 197, 199
  catalytic antibodies 212–20
  dendrimers 202–9
  improving protein enzymes 231–46
  molecular imprinting 209–12
  nucleic acids as catalysts *198*, 191–3,
    220–31
  synzymes *198*, 199–202

Jencks' libido rule 12

k_2 36, *39*
k_{cat} 33, 34, *35, 40*
K_M 6–7, 19, 32–4, *35*, 37–41, 49–52
k_{obs} 30–1, 35–7
k_0 34–7
K_S 34–5
K_{TS} 34–5, 37–9, 51
k_{uncat} 31–2, 34–41, 51–2
Kemp elimination 44, *46*, 200–1,
  216–19, *246*
  protein design 244–6
ketone enolization 9, 13
kinetic equivalence 13–14
kinetics 6–9, 30, 39, 44, 47, 49
  see also catalytic efficiency

β-lactamases 91–2
  models 93–6
lactate dehydrogenase 161, *163*
lactonization 14, 67
lanthanum cation 125
leadzyme 228
Lehn's cryptand 102–3, 106
L-leucine *p*-nitroanilide 88, 93
libido rule 12
libraries 195–7
  catalytic antibodies 212–20
  molecularly imprinted polymers
    209–12
  nucleic acids 226–7
  peptide dendrimers 202–9
  proteins 212–20, 232–5, 243–7
liver alcohol dehydrogenase 161, *164*
lock-and-key principle 4
lysozyme 133–4, *135*

magnesium cation 103–4, 118, 157–8,
  164–6, 222–4, 237
  and enolases 236–8
magnesium-bound hydroxide 103
maleamic acid derivatives 8, 80
malonate esters 18
mandelate racemase 157–9, *237*
medium effect 43–6

metal coordination 83–4, 118
  and enolases 151–2, 236–8
  in molecular imprinting 210–12
  and ribozyme catalysts 221, *223*
  see also individual metals
metal ion catalysis 83–96, 102–6, 118–26,
    157–8, 168–71, 164, *223, 237, 239,*
metallopeptidases/amide hydrolases
    83–6
  common enzymes 86
  models with one metal centre
    intermolecular reactions 86–8
    intramolecular reactions 88–90
    supramolecular models 90
  models with two metal centres 88–95,
    121, 125
metallophosphatases
  diesterases 118–20
  monoesterases 103–5, 118–20
    models 105–6
metalloproteinases 86
methyl acetate 87–8
Michael reaction 229, *230*
Michaelis constant 32, *33*, 34, 39
Michaelis–Menten kinetics 6–7, 19,
    31–2
microenvironments 43, 46, 69, 53
minimalist protein redesign 56–7
MIP *see* molecularly imprinted polymers
model enzyme, definition of 27
models *see* enzyme models
modified enzymes *see* enzyme models
modular design 55
molecularly imprinted polymers 52,
    209–12
*m*RNA display 234
muconate lactonizing enzyme *237*
mutant enzymes 52, 56–7

NAD$^+$/NADH system 25, 26, 159–65
naphthyl phosphate 102
nitrocefin 94
p-nitrophenol esters 48–9, 69, *72*, 74–5,
    88–9, 199–200, 202
m-nitrophenyl acetate 48–9
p-nitrophenyl β-glucoside 143–4
p-nitrophenyl phosphates 105, 124–5

p-nitrophenyl ribosides 138–9
no-mechanism reactions *see*
    pericyclic reactions
noncovalent interactions
  ground state binding 43–5, 47–50,
    129, 181
  increased transition state binding
    43–5, 50–2, 116, 139, 153,
    183–5, 188, 191, 209, 214
nucleases 119–20
  see also phosphoryl transfer
  site-specific 55
  zinc-finger 126
nucleic acid catalysts *198*, 220, *222*
  mechanisms 220–5
  new catalyst by SELEX *198*, 225–8
    DNAzymes *198*, 228–9
    non-natural reactions 229–30
    RNAzymes *198*, *222*, 220–31
    see also ribozymes
nucleophilic catalysis
  by enzymes 62, 64, 77, 100–1, 104,
    113–4, 120, 133–7, 242
  by models 74–6, 78, 80–1, 102–3,
    105, 115, 123, 127–8, 214, 223
  general hydrolysis mechanisms 9–13,
    81, 242
  intramolecular reactions 14–18
  metal cation enhanced 83–5

oxacillin 95–6
oxaloacetate tautomerase 56–7
oxyanion hole 64, 68, 76, 82

P1 nucleases 119–20
PAMAM dendrimers 203–4
papain 56, 76
penicillin G 94
peptide dendrimers 204–9
pericyclic reactions 180–1
  chorismate mutase reaction 181–2
    catalysis by antibodies 182–5
  Diels–Alder reaction 180, 183
    catalyzed by antibodies 185–6
    catalyzed by RNA 191–3
    supramolecular catalysis 187–91
  RNA catalysis 191–3

pH
  pH–rate profiles 7–8, 16, 78, 81, 114,
     117, 140, 142
  and rate constants 36–7
  side-chain ionisation 4–5
phage display 233, *234*, 235
phenotypes 196–7, 233, 235
phenyl α-D-mannopyranoside *209*
phenyl salicyl phosphate *107*, 108, *109*
phosphatase enzymes and models *see*
     alkaline phosphatase superfamily,
     phosphoryl transfer, nucleases
phosphate esters
  diesters 31–2, 97–8, 106–12
    and DNAzyme 228, *229*
  monoesters 97–100
  see also phosphoryl transfer
phosphatidylcholine 119
2-phospho-D-glycerate 155–7
phosphoenolpyruvate 155–7
phospholipase C 116, 119–20
phospholipase D superfamily 113,
     114, 118
phosphonamidates 50–2
phosphonate monoester hydrolase
     239, *240*, 241, 242
phosphoric acid 97
phosphoryl transfer 96–8
  from diesters 106–7
    intramolecular reactions 108–13
    metalloenzymes 118–26
    non-metalloenzymes 113–18
  from monoesters 98–100
    metalloenzymes 103–6
    non-metalloenzymes 100–3
phthalamic acid 80
pK$_a$ values 9–10, 12–13, 16, 21, 36–7, 49
  of amino-acids 4–5, 44–5
  of carbon acids/enolization 146–8
  of ionizable active site groups 43–5,
     221, 223
poly 4(5)-vinylimidazole 199
polyamidoamine dendrimers 203–4
polyethylenimine 199–201
polypeptides 2–3
polyreactivity/promiscuity *see*
     catalytic promiscuity

positioning of functional groups 46–50
protein catalysts
  de novo approaches 243–6
  development, challenges 231–3
  diversity space and technologies
     233–5
  protein superfamilies/enzyme function
     alkaline phosphatases *58*, 103–6,
       238–43
  web resources 236
protein structure 2–3, 236
protein tyrosine phosphatases 100–1
proton transfer (enolization) 145–8
  enzymes 152
    citrate synthase 155
    enolase superfamily 155–7
    mandelate racemase 157–9
    triose phosphate isomerase 152–5
  metal ion catalysis 151–2
  models
    intramolecular 149–51
    simple 148–9
proton transfer (from carbon) 200,
     216–19
proton-coupled electron transfer 167–8,
     177
pyrenesulfonate esters 206–8
pyridoxal 25, 147, 203–4
pyruvate-formate lyase 170–2

quinolinium esters 204–5

racemases 157–9, *237*
radical reactions 168
  adenosylcobalamin-based initiators
     168–71
  models 174, 176
    hydrogen atom transfers 176–80
    initiation stages 176
  radical centres/active sites
    pyruvate-formate lyase 171–2
    ribonucleotide reductase 172–4
rate accelerations 4, 18, 35–41, 44, 46,
     49–50, 53, 56, 59, 66, 71, 79, 90, 106,
     123, 125, 138–42, 181–2, 189, 192,
     200–2, 205, *207*, 210, *212*, *215*, 219–21,
     222, 231, 239–41, 243, 245, *246*

rate constants 29–40
  background 9, 36–7
  catalyzed reactions 32–4
  uncatalyzed reactions 30–2
  see also k_{cat}, *etc*
rate determining step 6–7, 9, 11, 13,
    18–19, 21, 32, 66, 79, 108, 110–12,
    146, 148–50, 154, 165, 176, 187
  intramolecular proton transfer 18
  limiting diffusion 4, 6, 32
rate profiles *see* pH–rate profiles
rates of reaction *see* catalytic efficiency
reactions of enzymes/enzyme models *58*
  acyl transfer 61–96
  glycosyl transfer 126–45
  hydrogen transfer 145–68
  pericyclic reactions 180–93
  phosphoryl transfer 96–126
  radical reactions 168–80
recognition and binding 20–1
recombinant DNA technology 56–7
ribonucleases 114–16, 118–20
  see also phosphoryl transfer
ribonucleosides 108, 110–12
ribonucleotide reductase 172–4, 180
ribosome display 233, *234*, 235
ribozymes 191–3, 220–1
  mechanism of catalysis 220–5
  new catalysts using SELEX 227–31
RNA
  catalysis of pericyclic reactions 191–3
  cleavage reactions 108, 110–12
    bait and switch tactic 214, *216*
    supramolecular models 120–3
  ligase reaction 243, *244*
  see also nucleic acid catalysts,
      ribozymes
RNAse A 116
  models 117–18
RNAzymes *see* ribozymes

salicyl glucosides 140
salicyl phosphates 102, *107*, 108, *109*
salicylamides 68–9
selective binding *see* binding
SELEX methodology *198*, 225–6
  applications 227–31

serine proteases 14–15, 26
  carboxylic acyl transfer
    active sites 62–4
    intramolecular models 66–9
    mechanism and key steps 64–6
    supramolecular models 69–75
SH hydrolases, acylation of 75–6
  models 76–9
sequence space 233
site-directed mutation 56–7
size of enzymes 3–4
solvent catalysis 43–6
specific acid-base catalysis 8
streptavidin 55
structure of enzymes 2–4
substrate binding *see* binding
substrates
  activated 31, 199, 210, 213, *218*, 219
  non-natural/nonspecific 53, 57–9
  positioning by covalent design 47
subtilisin 63, 68
o-succinylbenzoate synthase 156, 238
superfamily *58*, 236–43
synzymes 199–202

templating 209–12
tetralactams 68–9, *70*
tetraoxaparacyclophane derivatives 74
thermodynamics 13, 18–20, 34–5
thermolysin 50, 86
thiolysis 75–8
  intramolecular models 78–9
thiyl radical 172, 174, 177, *178*
TIM barrels 244
transition-state
  dissociation constants 34–5
  energy differences in 35–7
  noncovalent interactions 50–2
  and solvent effects 43–4
  stabilization of 20, 44–5, 50–2, 139–40,
      150, 152, 182, 200–1, 217–18
transition-state analogues 50, 190
  and catalytic antibodies 212–20
  with charged group (bait and switch)
      model RNA cleavage 214–16
      proton transfer from carbon 216–19

templates for molecular imprinting 209–10
trehalose-6-phosphate synthase 137, *138*
triose phosphate isomerase 152–5
turnover numbers 33, *38*, *40*, 41
  and sensitivity to inhibitors 50
tyrosyl radical 172 *173*

uncatalyzed reactions 19
  range of rates 39–40
  rate constants *see* k_{uncat}
uridine diphosphate-galactose-4-epimerase 160–1, *162*

4-vinylbenzeneboronic acid *209*
virtual screening 245

water as nucleophile 9–10
  enhanced nucleophilicity 81–6

zinc cation 83–4
  dizinc complexes 90–6, 121–2
  enolization catalysis 151–2
  liver alcohol dehydrogenase 161, *164*
  in molecular imprinting 210–12
  phosphatases 103–6
    see also phosphoryl transfer
  trizinc complexes 121–2
  see also metal ion catalysis, metallopeptidases
zinc-bound alkoxide 89
zinc-bound hydroxide mechanism 84
zinc-finger nucleases 126